Cell Therapy, Stem Cells, and Brain Repair

Contemporary Neuroscience

Cell Therapy, Stem Cells, and Brain Repair

Edited by

Cyndy Davis Sanberg, PhD

Saneron CCEL Therapeutics, Tampa, FL

and

Paul R. Sanberg, PhD, DSc

University of South Florida College of Medicine, Tampa, FL

HUMANA PRESS ✷ TOTOWA, NEW JERSEY

© 2006 Humana Press Inc.
999 Riverview Drive, Suite 208
Totowa, New Jersey 07512

www.humanapress.com

For additional copies, pricing for bulk purchases, and/or information about other Humana titles, contact Humana at the above address or at any of the following numbers: Tel.: 973-256-1699; Fax: 973-256-8341; E-mail: orders@humanapr.com, or visit our Website: www.humanapress.com

This publication is printed on acid-free paper. ⬭∞
ANSI Z39.48-1984 (American Standards Institute) Permanence of Paper for Printed Library Materials.

Cover art: figure 1, chapter 8

Cover design by Patricia F. Cleary

Printed in the United States of America. 10 9 8 7 6 5 4 3 2 1
eISBN 1-59745-147-9
Library of Congress Cataloging-in-Publication Data

Cell therapy, stem cells, and brain repair / edited by Cyndy Davis Sanberg and Paul R. Sanberg.
        p. cm.
     Includes bibliographical references and index.
     ISBN 1-58829-502-8 (alk. paper)
     1. Cellular therapy. 2. Neural stem cells. 3. Brain--Diseases--Treatment. I. Sanberg, Cyndy Davis. II. Sanberg, Paul R. III. Series
     [DNLM: 1. Brain Diseases--therapy. 2. StemCell Transplantation. 3. Cell Differentiation--physiology. 4. Stem Cells--physiology. WL359 C393 2006]
     RM287.C3852 2006
     616.8'0427–dc22   2005034441

Preface

As our world continues to evolve, the field of regenerative medicine follows suit. Although many modern day therapies focus on synthetic and natural medicinal treatments for brain repair, many of these treatments and prescriptions lack adequate results or only have the ability to slow the progression of neurological disease or injury.

Cell therapy, however, remains the most compelling treatment for neurodegenerative diseases, disorders, and injuries, including Parkinson's disease, Huntington's disease, traumatic brain injury, and stroke, which is expanded upon in more detail in Chapter 1 by Snyder and colleagues. Cell therapy is also unique in that it is the only therapeutic strategy that strives to replace lost, damaged, or dysfunctional cells with healthy ones. This repair and replacement may be due to an administration of exogenous cells itself or the activation of the body's own endogenous reparative cells by a trophic, immune, or inflammatory response to cell transplantation. However, the precise mechanism of how cell therapy works remains elusive and is continuing to be investigated in terms of molecular and cellular responses, in particular. Moreover, Chapter 11 by Emerich and associates, discusses some of the possibilities of cell immunoisolation and the potential for treating central nervous system diseases.

During the past 20 years most investigations have utilized cells derived from fetal tissue as a source of transplantable cells for cell therapy, which have demonstrated an underlying proof of principle for current cell transplants for a treatment of a variety of neurological diseases and injuries, including Huntington's disease which are discussed in Chapter 4 by Dunnett and colleagues. Chapter 4 also reviews challenges in harvesting the tissue, the analogy of developmental stages between species, clinical trials, alternative tissue sources, as well as specific xenogenic issues. In addition, stem cells have emerged as the leading topic regarding cell therapy. According to the National Institutes of Health, "a stem cell is a cell that has the ability to divide (self replicate) for indefinite periods-often throughout the life of the organism. Under the right conditions, or given the right signals, stem cells can give rise (differentiate) to the many different cell types that make up the organism. That is, stem cells have the potential to develop into mature cells that have characteristic shapes and specialized functions, such as heart cells, skin cells, or nerve cells."

Previous studies in fetal tissue also contributed a great deal to the discovery of neural stem cells. Neural stem cells are derived from fetal and adult brain and have the ability to divide and give rise to more stem cells or to several types of precursor cells, which can then become neurons and glia. Neural stem cells in the mammalian fetal brain have been located in the subventricular zone, ventricular zone, hippocampus, olfactory bulb, cerebellum, and cerebral cortex. Chapter 1 by Snyder and colleagues describes the promising potential of neural stem cells for therapeutic use. In addition, studies using neurospheres also help to identify the subependymal zone as another source of stem cells in the brain. Another source of neural stem cells, which are quite different than the fetal neural stem cells, is neural crest cells. During development, the neural crest cells migrate from the sides of the neural tube as it closes, and the cells differentiate into a variety of tissues, which are not all part of the nervous system. Many neural crest cells are responsible for comprising most of the peripheral nervous system, including hormone-producing glands, as well as skin, cartilage, bone, and many connective tissues within the body. Neural stem cells and neural transplantation in primates are discussed in Chapter 3 by Bjugstad and Sladek. This chapter also elaborates on direct comparisons between successful rodent studies and marginal human studies and the limitations of a rodent Parkinson's disease (PD) model; thereby, demonstrating the proof of principle for a primate PD model.

A new era in stem cell research began in 1998 with the derivation of embryonic stem cells. Techniques involving embryonic stem cells have developed greatly since 1998, when James Thomson and his colleagues reported methods for deriving and maintaining these cells. Stem cells derived from embryos have also been extensively studied and have demonstrated the remarkable ability to differentiate into neurons, glia, and numerous cell types in animals, which is summarized in Chapter 1. Despite current negative views regarding the use of these tissue types, this previous research has paved the way for many new types of stem cell research which follow similar experimental paths of the original embryonic stem cell research. This proof of principle involving embryonic stem cells, as well as an overview of various types of stem cells suitable for transplantation, particularly in Parkinson's disease, is reviewed in Chapter 2 by Brundin and colleagues.

Owing to the heightened ethical concerns and governmental issues regarding embryonic and fetal tissue research, cellular research has continued to expand its search for alternative sources of stem cells. More recently, adult stem cells, which are cells obtained post-birth, have made a breakthrough in the field of stem cell research. Adult stem cells make identical copies of themselves for long periods of time (self-renewal), and can produce mature cell types that have specific morphologies and functions. Their primary functions are to maintain the steady state of a cell and to replace cells that die due to injury or disease. Adult stem cells usually generate an intermediate cell type or types before they become fully differentiated. The intermediate cell type is commonly called a precursor or progenitor cell. This progenitor cell has the capacity to produce cells of the original tissue or organ (multipotent). For instance, stem cells isolated from the brain will give rise to neural cells, stem cells from the heart will give rise to cardiac cells, or stem cells from the bone marrow will give rise to blood cells. In addition, the adult stem cells also have the capacity to produce cells giving rise to many different cell types, tissues, and organs regardless of the origin of the stem cell (pluripotent). For example, stem cells from umbilical cord blood may give rise to neural cells, cardiac muscles, or other blood cells depending upon the condition or environment of the stem cells themselves. The ability of the adult stem cells to display pluripotency is quite similar to embryonic stem cells, thus expanding our resources for stem cells for cell therapy. Adult stem cells may be obtained from many different types of tissues, however, they retain the ability to produce many tissue types as well. These cells can be harvested from donors and isolated within the laboratory, where scientists culture and grow these cells for transplantation.

Bone marrow has also been found to be rich in adult stem cells. This idea, however, is not novel; hematopoietic stem cells were recognized as stem cells more than 40 years ago. However, more recent research has shown that these stem cells have exercised enormous potential for cellular therapy by demonstrating the capability of neuronal and astrocytic differentiation following transplantation. Thus, studies in bone marrow have advanced cell therapy to now include brain repair as well. Bone marrow actually contains three specific stem cell populations-hematopoietic stem cells, stromal cells, and endothelial progenitor cells, although more specifics on bone marrow stem cell types and classifications are discussed in Chapter 7 by Low and colleagues. In addition, Chapter 7 also includes a review of the experimental progress toward a therapeutic for each type of bone marrow stem cell, as well as the concepts and studies necessary to translate bone marrow stem cell research into clinic. Moreover, Chapter 10 by Emerich and colleagues covers the therapeutic potential of transplanted bone marrow stem cells into the choroids plexus (CP), in particular, as well as the future potential for using transplantable CP cells as a means of delivering neurotrophic factors to the brain and spinal cord. Chapter 5 by Dunbar and associates elaborates upon the specific use of autologous whole bone marrow and mesenchymal stem cell transplants in a model of Huntington's disease, and a comprehensive comparison between autologous and heterologous marrow stem cell transplants.

Another hematopoietic source that is rich in adult stem cells includes umbilical cord blood. The umbilical cord which supports the fetus during pregnancy, is delivered with the baby, and is typically discarded. Since the first successful umbilical cord blood transplants in children with Fanconi anemia, the collection of cord blood and cellular therapeutic use has grown rapidly. Moreover, there are none of the ethical issues regarding the use of cord blood stem cells compared to embryonic stem cells, and the method of harvesting the stem cells from the umbilical vein poses no risk to the mother or baby, since the cord is removed and set aside prior to the blood collection. From a cellular therapeutic perspective, umbilical cord blood offers many advantages. Like bone marrow it is rich in stem cells, but is much easier to obtain than bone marrow. Fortunately, both bone marrow and umbilical cord blood stem cells have been shown to migrate and engraft to neurological sites of injury, following non-invasive intravenous injection, and amazingly produce recovery of function resulting from stroke and other forms of neurological injury, which offers an extreme advantage for cell therapy with these cells. Chapter 13 by Vendrame and Willing, comprises an overview of human cord blood cells, their phenotype, functional characteristics, and potential as a therapy for neurodegenerative diseases and disorders. This chapter also discusses other hematopoietic stem cells, including G-CSF stimulated peripheral blood, and its therapeutic potential for brain repair.

Although the field of stem cell research has evolved into a promising therapy for brain repair many challenges still exist. The process of identifying the desired type of stem cell in culture will involve tedious research, while developing the right biochemical environment or media is essential to ensure that the stem cell differentiates into the desired cell type. Also, once the stem cells have been transplanted the cells must be integrated within the body's own tissue and organs and function correctly. Yet another challenge is tissue rejection. The body's immune system must not recognize the transplanted cells as foreign. Fortunately, cord blood-derived stem cells are considered to be more immune immature cells, thus making the incidence of tissue rejection much less than other types of transplantable cells. In addition, Sertoli cells are described in Chapter 9 in more detail for their potential role in immune system modulation and their capability to reduce rejection for cell transplants. Another concern is the possible risk of cancer. Cancer results when the cells continue to proliferate and keep further dividing beyond the desired point. This point is a delicate balance once the cells have been transplanted, fostering the growth of the new cells without them dividing out of control. Interestingly, however, much evidence has been presented that cells isolated from a specific human neuroteratocarcinoma (NT2N cells) have the ability to generate neurons once transplanted into stroke patients, which is outlined in Chapter 6 by Borlongan and associates. Thus, continued efforts are being made to address the positive and negative issues in order for these cell therapies to complete human clinical trials.

However, with these challenges in mind, stem cell therapy remains one of the best "natural" candidates to help heal the human body. Despite the many challenges, many scientists believe that cell therapy will revolutionize medicine. These cell therapies may one day offer cures for cancer, Parkinson's disease, diabetes, kidney disease, multiple sclerosis, cardiovascular disease, and symptoms of stroke. Cell therapy may also fill a tremendous need for chronic pain management and traumatic brain injury (TBI), which is examined in more detail using several intervention strategies in Chapter 8 by Eaton and Sagen. A variety of potential cell sources for chronic pain and TBI are elaborated upon in Chapter 8 as well. Stem cell therapies have also shown encouraging results in helping to repair spinal cord injuries, and helping to regain movement resulting from paralysis. It is also possible that the human life span could be increased due to the regeneration and repair of tissue and organs by stem cells. Stem cells also seem to be in the forefront in providing a treatment for brain repair, in general, as the incidence of neurological injuries and disease increase in our world today. While our knowledge of cell therapy continues to develop, so does our revolutionary precision in how to design a better therapy to treat disease. Chapter 12 by Polgar, identifies recent developments in health research methodology that may be useful for ensuring progress in cellular therapy for brain repair, goals for cellular therapy, best practices, and some critical analysis and ethics. In addition, the commercial and pharmaceutical implications of stem cells and their role in regenerative medicine are discussed in Chapter 14 by Cruz and Azevedo.

This compilation attempts to explicate previous cornerstones and milestones of neurological cellular therapy, which have provided a foundation for modern stem cell research. Ongoing challenges are discussed, as well as many obstacles that have been overcome already. The current direction of cellular research is described, and modern techniques involving certain subsets of cell populations explained. In addition, the ongoing discovery of stem cell sources for cell therapy is discussed, while expounding upon clinical applications for cell therapeutic brain repair as they become increasingly promising. The clinical applications include potential cell therapy for

Parkinson's disease, traumatic brain injury, and ischemic stroke. We hope to provide a good understanding of the stem cell research field by presenting literature from renowned scientists and clinicians in the field of cell therapy today, and share their data, conclusions and future investigations, and the challenges that they overcame to reach their results. Also, varying methods of cell transplantation are revealed and how the method or route of administration affects the behavioral outcome in animal injury models.

Scientists have begun to recognize the amazing versatility of these primitive cells, which exist for only a short period of time prior to differentiating into other cell types and tissues within the body. Since cells are the basic building blocks of the human body, it would only stand to reason that we should harness the power of these stem cells to sustain and repair the body's tissues and organs and with the appropriate research, as demonstrated here, the many obstacles of stem cell research that can be overcome. It is by sharing knowledge with reputable scientists and clinicians that enables the field of cellular research to continue to thrive and move forward.

Cyndy Davis Sanberg, PhD
Paul R. Sanberg, PhD, DSc

Contents

Contributors

SILVIA P. AZEVEDO • *Cryopraxis Criobiologia Ltd., Rio de Janeiro, Brasil*

KIMBERLY B. BJUGSTAD • *Department of Psychiatry, University of Colorado Health Sciences Center, Aurora, CO*

CESARIO V. BORLONGAN • *Neurology/Institute of Molecular Medicine & Genetics, Research and Affiliations Service Line, Augusta VAMC, Medical College of Georgia, Augusta, GA*

PATRIK BRUNDIN • *Section for Neuronal Survival, Wallenberg Neuroscience Center, Lund University, Lund, Sweden*

NICOLAJ S. CHRISTOPHERSEN • *NsGene A/S, Ballerup, Denmark, and Neuronal Survival Unit, Wallenberg Neuroscience Center, Lund University, Lund, Sweden*

ANA SOFIA CORREIA • *Section for Neuronal Survival, Wallenberg Neuroscience Center, Lund University, Lund, Sweden*

L. EDUARDO CRUZ • *Cryopraxis Criobiologia Ltd., Rio de Janeiro, Brasil*

GARY L. DUNBAR • *Brain Research and Integrative Neuroscience Center, Department of Psychology, Central Michigan University, Mt. Pleasant, MI*

STEPHEN B. DUNNETT • *Brain Repair Group, School of Biosciences, University of Cardiff, Cardiff, United Kingdom*

MARY EATON • *The Miami Project to Cure Paralysis, VA RR&D Center of Excellence in Functional Recovery in Chronic Spinal Cord Injury, University of Miami School of Medicine, Miami, FL*

ROBERT B. ELLIOTT • *Diatranz NZLtd/Living Cell Technologies, New Zealand*

DWAINE F. EMERICH • *LCTBioPharma, Inc., Providence, RI*

CHRISTINE FOURNIER • *Neurology/Institue of Molecular Medicine & Genetics, Research and Affiliations Service Line, Augusta VAMC, Medical College of Georgia, Augusta, GA*

CRAIG R. HALBERSTADT • *The Transplant Center, Carolinas Medical Center, Charlotte, NC*

DAVID C. HESS • *Neurology/Institute of Molecular Medicine & Genetics, Research and Affiliations Service Line, Augusta VAMC, Medical College of Georgia, Augusta, GA*

YUEHUA JIANG • *Stem Cell Institute, University of Minnesota Medical School, Minneapolis, MN*

C. DIRK KEENE • *Department of Neurosurgery, University of Minnesota Medical School, Minneapolis, MN*

CLAIRE M. KELLY • *Brain Repair Group, School of Biosciences, University of Cardiff, Cardiff, United Kingdom*

LAURENT LESCAUDRON • *Institute de Transplantation et de Recherché en Transplantation, CHU Hotel Dieu, Nantes and Service de Physiologie Animale et Humaine, Faculte des Sciences et des Techniques, University de Nantes, Nantes, France*

JIA-YI LI • *Section for Neuronal Survival, Wallenberg Neuroscience Center, Lund University, Lund, Sweden*

WALTER C. LOW • *Department of Neurosurgery, , Graduate Program in Neuroscience, University of Minnesota Medical School, Minneapolis, MN*

JUSTIN D. OH-LEE • *Brain Research and Integrative Neuroscience Center, Department of Psychology, Central Michigan University, Mt. Pleasant, MI*

XILMA R. ORTIZ-GONZALEZ • *Department of Neurosurgery, University of Minnesota Medical School, Minneapolis, MN*

JITKA OUREDNIK • *Department of Anatomy, School of Veterinary Medicine, Iowa State University, Ames, IA*

KOOK IN PARK • *Department of Pediatrics, Yonsei University School of Medicine, Seoul, Korea*

STEPHEN POLGAR • *School of Public Health, Faculty of Health Sciences, La Trobe University, Victoria, Australia*

ANNE E. ROSSER • *Department of Neurology, University of Wales College of Medicine, Cardiff, United Kingdom*

LAURENT ROYBON • *Section for Neuronal Survival, Wallenberg Neuroscience Center, Lund University, Lund, Sweden*

ROYA SABETRASEKH • *The Burnham Institute, La Jolla, CA, and Centre for Molecular Biology and Neuroscience, Institute of Basic Medical Sciences, University of Oslo, Oslo, Norway*

JACQUELINE SAGEN • *The Miami Project to Cure Paralysis, University of Miami School of Medicine, Miami, FL*

CYNDY DAVIS SANBERG • *Saneron CCEL Therapeutics, Inc., Tampa, FL*

PAUL R. SANBERG • *Center of Excellence for Aging and Brain Repair, University of South Florida, Tampa, FL*

STEPHEN J. M. SKINNER • *Diatranz NZLtd/Living Cell Technologies, New Zealand*

JOHN R. SLADEK, JR. • *Department of Psychiatry, University of Colorado Health Sciences Center, Aurora, CO*

EVAN Y. SNYDER • *The Burnham Institute, La Jolla, CA*

YANG D. TENG • *Department of Neurosurgery, Harvard Medical School, Boston, MA, and Veterans Administration Hospital, Roxbury, MA*

CHRISTOPHER G. THANOS • *Department of Molecular Pharmacology, Physiology and Biotechnology, Brown University, Providence, RI*

ALFRED VASCONCELLOS • *LCTBioPharma, Inc, Providence, RI*

MARTINA VENDRAME • *Department of Neurology, Temple University School of Medicine, Philadelphia, PA*

CATHERINE M. VERFAILLIE • *Graduate Program in Neuroscience, Stem Cell Institute, University of Minnesota Medical School, Minneapolis, MN*

ALISON E. WILLING • *Center of Excellence for Aging and Brain Repair, University of South Florida, Tampa, FL*

Color Plates

A color insert appears after page 240, and contains the following illustrations:

Current Views of the Embryonic and Neural Stem Cell

Cell Replacement or Molecular Repair?

Roya Sabetrasekh, Yang D. Teng,
Jitka Ourednik, Kook In Park, and Evan Y. Snyder

ABSTRACT

Stem cell biology, construed in its broadest sense, has forced Medicine to view development and disease, and subsequent potential therapies, from an entirely different perspective (1–3). We have learned that there is an inborn plasticity and flexibility "programmed" into the organism and its organ systems (1). The repository of this plasticity is thought to be the stem cell—the most primordial cell in the body and in any given structure. Nearly two decades ago, investigators began to identify cells with surprising plasticity and a propensity for dynamically shifting their fates within cultures obtained from developing and mature organs (1). The existence of such cells challenged the prevailing dogma that organs were rigidly and immutably constructed. Stem cells, as these plastic cells came to be termed, began to garner the interest of the developmental community, as well as that of the repair, gene therapy, and transplant communities. This interest arose when it was recognized that stem cells could be expanded in number and reimplanted into organs, where they would reintegrate appropriately and seamlessly, shift their fate in response to local cues to compensate for the absence of cells, express new genes, and in some cases, help promote functional improvement in disease models (4–13) (Fig. 1). Exploiting the power of a cell that presumably had a pivotal role in development for repair purposes is somewhat analogous to rebooting a computer or reseeding a lawn. Optimizing these natural processes is the primary focus of today's regenerative medicine. There have been a wide range of compelling studies conducted in animal models using various stem cells, including models of aging, spinal cord injury, stroke, parkinsonism, amyotrophic lateral sclerosis (ALS), cancer, multiple sclerosis, blood

From: *Contemporary Neuroscience: Cell Therapy, Stem Cells, and Brain Repair*
Edited by: C. D. Sanberg and P. R. Sanberg © Humana Press Inc., Totowa, NJ

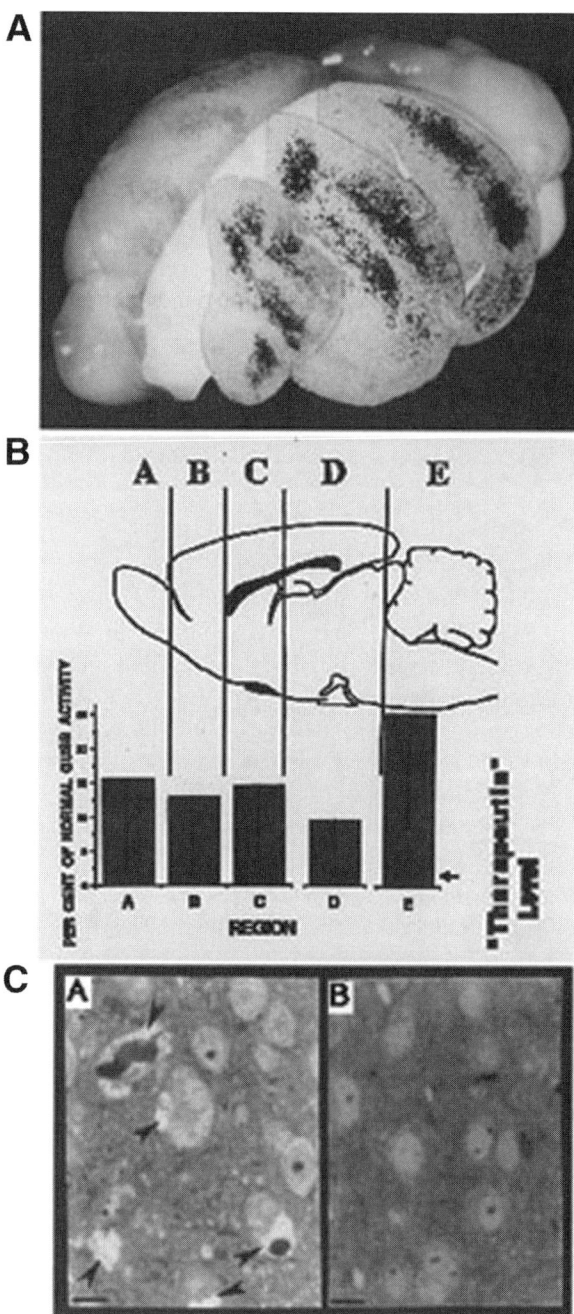

Fig. 1.

diseases, immunodeficiencies, enzyme deficiencies, myocardial infarcts, and diabetes.

This chapter outlines investigations that are capturing the most attention and considers the current state of scientific understanding and controversy regarding the properties of embryonic and neural stem cells. Its objective is to provide a framework to appreciate the medical promise of this research and also to describe the challenges of translating fundamental stem cell biology into novel clinical therapies.

Key Words: Embryonic; neural stem cells; MPS, tumor.

THERAPEUTIC POTENTIAL OF EMBRYONIC STEM CELLS

Most of the enthusiasm relating to embryonic stem (ES) cells results directly from the perceived need for cell replacement therapy for a host of degenerative diseases. Indeed, disorders of organ failure are not reversible, and organ transplantation cannot meet the needs of an ever-aging population. Primary pump failure in the heart, alcoholic or viral liver failure, β-cell–deficient type 1 diabetes, and Parkinson's disease (PD) are frequently cited as examples of monocellular deficiency states that might be amenable

---

Fig. 1. *(previous page)* Widespread engraftment of NSCs expressing GUSB throughout the brain of the MPS VII mouse. (Adapted from ref. 6). **(A)** Brain of a mature MPS VII mouse after receiving a neonatal intraventricular transplant of murine NSCs expressing GUSB. Donor NSC-derived cells, identified by their X-gal histochemical reaction (blue precipitate) for expression of the LacZ marker gene, have engrafted throughout the recipient mutant brain. Representative coronal sections (placed at their appropriate level by computer) show these cells to span the rostral-caudal expanse of the brain. **(B)** Distribution of GUSB enzymatic activity throughout the brains of MPS VII NSC transplant recipients. Serial sections were collected throughout the brains of transplant recipients and assayed for GUSB activity. Sections were pooled to reflect the activity present within the regions demarcated in the scheme. The regions were defined by anatomical landmarks in the anterior-to-posterior plane to permit comparison among animals. The mean levels of GUSB activity ($n = 17$) are presented as the percentage of average normal levels for each region. Untreated MPS VII mice show no GUSB activity biochemically or histochemically. Enzyme activity 2% of normal is corrective based on data from the liver and spleen. **(C)** Decreased lysosomal storage in a treated MPS VII mouse brain at 8 mo of age. (A) Extensive vacuolation representing distended lysosomes (arrowheads) in both neurons and glia in the neocortex of an 8-mo-old, untransplanted control MPS VII mouse. (B) Decrease in lysosomal storage in the cortex of an MPS VII mouse treated at birth from a region analogous to the untreated control section in part A. The other regions of this animal's brain showed a similar decrease in storage, compared to untreated, age-matched mutants in regions where GUSB was expressed. Scale bars: 21 μm.

to cell replacement strategies if a suitable and inexhaustible cell source could be found. Human ES cells may represent such a source, but the overriding challenge is to achieve efficient, directed differentiation of ES cells into therapeutically relevant cells, followed by proof-of-principle for effective restoration of tissue function in animal models. Once success is achieved in these areas of cell engineering many obstacles will remain, especially the immune barrier to tissue transplantation. Current strategies for confronting these challenges are outlined below.

ES cells are derived from the inner cell mass of the preimplantation embryo. When placed in culture, ES cells proliferate indefinitely and yet retain their potential to form all the tissues of the developing organism. Murine ES cells have been intensively studied for over 20 yr; yet, the first derivation of ES cells from the human embryo was reported in 1998 *(14)*. Although advances in ES cell biology have revolutionized the creation of mouse disease models, generating and breeding mice is time-consuming and costly. To address questions in cell and developmental biology, ES cells represent an excellent in vitro model system.

To maintain their undifferentiated, pristine state, ES cells are typically grown on feeder cell layers of mouse embryonic fibroblasts, and cultures are supplemented by the antidifferentiation cytokine leukemia inhibitory factor. When ES cells are removed from these conditions, they undergo spontaneous differentiation and initiate the diverse programs of tissue and cell specification *(15)*. ES cells recapitulate many developmental programs of the early embryo in culture, including cell generation from all three classical embryonic germ layers: ectoderm, mesoderm, and endoderm.

Because the mouse embryo remains microscopic and inaccessible during most of the first gestation week—a time when many central developmental programs are laid down—in vitro differentiation of ES cells provides a tractable model system to investigate the genetic and cell biological regulation of early development. Moreover, genes can be altered in ES cells through ectopic transgene expression or homologous recombination (HR), a process enabling specific sites in the genome to be altered directly. HR can be designed to produce gene deletion, mutation, or substitution. Using these techniques in ES cells, the genetic programs responsible for directing the development of blood, neurons, hepatocytes, cardiomyocytes, and a host of other tissues have been extensively explored.

Although ES cells have been touted as an inexhaustible resource for cell replacement therapies, they have already been proven as highly valuable research and discovery tools. By analyzing the effect of targeted gene deletions on the formation of specific cell lineages, ES cells can help validate potential therapeutic targets for small-molecule drug development. ES cells are emerging as a platform technology, around which chemical screens can be built, leading to the identification of compounds that promote or block cell differentiation. Schultz and colleagues *(16)* recently performed a chemical screen to identify agents that induce neurogenesis in ES cells, thereby establishing the proof-of-principle for using stem cell differentiation in assays for drug discovery. Kamp and colleagues *(17)* have shown that human ES cells differentiate into a number of cardiomyocyte classes, including embryonic atrial, ventricular, and nodal subtypes, each recapitulating their respective electrophysiologic properties and pharmacologic responses. Gauging the effects of compounds on the differentiation of specific cell populations from ES cells would provide a screen for potential drug toxicities prior to clinical development. The assembly of a genetically diverse bank of human ES cells, together with detailed knowledge of human genetic variation from the international haplotype mapping project ("the hap map"), could translate into a discovery platform for pharmacogenomics.

ES Cells as a Source of Neurons for Neurodegenerative Diseases

Despite inadequate knowledge about disease etiology and pathogenesis, neurodegenerative diseases (e.g., PD, Alzheimer's, Huntington's, amyotrophic lateral sclerosis, stroke, anoxic brain injury, and a host of lysosomal storage diseases with central nervous system (CNS) pathology) represent poorly managed diseases that are worthy targets for cell replacement therapy. The hippocampus and olfactory bulb maintain self-renewing populations of neural stem/progenitor cells, but there is scant evidence for cell renewal beyond these limited regions of the CNS. Given the likelihood that many classes of highly specialized neurons develop only during critical periods of embryogenesis, ES cells might, in principle, be directed to differentiate into specialized neuronal subtypes for use in cell replacement therapy. Several groups have reported success in differentiating specific neuronal subtypes from mouse and human ES cells, and some have reported positive data from transplantation of such cells into animal models of disease.

The most compelling reports of directed ES cell differentiation have derived from the laboratories of Jessell *(18)* and McKay *(19)*. Both have exploited knowledge of the morphogens and transcription factors that program neuronal development during embryogenesis and recapitulated the timing and sequence of exposure to direct neuronal patterning during in vitro ES cell differentiation. They initially used retinoic acid to program ectodermal commitment, then followed with exposure to the morphogen sonic hedgehog, which acts to "ventralize" neuronal subtypes. With these methods, Jessell and colleagues showed that they could pattern the formation of spinal motor neurons that successfully engrafted the embryonic spinal cord of the chick, extended axons, and formed synapses with target muscles. McKay's group exploited the instructive effects of sonic hedgehog and FGF8 to drive the commitment of ES cells to ventral midbrain fates and, ultimately, to tyrosine hydroxylase–positive dopaminergic neurons. These cells functioned after transplantation into a rodent model of PD. Isacson and colleagues *(20)* have similarly demonstrated improvement in a rodent model of PD from transplantation of undifferentiated ES cells into the striatum, suggesting that the local environment is capable of inducing proper development of dopaminergic neurons. Introduction of neuronal populations of differentiated murine ES cells into a rat model of spinal cord injury has shown improved motor function *(21)*, but it is by no means clear that the mechanism was direct neuronal reconstitution, as opposed to modulation of the repair process in the host through remyelination. Several research groups have demonstrated neuron formation from human ES cells *(22–25)*, presaging future human applications.

Despite these apparent successes, there may be as many as 200 distinct neuronal subtypes in the adult brain, and even within a given subtype, neurons show remarkable degrees of regional specificity. A great leap of faith is required to believe that neurons produced from ES cells in culture will recapitulate the differentiated features of specific neuronal subtypes and reestablish relevant neural networks produced during the formation of the embryonic brain. An alternative strategy is to differentiate ES cells into neural stem cells (NSC) and progenitors in vitro, then coax local environments in the diseased region of the brain or spinal cord to direct further differentiation and accommodation of neural cells to their new niche. Whether this will occur is a matter of pure speculation and is subject to hyperbolic claims. The applications of NSCs are described in detail below.

THERAPEUTIC POTENTIAL OF NSCS

The recognition that NSCs propagated in culture could be reimplanted into mammalian brain, where they could reintegrate appropriately and stably express foreign genes *(4,26)* made this strategy an appealing alternative for CNS gene therapy and repair. Numerous subsequent studies over the past decade *(27)* reaffirmed that neural progenitors from many regions and developmental stages could be maintained, perpetuated, and passaged in vitro by epigenetic, and genetic methods. Examples include the transduction of genes interacting with cell cycle proteins (e.g., *vmyc*) and by mitogen stimulation (e.g., epidermal growth factor and/or basic fibroblast growth factor; *(4,26,28–32)*. Some of these methods may operate through common cellular mechanisms. This speculation is supported by the observation that many progenitor cell lines behave similarly in their ability to reintegrate into the CNS, despite that they were generated by different methods, obtained from various locations, and reimplanted into various CNS regions. Some NSC lines appear sufficiently plastic to participate in normal CNS development from germinal zones of multiple regions along the neuraxis and at multiple stages of development from embryo to old age *(4,6,9,33–37)*. In addition, they appear to model the in vitro and in vivo behavior of some primary fetal and adult neural cells *(38–43)*, suggesting that insights gleaned from these NSC lines may legitimately reflect the potential of CNS progenitor or stem cells.

The inherent biologic properties of NSCs may circumvent limitations of other techniques for treating metabolic, degenerative, or other widespread lesions in the brain. They are easy to administer (often directly into the cerebral ventricles), are readily engraftable, and circumvent the blood–brain barrier. Unlike BMT, a preconditioning regime is not required before administration (e.g., total-body irradiation). One important property of NSCs is their apparent ability to develop into integral cytoarchitectural components *(47)* of many regions throughout the host brain as neurons, astrocytes, oligodendrocytes, and even incompletely differentiated, but quiescent, progenitors. Therefore, they may be able to replace a range of missing or dysfunctional neural cell types. A given NSC clone can give rise to multiple cell types within the same region. This is important in the likely situation where return of function may require the reconstitution of the whole milieu of a given region, e.g., not just the neurons but also the glia, and support cells required to nurture, detoxify, and/or myelinate the neurons. They appear to respond in vivo to neurogenic signals not only when they occur appropriately during development, but even when induced at later stages by certain neurodegenerative processes, like during apoptosis *(4,45)*. NSCs may be attracted to regions of neurodegeneration in the young, as well as in the elderly *(12,46–48*; Fig. 2).

NSCs also appear to accommodate to the engraftment region, perhaps obviating the necessity to obtain donor cells from many specific CNS regions or the imperative for precise targeting during reimplantation. The cells might express certain genes of interest intrinsically (e.g., many neurotrophic factors), or they can be engineered ex vivo to do so because they are readily transduced by gene transfer vectors. These gene products can be delivered to the host CNS in a direct, immediate, and stable manner *(6,7,47,49)*. Although NSCs can migrate and integrate widely throughout the brain

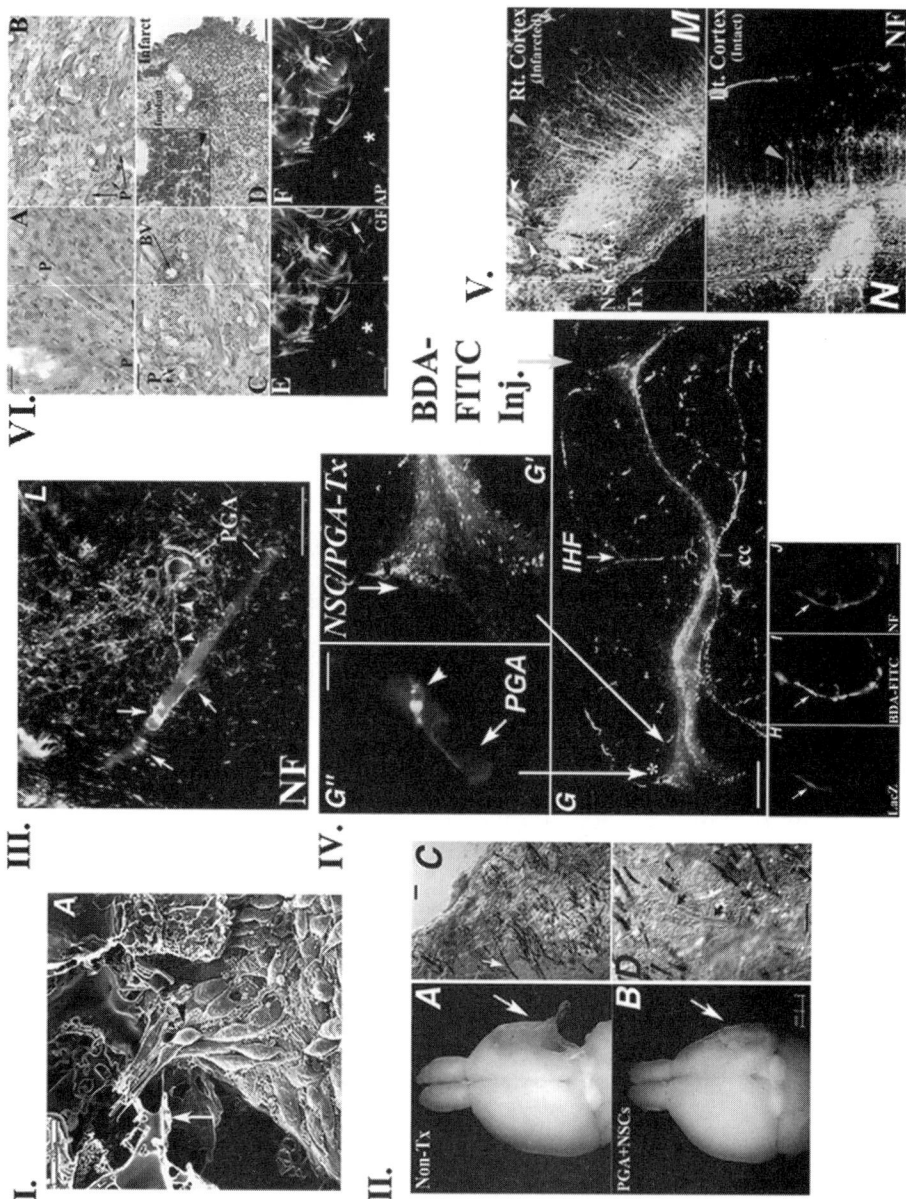

Fig. 2. The injured brain interacts reciprocally with NSCs supported by scaffolds to reconstitute lost tissue—evidence from hypoxic-ischemic (HI) injury. (Modified from ref. 8.)

(I) Characterization of NSCs in vitro when seeded upon a PGA scaffold. Cells seen with scanning electron microscopy at 5 d after seeding were able to attach to, impregnate, and migrate throughout a highly porous PGA matrix (arrow). The NSCs differentiated primarily into neurons (>90%) that sent out long, complex processes that adhered to, enwrapped, and interconnected the PGA fibers.

(II) Implantation of NSC–PGA complexes into a region of cavity formation following extensive HI brain injury and necrosis. (A) Brain of an untransplanted (non-Tx) mouse subjected to right-HI injury with extensive infarction and cavitation of the ipsilateral right cortex, striatum, thalamus, and hippocampus (arrow). In contrast with part B, the brain of a similarly injured mouse implanted with an NSC–PGA complex (PGA+NSCs) (generated in vitro as per part I. into the infarction cavity 7 d after the induction of HI (arrow) (n = 60). At maturity (age-matched to the animal pictured in part A), the NSC–scaffold complex appears in this whole-mount to have filled the cavity (arrow) and become incorporated into the infracted cerebrum. Representative coronal sections through that region are seen at higher magnification in parts C and D, in which parenchyma appears to have filled in spaces between the dissolving black polymer fibers (white arrow in part C) and, as seen in part D, even to support neovascularization by host tissues. (Blood vessel is indicated by closed black arrow in part D; open arrow in part D points to degrading black polymer fiber.) Scale bars (C and D): 100 μm.

(III) Characterization in vivo of the neural composition of NSC–PGA complexes within the HI-injured brain. At 2 wk following transplantation of the NSC–PGA complex into the infarction cavity, donor-derived cells showed robust engraftment within the injured region. An intricate network of multiple long, branching NF+ (green) processes were present within the NSC–PGA complex and its parenchyma enwrapping the PGA fibers (orange autofluorescent tube-like structures under a Texas Red filter), adherent to and running along the length of the fibers (arrows), often interconnecting and bridging the fibers (arrowheads). Those NF+ processes were of both host and donor derivation. In other words, not only were donor-derived neural cells present, but host-derived cells also seemed to have entered the NSC–PGA complex, migrating and becoming adherent to the PGA matrix. In a reciprocal manner, donor-derived (lacZ+) neurons (NF+ cells) within the complex appeared to send processes along the PGA fibers out of the matrix into host parenchyma, as seen in part IV. Scale bars: 100 μm. (*Figure 2 caption is continued on the next page.*)

9

Fig. 2. (*continued*) (**IV**) Long-distance neuronal connections extend from the transplanted NSC–PGA complexes in the HI-injured brain toward presumptive target regions in the intact contralateral hemisphere. By 6 wk following engraftment, donor-derived *lacZ*+ cells appeared to extend many exceedingly long, complex NF+ processes along the length of the disappearing matrix, apparently extending into host parenchyma. To confirm the suggestion that long-distance processes projected from the injured cortex into host parenchyma, a series of tract-tracing studies were performed. [G–G″] BDA-FITC was injected (G) into the contralateral intact cortex and external capsule (green arrow) at 8 wk following implantation of the NSC–PGA complex into the infarction cavity (NSC/PGA-Tx). Axonal projections (labeled green with fluorescein under an FITC filter) are visualized (via the retrograde transport of BDA), leading back to (across the interhemispheric fissure (IHF) via the corpus callosum ['cc″]), and emanating from, cells in the NSC–PGA complex within the damaged contralateral cortex and penumbra (seen at progressively higher magnification in parts G′ (region indicated by arrow to part G) and G″ (region indicated by an arrow and asterisk in part G). In part G″, the retrogradely BDA-FITC–labeled perikaryon of a representative neuron adherent to a dissolving PGA fiber is well visualized. The fact that such cells are neurons of donor derivation is supported by their triple labeling (H–J) for *lacZ* (H) (βgal), BDA-FITC (I), and the neuronal marker NF (J); arrow in (H–J), indicates the same cell in all three panels). Such neuronal clusters were never seen under control conditions, i.e., in untransplanted cases or when the vehicle, or even an NSC suspension unsupported by scaffolds, was injected into the infarction cavity. Scale bars: (G) 500 μm; (G″) 20 μm; (H–J) 30 μm.

Fig. 2. (*continued*) (**V**) Adverse secondary events that typically follow injury (e.g., monocyte infiltration and astroglial scar formation) are minimized by and within the NSC–PGA complex. (A–D) Photomicrographs of H&E-stained sections prepared to visualize the degree of monocyte infiltration in relation to the NSC–PGA complex and the injured cortex 3 wk following implantation into the infarction cavity. Monocytes are classically recognized under H&E as very small cells with small round nuclei and scanty cytoplasm (e.g., inset in part D, arrowhead). Although some localized monocyte infiltration was present immediately surrounding a blood vessel (BV in part C, arrow) that grew into the NSC–PGA complex from the host parenchyma, there was little or no monocyte infiltration either in the center of the NSC–PGA complex (B) or at the interface between the NSC–PGA complex and host cortical penumbra (A). This is in stark contrast to the excessive monocyte infiltration seen in an untransplanted infarct of equal duration, age, and extent (D), the typical histopathologic picture otherwise seen following HI brain injury (see inset, a higher magnification of the region indicated by the asterisk in part D; a typical monocyte is indicated by the arrowhead). Whereas neural cells (nuclei of which are seen in A–C) adhere exuberantly to the many polymer fibers (P in parts A–C), monocyte infiltration was minimal, compared to that in part D. (E,F) Astroglial scarring (another pathological condition confounding recovery from ischemic CNS injury) is also kept among the cell types into which NSCs differentiated when in contact with the PGA fibers, there was minimal astroglial presence either of donor or host origin *away* from the fibers (*). (E) GFAP immunostaining that is recognized by a fluorescein-conjugated secondary antibody (green) is observed. Note little scarring in the regions indicated by the asterisk. Under a Texas red filter (F) (merged with the fluorescein filter image), the tube-like PGA fibers (arrowhead in both panels) become evident (as autofluorescent orange), and most of the donor-derived astrocytes (arrows) (yellow because of their dual *lacZ* and GFAP immunoreactivity) are seen to be associated with these fibers, again leaving most regions of the infarct (*) astroglial scar-free. (Arrows in parts E and F point to the same cells.) Far from creating a barrier to the migration of host- or donor-origin cells, or to the ingrowth/outgrowth of axons of host- or donor-origin neurons (as per parts III and IV), NSC-derived astrocytes may have helped provide a facilitating bridge. Scale bars: (A) 10 μm; (C,D) and (E,F) 20 μm.

11

particularly well when implanted into germinal zones, allowing reconstitution of enzyme or cellular deficiencies in a global manner *(6,7,47)*, this extensive migratory ability is present even in the parenchyma of the diseased adult and aged brain *(6,7,50)*. Despite their extensive plasticity, NSCs never give rise to cell types inappropriate to the brain, such as muscle, bone, teeth, or yield neoplasms.

These attributes of NSCs may provide multiple strategies to treat CNS dysfunction. As proof of principle, they were first tested experimentally in mouse models of genetically based neurodegeneration. Their ability to mediate gene therapy was affirmed in a model of the neurogenetic lysosomal storage disease (LSD)—mucopolysaccharidosis type VII (MPS VII; *6*). Mice homozygous for a frameshift mutation in the β-glucuronidase gene are devoid of the secreted enzyme β-glucuronidase (GUSB). The enzymatic deficiency results in lysosomal accumulation of undegraded glycosaminoglycans in the brain and other tissues, causing a fatal progressive degenerative disorder. Treatments for MPS VII and most other LSDs are designed to provide a source of normal enzymes for uptake by diseased cells—a process termed *cross-correction (51)*. The goal of ex vivo gene therapy is to engineer donor cells to express the normal GUSB protein for export to other host cells. The engraftment and integration of GUSB overexpressing NSCs throughout the newborn MPS VII mutant brain succeeded in providing a sustained, lifelong, widespread source of cross-correcting enzyme in a manner not previously achieved *(6)*.

A rapid intraventricular injection technique was devised for the diffuse engraftment of the NSCs. Injecting the progenitors into the cerebral ventricles presumably allowed them to gain access to most of the subventricular germinal zone (SVZ), as well as to networks of cerebral vasculature, along the surface of which they would also migrate. This approach worked equally well in the fetus, where donor NSCs gained access to the ventricular germinal zone *(47)*, migrating into the parenchyma within 24–48 h. This engraftment technique, exploiting many of the inherent properties of NSCs, permitted missing gene products to be delivered without disturbing other neurobiological processes and was a potential strategy for gene therapy of a class of neurogenetic diseases that had not been adequately treated thus far (Fig. 3A). Although MPS VII may be regarded as "uncommon," the broad category of diseases that it models (neurogenetic conditions) afflicts as many as 1 in 1500 children and serves as a model for many adult neurodegenerative processes of genetic origin. (Alzheimer's disease could broadly fall into this category.) Therapy instituted early in life might arrest disease progression and prevent irreversible CNS alterations. Even in the adult brain, there are routes of relatively extensive migration followed by both endogenous and transplanted NSCs *(52,53)*. If injected into the cerebral ventricles of normal adult mice, NSCs (including those expressing transgenes) will integrate into the SVZ and migrate long distances, e.g., to the olfactory bulb, where they differentiate into interneurons, and occasionally into subcortical paren-chyma, where they become glia *(9,34,35,54,55)*. These migratory paths are still relatively restricted and stereotyped, compared to that seen in the fetal or newborn brain. However, in the degenerating, abnormal, or injured adult brain (as discussed below), migration by foreign gene-expressing NSCs can be extensive and directed specifically to regions of pathology—a phenom-enon observed with stroke, head injury, dopaminergic dysfunction, brain tumors, and amyloid plaques.

The therapeutic paradigm described above can be extended to other untreatable neurodegenerative diseases that are characterized by an absence of gene products and/or the accumulation of toxic metabolites. In almost all cases, NSCs, because they are normal cells, constitutively express normal amounts of the particular enzyme in question. The extent to which this amount needs to be augmented may vary from model to model and enzyme to enzyme. Reassuringly, in most inherited metabolic diseases, the amount of enzyme required to restore normal metabolism and forestall CNS disease may be quite small. It is significant to note that while the histograms in Fig. 1B illustrate the widespread distribution of a lysosomal enzyme, they could similarly reflect the NSC-mediated distribution of other diffusible (e.g., synthetic enzymes, neurotrophins, viral vectors; *56,57*) and nondiffus-ible (e.g., myelin, extracellular matrix) factors, as well as the distribution of "replacement" neural cells (see the following section). For example, neural progenitors and stem cells have been used for the local expression of NT-3 within the rat spinal cord, nerve growth factor and brain-derived neu-rotrophic factor within the septum, and tyrosine hydroxylase, Bcl-2, and glial cell–derived neurotrophic factor to the striatum *(58–65)*. These earlier stud-ies helped to advance the idea that NSCs, as a prototype for stem cells from any solid organ, might aid in reconstructing both the molecules, along with the cells of a maldeveloped or damaged organ. A further complexity, how-ever, is the recognition that the same NSC may not be able to be engineered to express certain neurotrophic agents simultaneously, because they may be processed antagonistically within the cell and/or within the environment. Therefore, a greater knowledge of the NSC processing of certain molecules is a prerequisite *(66)*.

I.

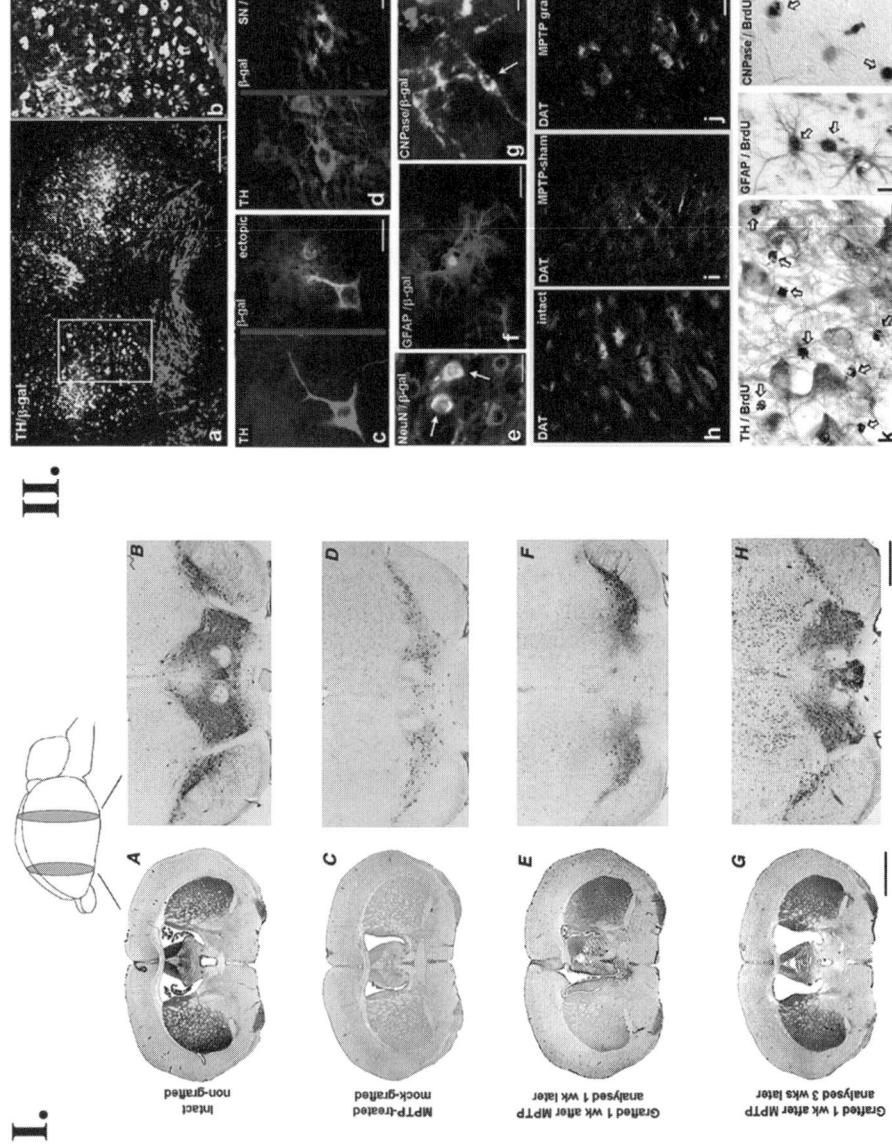

II.

14

Fig. 3. NSCs possess an inherent mechanism for rescuing dysfunctional neurons: evidence from the effects of NSCs in the restoration of mesencephalic dopaminergic function. (Modified from ref. 12.)

(**I**) TH expression in mesencephalon and striatum of aged mice following MPTP lesioning and unilateral NSC engraftment into the substantia nigra/ventral tegmental area (SN/VTA). A model that emulates the slow dysfunction of aging dopaminergic neurons in the SN was generated by giving aged mice high doses of MPTP repeatedly. Scheme (top) indicates the levels of analyzed transverse sections along the rostrocaudal axis of the mouse brain. Representative coronal sections through the striatum are presented in the left column (A,C,E,G) and through the SN/VTA area in the right column (B,D,F,H). (A,B) Immunodetection of TH (black cells) shows the normal distribution of DA-producing TH+ neurons in coronal sections in the intact SN/VTA (B) and their projections to the striatum (A). (C,D) Within 1 wk, MPTP treatment caused extensive and permanent bilateral loss of TH immunoreactivity in both the mesostriatal nuclei (C) and the striatum (D), which lasted lifelong. Shown in this example, and matching the time point in (G,H), is the result of 4-wk MPTP treatment in a mock-grafted animal. (E,F) Unilateral (right side) stereotactic injection of NSCs into the nigra is associated (within 1 wk after grafting) with substantial recovery of TH synthesis within the ipsilateral DA nuclei (F) and their ipsilateral striatal projections (E). By 3 wk posttransplant, however (G,H), the asymmetric distribution of TH expression disappeared, giving rise to TH immunoreactivity in the midbrain (H) and striatum (G) of both hemispheres that approached that of intact controls (A,B) and gave the appearance of mesostriatal restoration. Similar observations were made when NSCs were injected 4 wk after MPTP treatment (not shown). Bars: 2 mm (left), 1 mm (right). Note the ectopically placed TH+ cells in part H. These are analyzed in greater detail, along with the entire SN in part **II**. (*Figure 3 caption is continued on the next page.*)

Fig. 3. (*continued*) **(II)** Immunohistochemical analyses of TH, DAT, and BrdU-positive cells in MPTP-treated and grafted mouse brains. The presumption was initially that the NSCs had replaced the dysfunctional TH neurons. However, examination of the reconstituted SN with dual β-gal (green) and TH (red) ICC showed that (a,c) 90% of the TH+ cells in the SN were host-derived cells, which had been rescued and were only 10% donor-derived (d). Most NSC-derived TH+ cells were just above the SN ectopically (blocked area in part a, enlarged in part b). These photomicrographs were taken from immunostained brain sections from aged mice exposed to MPTP, transplanted 1 wk later with NSCs, and sacrificed after 3 wk. The following combinations of markers were evaluated: TH (red) with β-gal (green) (a–d); NeuN (red) with β-gal (green) (e); GFAP (red) with β-gal (green) (f); CNPase (green) with β-gal (red), as well as TH (brown) and BrdU (black) (k); GFAP (brown) with BrdU (black) (l); and CNPase (brown) with BrdU (black) (m). Anti-DAT–stained areas are revealed in green in the SN of intact (h), mock-grafted (i), and NSC-grafted (j) brains. Three different fluorescence filters specific for Alexa Fluor 488 (green), Texas Red (red), and a double filter for both types of fluorochromes (yellow) were used to visualize specific antibody binding. Parts c, d, and h–j are single-filter exposures; a,b, and e–g are double-filter exposures. Part a shows a low-power overview of the SN+VTA of both hemispheres, similar to the image in part H of Fig. 2. The majority of TH+ cells (red cells in part a) within the nigra are actually of host origin (~90%), with a much smaller proportion of donor derivation (green cells) (~10%) (representative close-up of such a donor-derived TH+ cell in part d). Although a significant proportion of NSCs did differentiate into TH+ neurons, many of these resided ectopically, dorsal to the SN (boxed area in part a, enlarged in b; high-power view of donor-derived (green) cell that was also TH+ (red) in part c), where the ratio of donor-to-host cells was inverted: ~90% donor-derived vs ~10% host-derived. Note the almost complete absence of a green β-gal–specific signal in the SN/VTA whereas ectopically, many TH+ cells were double-labeled and thus NSC-derived (appearing yellow-orange in higher power under a red/green double-filter in panel b). (c–g) NSC-derived non-TH neurons (NeuN+) (e, arrow), astrocytes (GFAP+) (f), and oligodendrocytes (CNPase+) (g, arrow) were also seen, both within the mesencephalic nuclei and dorsal to them. (h–j) The green DAT-specific signal in panel j suggests that the reconstituted mesencephalic nuclei in the NSC-grafted mice (as in panel I–h were functional DA neurons, comparable to those seen in intact nuclei (h), but not in MPTP-lesioned sham-engrafted controls (i). This further suggests that the TH+ mesostriatal DA neurons affected by MPTP are indeed functionally impaired. (Note that sham-grafted animals (i) contain only punctate residual DAT staining within their dysfunctional fibers, whereas DAT staining in normal (h) and, similarly in engrafted (j) animals, was normally and robustly distributed within processes and throughout their cell bodies.) (k–m) Any proliferative BrdU+ cells after MPTP insult and/or grafting were confined to glial cells, whereas the TH+ neurons (k) were BrdU–. This finding suggested that the reappearance of TH+ host cells was not the result of neurogenesis, but rather the recovery of extant host TH+ neurons. Bars: 90 μm (a); 20 μm (c,d,e); 30 μm (f); 10 μm (g); 20 μm (h–j); 25 μm, (k); 10 μm, (l); and 20 μm (m).

16

Therapeutic Uses of NSCs Against Brain Tumors

In 2000, Aboody et al. reported that transplanted exogenous murine and human NSCs were capable of "homing in" over long distances onto intra-cerebral xenogeneic brain tumors deposited into rodent brains (67). The authors also demonstrated the ability of NSCs to "track" tumor cells escaping from the original inoculated tumor mass and invading the normal brain. Specially tagged NSCs were found adjacent to invading tumor cells that appeared to be infiltrating normal brain along white-matter tracts, tumor-related endot-helium, and interstitial spaces. There appeared to be a particular predilec-tion for tumor-associated endothelium. The NSCs could be introduced either intraparenchymally or into the lateral ventricles with equal homing ability. Whether introduced into the ipsilateral or contralateral cerebral ventricles, or into the ipsilateral or contralateral cerebral parenchyma, NSCs were able to migrate toward the implanted tumor and appose themselves intimately to escaping, infiltrating tumor cells, even at far distances from the main tumor mass. Indeed, NSCs injected into the tail vein demonstrated successful intracranial tumor "homing" (establishing a paradigm, since it is used by some investigators for other intracranial pathologies, e.g., models of mul-tiple sclerosis; 68). The blood–brain barrier normally acts as an efficient barrier to the successful delivery of many therapeutic agents and dictates the limited repertoire of brain tumor chemotherapeutic agents (69,70). In this instance, the blood–brain barrier did not appear to affect the migration and homing of NSCs toward intracranial pathology, as modeled by the neoplasms.

This "gliomatropic" capability of NSCs bodes well for their use in clini-cal applications. NSCs could serve as tumor-directed homing devices expressing a variety of antitumor genes, including those encoding cytotoxic, antiangiogenic, antimitotic, antimigratory, immunomodulatory, prodiffer-entiating, and/or proapoptotic agents. Ultimately, the goal is for NSCs to be used in conjunction with, and to optimize, other therapeutic modalities by providing the ability to track invading cells—the bane of most gene, radia-tion, surgical, and pharmacological strategies—and complete eradication of the glioma cells (71). Aboody et al. tested this hypothesis with the use of cytosine deaminase (CD)–expressing murine NSCs. This enzyme converts the nontoxic precursor 5-fluorocytosine (5-FC) into the toxic compound 5-fluorouracil. Injection of 5-FC into tumor-bearing mice that had been inoculated with NSCs engineered to express CD was followed by the dra-matic reduction of intracranial tumor burden. CD has a particularly impres-sive bystander effect. Barresi et al. replicated this phenomenon by showing in vivo regression of implanted C6 glioma cells after coinoculation of an immortalized neural progenitor cell line ST14A expressing CD, followed by administration of 5-FC *(72)*. This work gave further credence to the utility of exploiting NSCs to effect the well-established, but hitherto unsuccessful, "prodrug/prodrug–converting enzyme" strategy to destroy intracerebral malignancies.

The virtue of the prodrug approach is twofold. First, the NSC has to kill only a small percentage of tumor cells to have a large impact on other tumor cells as a result of the bystander effect. Like an exploding hand grenade, dying tumor cells send out toxic factors or signals that are inimical to many surrounding tumor cells. CD's action on 5-FC creates one the greatest bystander effects of all the prodrug-converting enzymes. Second, although no adverse effects or contributions to tumor growth from NSCs have ever been detected in these models, if such an unlikely situation arises, the CD within the NSC would cause it to self-eliminate if it were to become mitotic, providing a built-in safety mechanism.

Extending the use of NSCs to deliver cytolytic genes, and based on the report by Lynch et al. that NSCs could be engraftable, mobile intracranial viral-packaging lines *(73)*, Herrlinger et al. showed that murine NSCs, engineered to release replication-conditional herpes simplex virus (HSV) thymidine kinase (TK), were efficient in the destruction of an intracranial tumor mass, as well as isolated, escaping tumor microdeposits *(74)*. In preliminary studies, William Weiss and colleagues have validated the gliomatropism of murine NSCs in a transgenic oligodendroglioma-bearing mouse model previously described. This double transgenic mouse is the product of mating a transgenic mouse in which a mutant epidermal growth factor receptor is transcribed from the S100 promoter (S100 β-v-*erb*B) with an INK4aNull mouse, yielding progeny that succumb to tumors by 6 mo of age. Investigators showed that NSCs homed in on spontaneously developing tumors. With the use of CD-expressing NSCs (clone C17.2) in pilot studies, host survival improved. Because the NSCs seemed to "find" even small "subclinical" tumors, whose presence was unsuspected by the investigators' prehistological examination, NSCs may also be used in tumor diagnosis if armed with tags that can be imaged in the living state (see below). It also warrants mentioning that transgenic mice that spontaneously develop brain tumors appear to provide models for studying tumor biology and therapies, including the glioma-specific migratory potential of NSCs, which far more faithfully emulate the true clinical situation in humans and are therefore superior to older models that depend on the artificial implantation of glioma cell lines.

Immunomodulation through the judicious use of cytokines has been shown to be extremely effective in the destruction of experimental brain tumors via either direct cytotoxicity or the activation of immune-mediated antitumor effects *(75–78)*. Benedetti and colleagues showed that intra-tumoral injection of interleukin 4 (IL-4)–expressing murine NSCs effected radiographic tumor regression and prolongation of host survival *(79)*. In addition, the authors detected the presence of NSCs several weeks postinjection in the recipient animals. This evidence implied persistence of implanted NSCs and possibly persistence of antitumor effects in vivo. The authors also described an inherent tumor inhibitory effect exhibited by the injected stem cells. This phenomenon was first observed many years ago using murine NSC clone C17.2. Gliastatin, a membrane-associated factor isolated from these NSCs, was capable of converting C6 glioma cells to cells resembling phenotypically normal astrocytes simply via cell contact *(80)*.

Exploiting the specific homing capability of NSCs to intracranial tumors, Ehtesham et al. transfected murine neural progenitors ex vivo with the IL-12 gene using an adenoviral vector *(81)*. Stable expression of the IL-12 protein was demonstrated in vitro and in vivo. Implantation of IL-12 expressing NSCs into tumor-bearing syngeneic mice effected tumor destruction and improved host survival. In addition, the authors showed enhanced tumor infiltration by T lymphocytes as a result of IL-2 expression in close proximity to the tumor mass. In a separate study, the authors transfected similarly derived murine neural progenitors with the human tumor necrosis factor–related apoptosis-inducing ligand (TRAIL) gene using a replication-deficient adenoviral vector *(82)*. TRAIL belongs to the tumor necrosis factor superfamily of proapoptotic proteins, which has been previously shown to induce apoptosis in experimental tumor models *(83)*. Introduction of TRAIL-expressing neural stem/progenitor cells into nude mice bearing gliablastoma multiforme xenografts was followed by the eradication of tumor via induction of apoptotic cell death.

FUTURE DIRECTIONS AND APPLICATION OF NSCS IN BRAIN TUMORS

The Therapeutic Package: New Candidates

NSCs offer a way to deliver a broad range of therapeutic molecules to intracranial glioma cells in a specific fashion. Immunomodulatory factors hold promise. Ehtesham et al. have already shown the successful and efficacious delivery of IL-12 and TRAIL to implanted brain tumors *(80,81)*. Other cytokines with demonstrated promise in the treatment of glioma could be similarly delivered by NSCs *(84)*. For example, IL-24 or melanoma differentiation–associated-7 induces the expression of the proapoptotic protein BAX and suppresses the growth of glioma cells *(85)*. Aoki et al. showed that IL-7 reduced the tumorigenicity of glioma cells in vivo via a T-cell–mediated mechanism *(86)*. Parker et al. also showed immune-mediated suppression of in vivo glioma with IL-12 delivered by an HSV vector *(87)*. Utilizing NSCs as gliomatropic cytokine/drug delivery vehicles would likely increase the therapeutic potential of these agents significantly.

NSCs can also prolong the bioavailability of a therapeutic agent in vivo. For example, the potent antiangiogenic molecule endostatin has a relatively short half-life in vivo, and intracranial bioavailability is restricted by the blood–brain barrier. Encapsulation of cells from an embryonal kidney cell line, engineered to produce endostatin in alginate gel beads, significantly prolonged the sustained release of the protein intracranially, resulting in a measurable reduction in tumor-related neoangiogenesis *(88)*. NSCs transfected with the endostatin gene or genes for other antiangiogenic molecules can likely serve a similar role. Yet, they have the added advantage of a migratory delivery system that is specific for glioma and tumor-related endothelium, is transcriptionally and translationally active for several weeks postintroduction, and can track invading cells throughout the brain that might set up new tumor foci.

Herrlinger et al. reported the use of NSCs in the focused delivery of engineered HSV-expressing TK into implanted brain tumors *(56)*. The authors delayed cell cycle entry and hence viral replication in the NSCs with mimosine prior to transplantation into the tumor-bearing rodent host. However, this had no deleterious effects on the migratory potential of the NSCs. Once the cytostatic effect of mimosine had expired, viral production reinitiated in close proximity to the tumor bulk, as well as escaping glioma cells, with effective tumor destruction following the administration of gangcyclovir to trigger the TK-mediated cytolysis.

Oncolytic viruses have recently garnered much attention in neuro-oncology *(89)*. There are multiple naturally occurring and engineered viruses being considered for glioma-specific treatment *(90)*. Taking advantage of the prevalence of *ras* mutations in brain tumors, reovirus has been shown to affect selective oncolysis of glioma and medulloblastoma cells *(91,92)*. Clinical trials are ongoing to assess the effectiveness of the virus in patients with gliablastoma multiforme. Advances in the understanding of the molecular genetics of vesicular stomatitis virus (VSV) and isolation of unique attenuated strains (AV1 and AV2) have recently propelled the field of tumor oncolytic therapy forward *(92)*. VSV AV1 and AV2 affect tumor-specific lysis by taking advantage of the abrogation of the interferon α/β (interferon α/β)–signaling pathway in tumor cells that endow them with growth independence, yet make them susceptible to viral infection and oncolysis. These attenuated VSV strains have favorable therapeutic indices and demonstrate selective killing of cancer cells (including CNS tumors) from the NCI-60 tumor cell panel. Can we exploit NSCs as carriers of these lethal viruses and enhance their delivery to infiltrative glioma cells? This is an intriguing concept and has yet to be proven. Assuming NSCs have none of the genetic mutations that endow the glioma cells with growth advantage (but also susceptibility to viral-mediated oncolysis), then NSCs should be able to survive viral infection and deliver the lethal package to glioma cells. It is beyond the scope of this chapter to provide a comprehensive and up-to-date review on novel glioma therapeutics, but importantly, NSCs offer a way to effectively deliver therapeutically relevant molecules and drugs to diffuse glioma in vivo.

A ROADMAP TO THE CLINIC

What is preventing imminent translation to the clinic? The answer, most simply put, is that the safest, most practical, most efficacious methods for exploiting stem cell biology need to be determined. With regard to applications to diseases of aging, we suggest the following cautions.

Understand the disease. It is not possible to "ask" the stem cell to "fix" a disease without truly understanding what that entails: what pathological process needs to be blunted and/or what cell type or gene needs to be replaced? Alzheimer's disease is the "poster child" in this regard—what the stem cell should do is not known. For many diseases, we have been unduly presumptuous in concluding what cell type is needed, where the true locus of the disease resides, and what's required to reconstitute a given region and to restore function. Diseases may actually need *multiple* cell types to reconstruct a milieu, not just the cells that have died but also the support cells that "chaperone" them by providing ongoing nutritional and detoxification support. This realization has emerged for ALS, PD, stroke, spinal trauma, diabetes, and myocardial infarction. Furthermore, abnormalities are dynamic; a given disease may have different needs at various times within the same patient. In addition, when investigating a disease, it is important to know if the underlying process is cell autonomous or cell nonautonomous, i.e., to be aware that even new cells may die unless adequately protected in advance.

Understand the stem cell. Is the regenerative task at hand within the biological "skill set" of the stem cell? This question takes on prominence when asking whether a stem cell from one organ system (e.g., blood) can transdifferentiate into a cell from another organ system (e.g., brain). The evidence, when examined *critically* to date, actually suggests that this probably *cannot* happen, or at least not with sufficient efficiency. To direct stem cells to yield cells of the right type(s) and number, in the right ratio, in the right location, with the right connections and the appropriate partners, without making any wrong connections, and to shield nontargeted cells and regions from such influences presupposes a level of understanding that still needs to be acquired in many cases.

Understand what the stem cell is doing. As noted above, because it is now becoming understood that stem cells intrinsically produce many growth, protective, and anti-inflammatory factors, the improvement observed in some recipients is likely not owing to replacement of a damaged cell alone (if at all), as had been the original assumption, but rather to the provision by stem cells of factors promoting the survival and enhanced function of *host* cells and their connections. The emerging recognition of a dynamic reciprocal stem cell–host interaction adds another level of complexity but also a powerful new avenue for stem cell–mediated recovery. Exploiting and optimizing it, however, requires a better understanding of exactly which mechanisms are mediating the observed effects.

Suit the type of stem cell to the purpose. Of the various putative stem cell sources—inner cell mass, gonadal ridge, marrow, blood, muscle, brain—which one is best for any particular therapeutic challenge? These decisions can be made only empirically, by comparing stem cells head-to-head under the same conditions, in the same experimental models, employing the same metrics, and balancing their respective advantages and disadvantages. A disease of a particular organ system is probably most efficiently treated with stem cells from that organ (e.g., for the CNS, a NSC or an ES cell directed to become a NSC). Given that, how does one then decide between using an ES cell directed down a given lineage and using a somatic stem cell derived directly from that structure? Again, the answer may vary from structure to structure and from disease to disease. Starting with an ES cell insures an unlimited supply of starting materials but requires a knowledge of developmental mechanisms (for molding its identity) that might not be required of stem cells derived *directly* from the tissue of interest where nature has done the instruction, e.g., the nervous system. Alternatively, for some organs, such as the heart and pancreas, one cannot identify young cells from the organ itself in sufficient numbers (if at all), hence suggesting the need for a more embryonic source. In dealing with ESCs, however, one must be certain to direct them invariantly toward a given cell type and create safeguards against the appearance of inappropriate cells. Furthermore, while stem cells derived from adults with Huntington's, Alzheimer's disease, PD, amyotrophic lateral sclerosis, multiple sclerosis, and other genetically based diseases have been touted as therapeutic options for such patients, this strategy suffers from the risk that these cells may already be flawed, harboring a genetic defect or predisposition to the disease in question. So, the choice of which cell for which disease might, in fact, vary from disease to disease, to be determined empirically.

Complex diseases require complex solutions. Stem cell–mediated therapy has never been viewed as isolated. The complexity of most disorders requires multifaceted solutions. Cell-mediated interventions must be intertwined with adjunctive pharmacological and molecular interventions, bioengineering, and rehabilitation. Restitution of function likely requires more than the replacement of a single cell type but rather reconstitution of the entire milieu. Stem cell combinations may be required, including various types at different developmental stages for diverse aspects of a given disease. While stem cells are just one weapon in an armamentarium, they may be the "glue" that bonds these multiple approaches.

Logistics. Ultimately, the logistical issues determine whether clinical translation is feasible and successful. For example, what is the proper time to intervene? How, where, how often, at what rate, and in what numbers should cells be delivered? What is the optimal degree of maturity of a stem cell for a particular disease—a mature and rigidly committed state or a less mature and more plastic state? What is the most efficacious and safest method for scaling up a given somatic stem cell to clinically relevant numbers? How much of an obstacle will immune barriers actually be in transplantation paradigms? The answer to this last question could critically influence how much emphasis is placed on using autologous cells or the need for somatic cell nuclear transfer.

"Low-Hanging Fruit" for Stem Cell–Mediated Therapeutics in the Aging Field

Despite the cautious and circumspect admonitions, there are disease processes that are within the purview of proven stem cell properties and might be approachable in the near future.

Cancer (particularly primary and metastatic brain tumors). Stem cells are rapidly drawn to such pathology and can effectively express antitumor genes. In such conditions, differentiation into a particular cell type is unimportant. The stem cells simply need to track abnormal cells and deliver a therapeutic agent while creating no abnormalities.

Diseases where the rescue and protection of cells and circuits or blunting of inflammation and scarring is therapeutic. Given the complexities of development, *preserving* preexisting cell types and their connections is as important, and probably safer and more tractable, than attempting to reconstruct proper functional new types. For stroke, spinal cord injury, head trauma, amyotrophic lateral sclerosis, PD, multiple sclerosis, and diabetes, these actions are attainable by stem cells, via their inherent regulated local release of a range of factors, in a manner unlikely to be achieved by drugs, pumps, or viral vectors. It would be optimal to approach these pathologies early in their onset before muscle atrophy, contractures, scarring, and various complications present additional barriers, and when the signals directing stem cell behavior appear to be at their peak. Furthermore, for such diseases as AD, where inflammation has a prominent role in pathophysiology, inhibiting the inflammatory process via a cell's constitutive expression of anti-inflammatory molecules may have a significant therapeutic impact in a disease whose necessary therapies are otherwise poorly understood.

Diseases where the integration of new cell types is unimportant. For example, a cell that faithfully mimics the function of a glucose-sensing insulin-producing β-cell may be therapeutic for diabetes, even if it did not reside in the pancreas.

CONCLUSIONS

In achieving modest goals—in "picking" this "low-hanging fruit"— the stem cell field will not only establish a track record of safety and efficacy, but its craft and knowledge will improve enormously, allowing it to move iteratively to more difficult challenges, such as cell replacement. Furthermore, although eradicating disease is clearly the goal, each incremental step in that direction may help relieve suffering.

When considering the practical application of stem cell biology to clinical situations, it is instructive to remember that the stem cell field emerged as the unanticipated byproduct of investigations by developmental biologists into fundamental aspects of organ generation and plasticity. Stem cell behavior is ultimately an expression of developmental principles, an alluring vestige from the more plastic and generative stages of organogenesis. In attempting to apply stem cell biology therapeutically, it is prudent to always note what role the stem cell has in *development* and what cues it was "designed" to respond to in trying to understand the "logic" behind its "behavior." If aging—and the pathological conditions that accompany aging—are regarded as just one aspect of a developmental arc, then it may be possible to intelligently figure out how stem cell biology and aging biology may dovetail for therapeutic and scientific ends.

REFERENCES

1. Imitola, J., Park, K. I., Teng, Y. D., et al. (2004) Stem cells: cross-talk and developmental programs. *Phil. Trans. R Soc. Lond. B* **359**, 823–837.
2. Snyder, E. Y., Sidman, R. L., Goodell, M. A., and Daley, G. Q. (2004) Realistic prospects for stem cell therapies for the nervous system *J. Neurosci. Res.* **76**, 157–168.
3. Park, K. I., Ourednik, J., Ourednik, V., et al. (2002) Global Gene and Cell Replacement Strategies via Stem Cells. *Gene Therapy* **9**, 613–624.
4. Snyder, E. Y., Yoon, C., Flax, J. D., and Macklis, J. D. (1997) Multipotent neural precursors can differentiation toward replacement of neurons undergoing targeted apoptotic degeneration in adult mouse neocortex. *Proc. Natl. Acad. USA* **94**, 11663–11668.
5. Rosario, C. M., et al. (1997) Differentiation of engrafted multipotent neural progenitors towards replacement of missing granule neurons in meander tail cerebellum may help determine the locus of mutant gene action. *Development* **124**, 4213–4224.
6. Snyder, E. Y., Taylor, R. M., and Wolfe, J. H. (1995) Neural progenitor cell engraftment corrects lysosomal storage throughout the MPS VII mouse brain. *Nature* **374**, 367–370.
7. Yandava, B. D., Billinghurst, L. L., and Snyder, E. Y. (1999) "Global" cell replacement is feasible via neural stem cell transplantation: evidence from the dysmyelinated shiverer mouse brain. *Proc. Natl. Acad. Sci. USA* **96**, 7029–7034.
8. Park, K. I., Teng, Y. D., and Snyder, E. Y. (2002) The injured brain interacts reciprocally with neural stem cells supported by scaffolds to reconstitute lost tissue. *Nat. Biotechnol.* **20**, 1111–1117.
9. Flax, J. D., Aurora, S., Yang, C., et al. (1998) Engraftable human neural stem cells respond to developmental cues, replace neurons, and express foreign genes. *Nat. Biotechnol.* **16**, 1033–1039.
10. Park, K. I., Liu, S., Flax, J. D., et al. (1999) Transplantation of neural stem-like cells: developmental insights may suggest new therapies for spinal cord and other CNS dysfunction, *J. Neurotrauma* **8**, 675–687.
11. Zlomanczuk, P., Mrugala, M., de la Iglesia, H. O., et al. (2002) Transplanted neural stem cells respond to remote photic stimulation following incorporation within the suprachiasmatic nucleus. *Exp. Neurol.* **174**, 162–168.
12. Ourednik, J., Ourednik, V., Lynch, W. P., et al. (2002) Neural stem cells display an inherent mechanism for rescuing dysfunctional neurons. *Nat. Biotechnol.* **20**, 1103–1110.
13. Imitola, J., Raddassi, K., Park, K. I., et al. (2004) Neuroinflammation's Other Face: Directed migration of human neural stem cells to site of CNS injury by the SDF1α/CXCR4-dependent pathway. *Proc. Natl. Acad. Sci. USA* **101**, 18117–18122.
14. Thomson, J. A., Itskovitz-Eldor, J., Shapiro, S. S., et al. (1998) Embryonic stem cell lines derived from human blastocysts. *Science* **282**, 1145–1147.

15. Doetschman, T. C., Eistetter, H., Katz, M., et al. (1985) The in vitro development of blastocyst-derived embryonic stem cell lines: formation of visceral yolk sac, blood islands and myocardium. *J. Embryol. Exp. Morphol.* **87,** 27–45.

16. Ding, S., Wu, T. Y., Brinker, A., et al. (2003) Synthetic small molecules that control stem cell fate. *Proc. Natl. Acad. Sci. USA* **100,** 7632–7637.

17. He, J. Q., Ma, Y., Lee, Y., et al. (2003) Human embryonic stem cells develop into multiple types of cardiac myocytes. action potential characterization. *Circ. Res.* **93,** 32–39.

18. Wichterle, H., Lieberam, I., Porter, J. A., and Jessell, T. M. (2002) Directed differentiation of embryonic stem cells into motor neurons. *Cell* **110,** 385–397.

19. Kim, J. H., Auerbach, J. M., Rodriguez-Gomez, J. A., et al. (2002) Dopamine neurons derived from embryonic stem cells function in an animal model of Parkinson's disease. *Nature* **418,** 50–56.

20. Bjorklund, L. M., Sanchez-Pernaute, R., Chung, S., et al. (2002) Embryonic stem cells develop into functional dopaminergic neurons after transplantation in a Parkinson rat model. *Proc. Natl. Acad. Sci. USA* **99,** 2344–2349.

21. McDonald, J. W., Liu, X. Z., Qu, Y., et al. (1999) Transplanted embryonic stem cells survive, differentiate and promote recovery in injured rat spinal cord. *Nat. Med.* **5,** 1410–1412.

22. Zhang, S. C., Wernig, M., Duncan, I. D., et al. (2001) In vitro differentiation of transplantable neural precursors from human embryonic stem cells. *Nat. Biotechnol.* **19,** 1129–1133.

23. Schuldiner, M., Eiges, R., Eden, A., et al. (2001) Induced neuronal differentiation of human embryonic stem cells. *Brain Res.* **913,** 201–205.

24. Reubinoff, B. E., Itsykson, P., Turetsky, T., et al. (2001) Neural progenitors from human embryonic stem cells. *Nat. Biotechnol.* **19,** 1134–1140.

25. Carpenter, M. K., Inokuma, M. S., Denham, J., et al. (2001) Enrichment of neurons and neural precursors from human embryonic stem cells. *Exp. Neurol.* **172,** 383–397.

26. Snyder, E. Y., et al. (1992) Multipotent neural cell lines can engraft and participate in development of mouse cerebellum. *Cell* **68,** 33–55.

27. Vescovi, A. L. and Snyder, E. Y. (1999) Establishment and properties of neural stem cell clones: plasticity *in vitro* and *in vivo*. *Brain Pathol.* **9,** 569–598.

28. Renfranz, P. J., Cunningham, M. G., and McKay, R. D. G. (1991) Region-specific differentiation of the hippocampal stem cell line HiB5 upon implantation into the developing mammalian brain. *Cell* **66,** 713–719.

29. Ray, J. (1994) Spinal cord neuroblasts proliferate in response to basic fibroblast growth factor. *J. Neurosci.* **14,** 3548–3564.

30. Reynolds, B. A. and Weiss, S. (1992) Generation of neurons and astrocytes from isolated cells of the adult mammalian central nervous system. *Science* **27,** 1707–1710.

31. Gritti, A., et al. (1996) Multipotential stem cells from the adult mouse brain proliferate and self-renew in response to basic fibroblast growth factor. *J. Neurosci.* **16,** 1091–1100.

32. Gage, F. H., et al. (1996) Survival and differentiation of adult neuronal progenitor cells transplanted to the adult brain. *Proc. Natl. Acad. Sci. USA* **92,** 11879–11883.
33. Ourednik, V., et al. (2001) Segregation of human neural stem cells in the developing primate forebrain. *Science* **293,** 1820–1824.
34. Suhonen, J. O., Peterson, D. A., Ray, J., and Gage, F. H. (1996) Differentiation of adult hippocampus-derived progenitors into olfactory neurons in vivo. *Nature* **383,** 624–627.
35. Fricker, R. A., et al. (1999) Site-specific migration and neuronal differentiation of human neural progenitor cells after transplantation in the adult rat brain. *J. Neurosci.* **19,** 5990–6005.
36. Zhang, S.-C., et al. (2001) In vitro differentiation of transplantable neural precursors from human embryonic stem cells. *Nature Biotech.* **19,** 1129–1133.
37. Rubinoff, B. E., et al. (2001) Neural progenitors from human embryonic stem cells: derivation, expansion, and characterization of their developmental potential in vitro and in vivo. *Nat. Biotech.* **19,** 1134–1140.
38. Park, K. I., Jensen, F. E., Stieg, P. E., and Snyder, E. Y. (1998) Hypoxic-ischemic (HI) injury may direct the proliferation, migration, and differentiation of endogenous neural progenitors. *Soc. Neurosci. Abstr.* **24,** 1310.
39. Magavi, S. S., Leavitt, B. R., and Macklis, J. D. (2000) Induction of neurogenesis in the neocortex of adult mice. *Nature* **405,** 951–955.
40. Gould, E., Reeves, A. J., Graziano, M. S., and Gross, C. G. (1999) *Science* **286,** 548–552.
41. Weiss, S., et al. (1996) Is there a neural stem cell in the mammalian forebrain? *Trends Neurosci.* **19,** 387–393.
42. Morshead, C. M., et al. (1994) Neural stem cell in the adult mammalian forebrain: a relatively quiescent subpopulation of subependymal cells. *Neuron* **13,** 1071–1082.
43. Davis, A. A. and Temple, S. (1994) A self-renewing multipotential stem cell in embryonic rat cerebral cortex. *Nature* **372,** 263–266.
44. Zlomanczuk, P., et al. (2002) Transplanted clonal neural stem-like cells respond to remote photic stimulation following incorporation within the suprachiasmatic nucleus. *Exp. Neurol.* **174,** 162–168.
45. Doering, L. and Snyder, E. Y. (2000) Cholinergic expression by a neural stem cell line grafted to the adult medial septum/diagonal band complex. *J. Neurosci. Res.* **61,** 597–604.
46. Park, K. I., et al. (1999) Transplantation of neural progenitor and stem-like cells: developmental insights may suggest new therapies for spinal cord and other CNS dysfunction. *J. Neurotrauma* **16/8,** 675–687.
47. Aboody, K. S., et al. (2000) Neural stem cells display extensive tropism for pathology in the adult brain: evidence from intracranial gliomas. *Proc. Natl. Acad. Sci. USA* **97,** 12846–12851.
48. Ourednik, J., et al. (1999) Massive regeneration of substantia nigra neurons in aged parkinsonian mice after transplantation of neural stem cells over-expressing L1. *Soc. Neurosci. Abstr.* **25,** 1310.

49. Lacorazza, H. D., Flax, J. D., Snyder, E. Y., and Jendoubi, M. (1996) Expression of human β-hexosaminidase α-subunit gene (the gene defect of Tay-Sachs disease) in mouse brains upon engraftment of transduced progenitor cells. *Nat. Med.* **4**, 424–429.

50. Ourednik, V., Ourednik, J., Park, K. I., and Snyder, E. Y. (1999) Neural stem cells: a versatile tool for cell replacement and gene therapy in the CNS. *Clin. Genet.* **46**, 267–278.

51. Neufeld, E. F. and Fratantoni, J. C. (1970) Inborn errors of mucopolysaccharide metabolism. *Science* **169**, 141–146.

52. Luskin, M. B. (1993) Restricted proliferation and migration of postnatally generated neurons derived from the forebrain subventricular zone. *Neuron* **11**, 173–189.

53. Lois, C. and Alvarez-Buylla, A. (1994) Long distance neuronal migration in the adult mammalian brain. *Science* **264**, 1145–1148.

54. Snyder, E. Y. (1998) Neural stem-like cells: developmental lessons with therapeutic potential. *Neuroscientist* **4**, 408–425.

55. Uchida, N., et al. (2000) Direct isolation of human central nervous system stem cells. *Proc. Natl. Acad. Sci. USA* **97**, 14720–14725.

56. Lynch, W. P., Sharpe, A. H., and Snyder, E. Y. (1999) Neural stem cells as engraftable packaging lines optimize viral vector-mediated gene delivery to the CNS: evidence from studying retroviral env-related neurodegeneration. *J. Virol.* **73**, 6841–6851.

57. Rubio, F. J., et al. (1999) BDNF gene transfer to the mammalian brain using CNS-derived neural precursors. *Gene Therapy* **6**, 1851–1866.

58. Himes, B. T., et al. (2001) Transplants of cells genetically modified to express neurotrophin-3 rescue axotomized Clarke's nucleus neurons after spinal cord hemisection in adult rats. *J. Neurosci. Res.* **65**, 549–564.

59. Liu, Y., et al. (1999) Intraspinal delivery of neurotrophin-3 using neural stem cells genetically modified by recombinant retrovirus. *Exp. Neurol.* **158**, 9–26.

60. Martinez-Serrano, A., et al. (1995) CNS-derived neural progenitor cells for gene transfer of nerve growth factor to the adult rat brain: complete rescue of axotomized cholinergic neurons after transplantation into the septum. *J. Neurosci.* **15**, 5668–5680.

61. Martinez-Serrano, A., Fischer, W., and Bjorklund, A. (1995) Reversal of age-dependent cognitive impairments and cholinergic neuron atrophy by NGF-secreting neural progenitors grafted to the basal forebrain. *Neuron* **15**, 473–484.

62. Anton, R., et al. (1994) Neural-targeted gene therapy for rodent and primate hemiparkinsonism. *Exp. Neurol.* **127**, 207–218.

63. Sabaate, O., et al. (1995) Transplantation to the rat brain of human neural progenitors that were genetically modified using adenoviruses. *Nat. Genet.* **9**, 256–260.

64. Anton, R., et al. (1995) Neural transplantation of cells expressing the anti-apoptotic gene bcl-2. *Cell Transplant.* **4**, 49–54.

65. Akerud, P., et al. (2001) Neuroprotection through delivery of GDNF by neural stem cells in a mouse model of Parkinson's disease. *J. Neurosci.* **21,** 8108–8118.
66. Lu, P., et al. (2000) Neural stem cells secrete BDNF and GDNF, and promote axonal growth after spinal cord injury. *Soc. Neurosci. Abstr* **26,** 332.
67. Aboody, K. S., Brown, A., Rainov, N. G., et al. (2000) Neural stem cells display extensive tropism for pathology in adult brain: evidence from intracranial gliomas. *Proc. Natl. Acad. Sci. USA* **97,** 12846–12851.
68. Pluchino, S., Quattrini, A., Brambilla, E., et al. (2003) Injection of adult neurospheres induces recovery in a chronic model of multiple sclerosis. *Nature* **422,** 688–694.
69. Pardridge, W. M. (2002) Drug and gene targeting to the brain with molecular Trojan horses. *Nat. Rev. Drug Discov.* **1,** 131–139.
70. Pardridge, W. M. (2002) Drug and gene delivery to the brain: the vascular route. *Neuron* **36,** 555–558.
71. Kabos, P., Ehtesham, M., Black, K. L., and Yu, J. S. (2003) Neural stem cells as delivery vehicles. *Expert Opin. Biol. Ther.* **3,** 759–770.
72. Barresi, V., Belluardo, N., Sipione, S., et al. (2003) Transplantation of prodrug-converting neural tumor therapy. *Cancer Gene Ther.* **10,** 396–402.
73. Lynch, W. P., Sharpe, A. H., and Snyder, E. Y. (1999) Neural stem cells as engraftable packaging lines can mediate gene delivery to microglia: evidence from studying retroviral env-related neurodegeneration. *J. Virol.* **73,** 6841–6851.
74. Herrlinger, U., Woiciechowski, C., Sena-Esteves, M., et al. (2000) Neural precursor cells for delivery of replication-conditional HSV-1 vectors to intracerebral gliomas. *Mol. Ther.* **1,** 347–357.
75. Jean, W. C., Spellman, S. R., Wallenfriedman, M. A., et al. (1998) Interleukin-12-based immunotherapy against rat 9L glioma. *Neurosurgery* **42,** 850–856; discussion 856–857.
76. Ehtesham, M., Samoto, K., Kabos, P., et al. (2002) Treatment of intracranial glioma with in situ interferon-gamma and necrosis factor-alpha gene transfer. *Cancer Gene Ther.* **9,** 925–934.
77. Rhines, L. D., Sampath, P., DiMeco, F., et al. (2003) Local immunotherapy with interleukin-2 delivered from polymer microspheres combined with interstitial chemotherapy: a treatment for experimental malignant glioma. *Neurosurgery* **52,** 872–879; discussion 879–880.
78. Benedetti, S., Pirola, B., Pollo, B., et al. (2000) Gene therapy of experimental brain tumors using neural progenitor. *Nat. Med.* **6,** 447–450.
79. Weinstein, D. E., Shelanski, M. L., and Liem, R. K. (1990) C17, a retrovirally immortalized neuronal cell line, inhibits the proliferation of astrocytes and astrocytoma cells by a contact- mechanism. *Glia* **3,** 130–139.
80. Ehtesham, M., Kabos, P., Kabosova, A., et al. (2002) The use of interleukin 12-secreting neural stem cells for the treatment of intracranial glioma. *Cancer Res.* **62,** 5657–5663.
81. Ehtesham, M., Kabos, P., Gutierrez, M. A., et al. (2002) Induction of glioblastoma apoptosis using neural stem cell-mediated delivery of tumor necrosis factor-related apoptosis-inducing ligand. *Cancer Res.* **62,** 7170–7174.

82. Walczak, H., Miller, R. E., Ariail, K., et al. (1999) Tumoricidal activity of tumor necrosis factor-related apoptosis- ligand in vivo. *Nat. Med.* **5,** 157–163.
83. Marras, C., Mendola, C., Legnani, F. G., and DiMeco, F. (2003) Immunotherapy and biological modifiers for the treatment of malignant brain tumors. *Curr. Opin. Oncol.* **15,** 204–208.
84. Yacoub, A., Mitchell, C., Lister, A., et al. (2003) Melanoma differentiation-associated 7 (interleukin 24) inhibits growth and enhances radiosensitivity of glioma cells in vitro and in vivo. *Clin. Cancer Res.* **9,** 3272–3281.
85. Aoki, T., Tashiro, K., Miyatake, S., et al. (1992) Expression of murine interleukin 7 in a murine glioma cell line results in reduced tumorigenicity in vivo. *Proc. Natl. Acad. Sci. USA* **89,** 3850–3854.
86. Parker, J. N., Gillespie, G. Y., Love, C. E., et al. (2000) Engineered herpes simplex virus expressing IL-12 in the treatment of experimental murine brain tumors. *Proc. Natl. Acad. Sci. USA* **97,** 2208–2213.
87. Bjerkvig, R., Read, T. A., Vajkoczy, P., et al. (2003) Cell therapy using encapsulated cells producing endostatin. *Acta Neurochir.* Suppl **88,** 137–141.
88. Chiocca, E. A. (2002) Oncolytic viruses. *Nature Rev. Cancer* **2,** 938–950.
89. Rainov, N. G. and Ren, H. (2003) Oncolytic viruses for treatment of malignant brain tumours. *Acta Neurochir.* Suppl **88,** 113–123.
90. Wilcox, M. E., Yang, W., Senger, D., et al. (2001) Reovirus as an oncolytic agent against experimental human malignant gliomas. *J. Natl. Cancer Inst.* **93,** 903–912.
91. Yang, W. Q., Senger, D., Muzik, H., et al. (2003) Reovirus prolongs survival and reduces the frequency of leptomeningeal metastases from medulloblastoma. *Cancer Res.* **63,** 3162–3172.
92. Stojdl, D. F., Lichty, B. D., tenOever, B. R., et al. (2003) VSV strains with defects in their ability to shutdown innate immunity are potent systemic anticancer agents. *Cancer Cell* **4,** 263–275.

2

Developing Novel Cell Sources
for Transplantation in Parkinson's Disease

Nicolaj S. Christophersen, Ana Sofia Correia, Laurent Roybon, Jia-Yi Li, and Patrik Brundin

ABSTRACT

Developing dopaminergic (DAergic) neurons that originate from aborted human embryos have been implanted into the brains of patients with Parkinson's disease (PD) and, in some cases, have successfully restored function. However, there are insufficient numbers of cells available to allow this therapy to become widely used. The limited amount of tissue from embryos may be circumvented by the use of cell lines that can be expanded in vitro for banking, then differentiated into DAergic neurons just prior to implantation into patients. Today, there are four main sources for such cell lines with future potential for banking and cell therapy for PD: human embryonic stem cells, human neural stem cells, human genetically immortalized stem/progenitor cells, and human adult-derived non-neural stem cells, such as bone marrow–derived stem cells. Currently, it is not possible to utilize these cell sources therapeutically for PD. The primary reasons are because it has not been feasible to effectively differentiate these cells into DAergic neurons and because the stability of phenotypic expression has been variable. This chapter describes methods to generate cells suitable for transplantation in PD in the future. The development of novel cell sources is described, along with an overview of the various types of stem cells that are suitable for grafting in PD.

Key Words: Parkinson; transplantation; dopamine; embryonic stem cells; neural progenitor; differentiation; immortalization.

INTRODUCTION

In the central nervous system, the most abundant source of dopaminergic (DAergic) cell bodies is located in the midbrain, mainly in the substantia nigra pars compacta (SNpc), the ventral tegmental area, and the retrorubral

From: *Contemporary Neuroscience: Cell Therapy, Stem Cells, and Brain Repair*
Edited by: C. D. Sanberg and P. R. Sanberg © Humana Press Inc., Totowa, NJ

field. These DAergic neurons extensively innervate the forebrain (e.g., projections to the caudate nucleus, putamen nucleus accumbens, olfactory tubercle) and several cortical and limbic regions, including the amygdala, lateral septum, and ventral hippocampus. The basic organization of midbrain DAergic neurons and their projections is consistent across most mammalians (1,2).

The DAergic projections originating in the midbrain and innervating the striatum are the most extensively studied catecholamine neurons, partly because the degeneration of DAergic neurons in the SNpc is one cause of the motor dysfunction present in Parkinson's disease (PD). This disorder is characterized by tremor, rigidity, hypokinesia, and postural instability. Levodopa treatment initially provides marked symptomatic relief; however, within 5–10 yr, most patients exhibit a gradual loss of efficacy taking levodopa that is associated with the appearance of involuntary movements (dyskinesias; 3).

Cell Therapy for PD: Proof of Concept

Both neural transplant studies performed in animal PD models and clinical grafting trials have suggested that cell replacement therapies may be effective in treating PD. Proof of this concept was observed in open-label trials more than a decade ago, when transplantation of human embryonic ventral mesencephalic tissue containing DA neurons was shown to be an effective therapy (4–6). The strategy has been to replace the population of degenerated DAergic neurons with neurons harvested from a donor at an early stage in development, when the cells are still dividing and able to grow processes to their appropriate targets in an adult host brain. This strategy is particularly suitable to explore in PD, since its main pathology is relatively focused on the nigrostriatal DAergic system (i.e., a specific neuronal population within a restricted area of the brain). The classical biochemical deficit in PD results from the loss of DAergic neurons in the SNpc. However, there is also degeneration of the acetylcholine and noradrenaline systems innervating the forebrain (7). This neuropathology may contribute to cognitive and other nonmotor features of this disease, some of which are relatively unresponsive to DAergic replacement therapy (8). Grafts specifically designed to replace the loss of DA in the striatum only treats the core pathology of PD. Nevertheless, this may be sufficient to alleviate the most disabling symptoms.

Thus far, over 300 patients worldwide with PD have received transplants of primary human embryonic tissue that is rich in DAergic neurons. Clinical improvements have been reported for up to 10 yr after transplantation surgery (9,10).

Challenges in the Cell Therapy Strategy

Currently, cell-based therapies for PD face three major problems. First, recent placebo-controlled trials have shown levodopa-independent dyskinesias in a minority of patients receiving embryonic mesencephalic tissue transplants *(11–13)*. The cause of the dyskinesias remains unknown, but the placement of transplants into potential hotspots in the dorsal and ventral part of the striatum may lead to the severe side effects *(14)*. In addition, the heterogeneous cellular composition of ventral mesencephalic grafts has also been suggested to underlie dyskinesias. Thus, selective transplantation of DAergic neurons from the SNpc has been thought to increase symptomatic relief and decrease motor side effects, compared to the grafting of DAergic neurons from the SNpc and adjacent ventral tegmental area *(15)*. However, this hypothesis has still not been addressed experimentally in an effective manner.

The second major problem facing scientists in the neural grafting field is that the two published studies on transplantation in PD following a double-blind placebo-controlled protocol did not show a significant improvement when compared with sham surgery *(11,13)*. For unknown reasons, the results differ from the earlier positive reports emanating from open-label trials *(9)*. Naturally, the differences between the two types of studies could be explained by a combination of the placebo effect and observer bias that occurred in the open-label trials. However, there are additional variations in surgical technique, patient selection, and immunosuppressive regimens that may have had equally important roles.

Third, there are significant practical and ethical problems associated with using embryonic/fetal tissues. Thus, one major restriction of the more widespread application of clinical transplantation is the limited availability of suitable human donor tissue, along with a poor survival of DAergic cells in the grafts. The human ventral mesencephalic tissue that is dissected from embryos and used for the grafts is only made up of 5–10% of cells that are destined to become DAergic neurons *(16)*; the remaining cells are other neurons, glia, or precursors. Furthermore, only a small fraction (approx 10%) of those cells that become DAergic neurons actually survive the grafting procedure *(17–19)*.

Due to the problems facing clinical neural grafting in PD, a number of alternative cell sources have been investigated. Cells for transplantation in PD should display the following characteristics:

1. Reliable viability and homogeneity.
2. Availability in large numbers at the planned time for surgery.
3. Ability to differentiate into DA-synthesizing neurons (particularly of the specific phenotype found in the SNpc).

4. Capacity for axonal growth.
5. Genetic stability and lack of tumor formation.
6. Capacity to survive in the host brain for the long term.
7. Long-lasting functional benefit without significant adverse effects.

Another source for implantation in PD is stem cells, defined as undifferentiated cells with high proliferative potential that can generate a wide variety of differentiated progeny *(20)*. They are particularly appealing for cell-based therapies because they can be made available in large numbers, they are relatively easy to maintain in culture and (by definition) exhibit a high differentiation potential, and they could consequently provide the missing DAergic neurons for grafting in PD.

STEM CELLS

Stem cells can be derived from either embryonic or adult tissue sources and are divided into different classes depending on when they are present and on their differentiation capacity. There are two categories based on when they appear during the lifetime of the organism: embryonic stem (ES) cells occur during early embryonic development, and somatic or adult-derived stem cells are present in different tissues in the fully developed organism. It is important to clarify that the term *embryonic stem cells* is usually reserved for cells derived from the inner cell mass of the preimplantation blastocyst (see below). However, other types of stem cells can be derived from later stage embryos that typically have greater lineage restriction. Based on their capacity to differentiate into various cell types, stem cells can be either *pluripotent*, giving rise to every cell of the organism (except the trophoblasts of the placenta), or *multipotent*, giving rise to all the cells of the organ in which the multipotent cell normally resides. As a potential cell source for PD therapy, the most commonly studied somatic stem cells are the neural stem cells derived from embryonic neural tissue.

Somatic Neural Stem Cells

Embryonic Neural Stem and Progenitor Cells

Embryonic neural stem and progenitor cells are obtained from embryonic neural tissue and are capable of self-renewing and generating neurons and glia. Over the past decade, research has shown that it is possible to select, epigenetically manipulate, and genetically engineer cells in culture prior to intracerebral transplantation *(21–25)*. Neural stem cells have been isolated from various parts of the brain, such as the midbrain and forebrain *(26,27)*. These cells can be grown as free-floating cell cultures in so-called *neurospheres*, where they will expand in number. These cells appear to retain

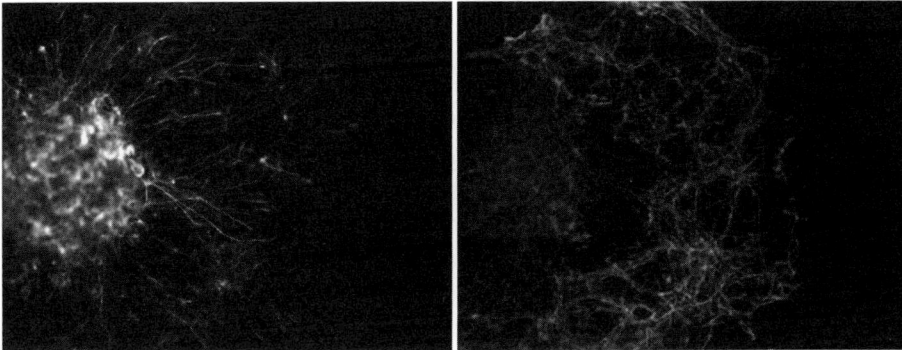

Fig. 1. Three-day differentiation of E14.5 rat ventral mesencephalic neurospheres. Neurons were stained for β-III-tubulin (red) and astrocytes for glial fibrially acidic protein (GFAP) (green), and they were counterstained using the nuclear marker Hoechst (blue).

their multipotency and the ability to develop along different progenitor, neuronal, and glial lineages under specific culture conditions *(28;* see Fig. 1).

However, it seems that neuronal differentiation potential differs between animal species and also because of the duration of cell propagation in culture. Human neurosphere cultures, which have been expanded for long periods (up to 150 d), can retain their ability to generate neurons with a high (40%) frequency *(23)*. In contrast, less than 5% of the cells from mouse neurospheres that have been passaged a few times (over 2 wk) appear capable of differentiating into neurons *(29)*. In the case of rat neurosphere cultures, the proportion of cells that differentiate into neurons decreases from around 63% to 17% following three passages over 3 wk *(30)*.

DAergic Differentiation by Epigenetic Factors

As mentioned above, the differentiation potential of an embryonic neural stem cell culture depends on the age of the donor tissue, the precise anatomical region dissected, and the length of time spent in culture *(31)*. This is also the case when differentiating these cells into DAergic neurons. Clearly, it is difficult to influence the differentiation of embryonic neural stem cells and coax them into adopting a DAergic neuronal phenotype. Thus far, the addition of various cytokines and erythropoietin to the culture medium and cell growth in low-oxygen tension have led to the most efficient production of DAergic neurons in vitro *(32,33)*. Expanded embryonic neural stem cells obtained from the ventral mesencephalon have also been transplanted successfully into hemiparkinsonian rats and have demonstrated functional improvement in reducing amphetamine-induced motor asymmetry in recipients *(25,27,34)*. This procedure could offer a powerful strategy for generating large numbers of DAergic neurons for transplantation, but the techniques to ensure good posttransplant survival are still not well established *(25,34)*. When comparing the numbers of surviving DAergic neurons when mesencephalic tissue is grafted with or without prior expansion in vitro, it appears that there are no major benefits of proliferating and differentiating the precursors in vitro before transplantation *(35)*. This may be explained by the cells' higher susceptibility to transplantation-related death after the period in culture *(25,35)*. Another study suggested that as many as around 75% of rat mesencephalic progenitors can be induced to express TH by treatment with a cocktail of cytokines, but following grafting, the majority of these cells appear to either die or suppress the DAergic phenotype *(34)*. The failure to maintain all the DAergic neurons after transplantation may indicate that when dealing with stem cell-derived neurons, this phenotype is not cell-autonomous but requires external cues or a more permissive environment.

DAergic Differentiation by Genetic Factors

Specific genes involved in the differentiation of DAergic neurons have been artificially expressed in embryonic neural stem cells to direct the differentiation into DAergic neurons. Most successfully, the orphan nuclear transcription factor Nurr1 has been used to engineer neural stem cells to differentiate into DAergic neurons *(36)*. Nurr1 is known to be implicated in the differentiation of mesencephalic DAergic neurons *(37)*. Thus, mice that are null-mutant for Nurr1 lack mesencephalic DAergic neurons *(38)*. Furthermore, ventral mesencephalic and cortical progenitors transduced with Nurr1 by viral vectors have been found to exhibit DAergic characteristics. Nurr1-transduced cells have shown to acquire immunoreactivity for several markers associated with DAergic neurons and were found to synthesize and release dopamine in vitro *(36)*. However, when these cells were transplanted into the striatum of rats with unilateral 6-hydroxydopamine lesions, there was no reduction of apomorphine-induced motor asymmetry. The lack of behavioral recovery was interpreted to be associated with immature neuronal morphologies and low survival of TH-immunopositive cells in vivo *(36)*.

Immortalized Neural Progenitor Cell Lines

Neural stem and progenitor cells are not ideal for cell banking because their mitotic competence is limited *(39)*. They exhibit a proliferative response to mitogens for only a limited time until they approach their natural senescence. Typically, a neural progenitor cell will undergo a number of divisions, then differentiate or undergo cell death *(40)*. As an adjunct to treatment with mitogens, it is possible to obtain large numbers of neural stem cells and progenitors by immortalizing them through genetic engineering *(41)*. The immortalizing gene arrests the cells at a certain developmental stage and prevents terminal differentiation *(42)*.

Most commonly, immature neural cells are immortalized using a retroviral vector that encodes the propagating-enhancing v-myc protein *(43–45)*. Interestingly, v-myc expression may be involved in telomerase activation, because telomerase is a direct target gene regulated by v-myc *(46)*. Maintaining telomere length is known to have an important role for continued cell proliferation *(39,47–51)*. V-myc propagated neural progenitor cell lines include C17.2, H6, HNSC.100, and MesII, which are derived from developing mouse cerebellum *(43,44)*, 15-wk-old human fetal telencephalon *(52)*, 10-wk-old human embryonic forebrain cultures grown as neurospheres *(53)*, and 8-wk-old human embryonic ventral mesencephalon *(54)*, respectively. By constitutively expressing such oncogenes as v-myc, cell lines proliferate indefinitely in culture; however, they still depend on mitogens (e.g., basic fibroblast growth factor [bFGF], epidermal growth factor, or serum) to divide. In the absence of mitogens, they exit the cell cycle and differentiate *(52,53,55)*. As an alternative to developing cell lines by constitutively expressing oncogenes, some groups have taken advantage of a tetracycline-controlled gene expression system *(56,57)*. This is how the human mesencephalic progenitor cell line MesII was generated *(54)*. MesII cells can differentiate and exhibit neurite extension, generate action potentials, express TH and the DA transporter, and produce DA (see Fig. 2A). This requires specific culture conditions that include the addition of tetracycline, dibuturyl cyclic adenosine monophosphate (cAMP), and glial cell line–derived neurotrophic factor (GDNF) to the growth medium (see Fig. 2B).

Although a tetracycline-regulated v-myc vector design is not safe for clinical application, MesII cells provide an interesting experimental tool and a proof of principle that human DAergic cell lines can be generated and illustrate that immortalized cell lines could be relevant for stem cell therapy for PD.

Another method to generate neural progenitor cell lines for PD is to overexpress Nurr1 in existing neural stem cell lines. Nurr1 overexpression in the C17.2 stem cell line has been shown to promote the differentiation of the stem cells into TH-positive neurons when they are cocultured with type 1 astrocytes from the ventral midbrain *(58)*. The identity of the astroglial factor is unknown, but it is evidently produced specifically by astrocytes in the developing ventral midbrain, because glial cells from other regions do not promote differentiation of the C17.2 cells into DAergic neurons. In a more recent study, cells resembling midbrain DAergic neurons were

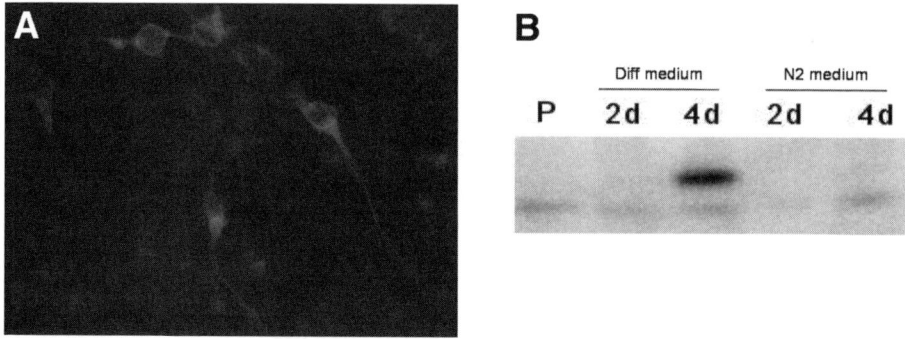

Fig. 2. Immortalized mesencephalic progenitor cells generate TH-positive neurons. (**A**) MesII cells expressing TH immunostaining (red) after differentiation. Cells were counterstained using the nuclear marker Hoechst (blue). (**B**) Western blot of MesII cells grown under specific culture conditions that include the addition of tetracycline, dibutyryl cAMP, and GDNF to the growth medium, which result in TH expression in MesII cells.

obtained from an immortalized multipotent neural stem cell line by overexpressing Nurr1, fibroblast growth factor-8 (FGF-8), and sonic hedgehog (Shh; *59*). Shh and FGF-8 are known to direct the differentiation of mesencephalic DA neurons during development *(60)*.

Several immortalized neural cell lines have been examined in transplantation studies. The oncogene must be sufficiently downregulated in the differentiating neural cells, to allow the differentiation programs to proceed and to avoid tumor formation after grafting. Ideally, proliferation should end, and differentiation should be initiated, before the cells are harvested for transplantation. Another option is to rely on the oncogene being spontaneously downregulated in conjunction with graft surgery. V-myc has been reported to be downregulated following intracerebral transplantation *(52,61)*, but detection of *v-myc* mRNA is still possible using reverse transcriptase-polymerase chain reaction *(62)*. This degree of downregulation is unlikely to be sufficient to meet the level of safety required in clinical trials. Nevertheless, in experimental animals, cell lines generated by the constitutive expression of immortalizing oncogenes have been shown as nontumorigenic *(63)*. Following transplantation into neurogenic regions (e.g., the hippocampus), HNSC.100 cells stop dividing after 2 d and spontaneously generate neurons, astrocytes, and oligodendrocytes *(62,64)*. However, results were different when the cells were transplanted into the striatum and substantia nigra of the adult intact rat brain. In these sites, which pertain to neural replacement in PD, cells from the HNSC.100 line displayed less neuronal differentiation in vivo than in vitro and did not generate DAergic neurons *(62)*.

An immortal source of cells can supply unlimited numbers of homogenous cells for research purposes. They are usually relatively easy to modify genetically and can be readily characterized in detail if they were ever candidates for clinical use. However, there is always the potential risk of creating transformed cells that completely lack growth control mechanisms. Such cells have no contact inhibition, can grow in soft agar, and give rise to tumors in the nude mouse. Their inability to respond to normal signals to withdraw from the cell cycle make them poor candidates for clinical use. To provide an extra level of assurance, it would be prudent to engineer cells with a CRE-loxP recombinase system to remove the immortalizing genes just prior to, or just following, implantation *(65)*. Even if genetically modified cell lines are never used in clinical studies, they can be valuable experimental tools. For example, genetically modified neural stem cells can be compared to neurospheres grown in a mitogen-enriched medium to provide more insight into the key genetic and molecular events that govern neural stem cell differentiation.

Embryonic Stem Cells

First described in 1980 *(66)*, ES cells are pluripotent and form the inner cell mass from the blastocyst stage of all mammalian embryos *(67)*. In addition to ES cells, there are other embryo-derived stem cell lines considered to be pluripotent and immortal: embryonic germ cells and embryonic carcinoma cells. Although embryonic germ and carcinoma cells can develop into a wide range of cell types in vitro and in vivo, ES cells have been shown to differentiate into the widest range of cell types.

To derive ES cells, the isolated inner cell mass is plated on embryonic fibroblasts and is grown in a culture medium supplemented with fetal bovine serum. Alternatively, human ES cells can be grown in a serum-free medium *(68,69)*. ES cells express high levels of telomerase activity, i.e., they are not prone to senescence and are therefore suitable for long-term culture *(70)*. To date, most work has been done on murine ES cells; however, in the last 6 yr, human ES cells have been isolated and cultured *(71–73)*. This chapter describes how ES cells can be differentiated into DAergic neurons and the specific scientific challenges that will be encountered before they can be considered for grafting in patients with PD.

DAergic Differentiation of ES Cells by Soluble Factors

Many studies have been aimed at directing the differentiation of ES cells into DAergic neurons in vitro. Essentially, there are three successful methods for the differentiation of mouse ES cells into DAergic neurons (see Table 1).

Table 1
Protocols Used to Differentiate ES Cells from Mouse and Human Sources into DAergic Neurons

Author and year	ES cell sources	Key features of cell differentiation method in vitro	TH+ neurons in vitro	Graft survival
Bjorklund, L. M., et al. (87)	Mouse ES cells	No special protocol. Cells dissociated prior to grafting	ND	2059 TH+ cells; no survival in 24% of rats[a]
Lee, S. H., et al. (74)	Mouse ES cells	Grown as EBs; exposed to bFGF, Shh, FGF-8, AA	5%[b]	ND
Kim, J. H., et al. (82)	Mouse ES cells overexpressing Nurr1	Transduced with the *Nurr1* gene; grown as EBs; exposed to bFGF, Shh, FGF-8	56%[c]	4% of total number of implanted cells
Shim, J. W., et al. (84)	Mouse ES cells overexpressing Bcl-XL	*Bcl-XL* gene; grown as EBs; exposed to bFGF, AA	30.9%[d]	3.6% of total number of implanted cells
Kawasaki, H., et al. (76,77)	Mouse or primate ES cells cocultured with PA6 stromal cells	Grown on PA6 cells	16% and 9%	Mouse: 3% of total number of implanted cells; Primate: ND
Barberi, T., et al. (78)	Mouse ES cells cocultured with MS5 stromal cells	Grown on MS5 cells; exposed to Shh, FGF-8, bFGF, BDNF, AA	?	10–20% of total number of implanted cells
Park, S., et al. (79)	Human ES cells	Grown as EBs; exposed to bFGF, TGF-α	20%[e]	ND
Perrier, A. L., et al. (80)	Human ES cells cocultured with MS5 stromal cells	Grown on MS5 cells; exposed to Shh, FGF-8, BDNF, GDNF, TGF-β3, dbcAMP, AA	[f]19.2–39.5%	ND

[a]No information about the number of implanted cells. [b]Percentage of TH+ neurons out of total number of cells (6.9% of 71.9%). [c]Percentage of TH+ neurons out of total number of cells (based on ref. 74: 78% of 71.9%). [d]Percentage of TH+ neurons as described in ref. 84. [e]Percentage of TH+ neurons as described in ref. 79. [f]Percentage of TH+ neurons out of total number of cells (64% of 30–79% of 50%). ND, not determined. For explanation of abbreviations, see main text. ?, not determined.

The first method is based on a five-step culture procedure *(74)* in which the in vivo neural development is mimicked by sequential exposure to epigenetic signals. The first step is to generate aggregates called *embryoid bodies* (EBs) from ES cells by dissociating the ES cells into single-cell suspension in a medium containing leukemia-inhibiting factor (LIF) and serum. The EB is an aggregate that can be maintained in a suspension culture and contains precursor cells from all three germ layers *(75)*. The EBs are formed after 4 d, then plated onto an adhesive tissue culture substrate in a medium containing LIF and serum. Neural progenitor cells are then selected by growing the cells in a serum-free medium, and these cells are grown in the presence of bFGF, Shh, and FGF-8. Withdrawal of these factors and the addition of ascorbic acid (AA) induce the final differentiation into DAergic neurons (approx 5% of the total cell number).

The second method designed to obtain DAergic neurons from ES cells involves the differentiation of ES cells on the stromal cell line PA6 *(76)*. The PA6 cell line has been shown to release and/or possess cell surface factors that direct the differentiation of mouse ES cells into DAergic neurons. These factors are still unknown and have collectively been named *stromal cell-derived inducing activity* (SDIA). After being grown on top of PA6 cells, about 16% and 9% of mouse and primate ES cells, respectively, develop into TH-expressing neurons *(76,77)*. These TH-immunopositive cells release DA upon depolarization by high-potassium stimulation.

The third technique also involves stromal cells; in this case, the MS5 cell line *(78)*. Using this cell line, mouse ES cells can be differentiated into neural stem cells. These progenitor cells are then differentiated into DAergic neuron progenitor cells by the addition of Shh and FGF-8. Following the generation of DAergic neuron progenitor cells, bFGF induces further cell proliferation. Subsequently, the withdrawal of these factors and the addition of ascorbic acid and brain-derived neurotrophic factor (BDNF) results in the differentiation into DAergic neurons.

Regarding directed differentiation of human ES cells into DAergic neurons, only two studies have been published thus far *(79,80;* see Fig. 3 and Table 1).

The first method involves the formation of EBs, followed by the selection of neuronal stem cells by incubation in a serum-free insulin/transferrin/selenium/fibronectin medium and expansion of neural stem cells in the presence of bFGF *(79)*. In the final step, DAergic neurons are enriched by the removal of bFGF and addition of transforming growth factor-α (TGF-α). Following 21 d in the presence of TGF-α, about 15% of the differentiated human ES cells in culture express TH and release DA.

A

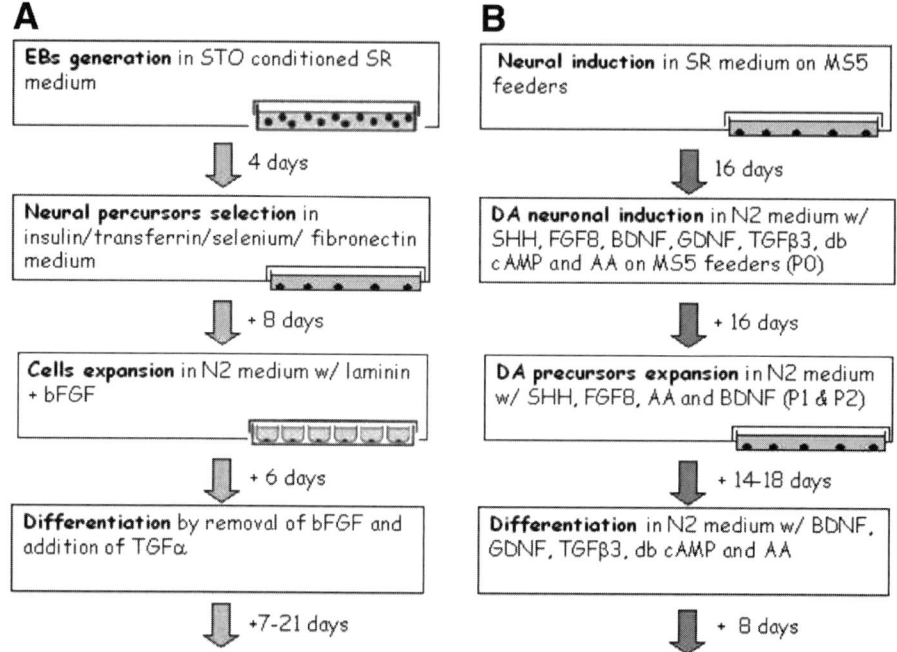

Fig. 3. The in vitro differentiation methods used to generate DAergic neurons from human ES cells. (**A**) ref. *79* and (**B**) ref. *80*. SR, serum replacement; P0, P1, P2, passages 0, 1, and 2.

The second method is an adaptation of the previously published method for the DAergic neuronal differentiation of mouse ES cells *(78)*. The first step of the 50-d differentiation procedure is an initial neural induction on stromal cells, followed by neural stem cell expansion and differentiation by sequential exposure to Shh, FGF-8, BDNF, GDNF, dibutyryl cAMP, AA, and TGF-β3. This factor is expressed in early embryonic structures in which midbrain DAergic neurons develop, and it appears to be essential for both the induction and survival of this type of neurons in vitro and in vivo *(81)*. This protocol has been tested on three different human ES cell lines. Interestingly, the cell proportions that differentiate into TH-immunopositive neurons, expressed as a percentage of the β-III-tubulin expressing cells, are both relatively high and similar: ~64%, ~70%, and ~79%.

DAergic Differentiation of ES Cells by Genetic Engineering

As described in an earlier section, knowledge of the genetic control of developing DA neurons is rapidly advancing and is now being applied to ES cell research. Stable Nurr1 overexpression in mouse ES cells has been shown to increase the proportion of TH-immunopositive neurons from 5% to 50%, expressed as a fraction of the β-III tubulin-immunopositive population. Treatment with Shh and FGF-8 at early stages of the differentiation increases this percentage further to 78% *(82)*. In a similar study, the combined therapy of Nurr1-overexpressing ES cells with AA, Shh, and FGF-8 increased the proportion of TH-positive neurons from 14–61% *(83)*.

To optimize DAergic differentiation and minimize cell death, anti-apoptotic genes (e.g., Bcl-XL) have been overexpressed in ES cells *(84)*. This overexpression of Bcl-XL in mouse ES cells increased the estimate of TH-immunopositive neurons from 18.4% to 30.9% *(84)*.

In another study, a human ES cell line was genetically modified to express TH and guanosine triphosphate cyclohydrolase I—two enzymes necessary for DA synthesis *(85)*. Thus, the endogenous capacity of the ES cells to differentiate into DAergic neurons was not utilized; instead, they were used as vehicles to express the DA-synthesizing machinery. These genetically modified ES cells could produce levodopa in vitro *(85)*. When they were transplanted into the striatum of hemiparkinsonian rats with unilateral 6-hydroxydopamine (6-OHDA) lesions of the nigrostriatal pathway, the cells were reported to express TH immunoreactivity and elicit a 46% reduction in the apomorphine-induced rotational behavior up to 6 wk after transplantation, compared to the rotational behavior before transplantation *(85)*. However, there was no evidence of DA production in vivo, and functional assays other than drug-induced rotation score were not performed.

DAergic Differentiation of ES Cells In Vivo

The most stringent approach to test whether ES cells have differentiated into true DAergic neurons is to graft them to the adult brain of animals with DA-depleting lesions. In this frequently employed paradigm, it is possible to determine if they can undertake behavioral functions normally associated with DAergic neurons.

Owing to their pluripotency, undifferentiated ES cells transplanted into immunosuppressed mice tend to generate lethal teratomas that contain a wide variety of somatic cells from all three germ layers *(86)*. Surprisingly, a significant proportion of mouse undifferentiated ES cells dissociated into single-cell suspensions has been reported to differentiate into DAergic neurons when grafted to the striatum of immunosuppressed rats *(87)*. Such grafts can even restore parkinsonian symptoms in hemiparkinsonian rats *(87)*. The mechanisms that cause these undifferentiated ES cells of mouse origin to adopt a DAergic neuronal phenotype when transplanted at low density are not understood. Neuronal differentiation has been suggested to represent a "default" pathway of differentiation for mouse ES cells *(88)*. The theory is that the lack of cell–cell contacts that might stimulate the ES cells to adopt other differentiation protocols would be absent in the single-cell suspensions. Consequently, the cells would only differentiate along the default pathway. However, among 25 rats receiving transplants of 1000–2000 ES cells, lethal teratomas were seen in five rats, and six rats had no surviving grafts at all *(87)*. Clearly, single-cell suspensions do not lead to exclusively DAergic neurons. In addition, although these results show that mouse ES cells can develop into DAergic neurons in the environment of the lesioned striatum, these data also demonstrate that a cell-based therapy for PD cannot rely on undifferentiated stem cells because of their tendency to form teratomas.

When mouse ES cells that had been grown on PA6 cells (i.e., exposed to SDIA) were implanted into the striatum of mice with 6-OHDA lesions, surviving DAergic neurons were observed 2 wk postgrafting. However, only a minority (3%) of the implanted cells were TH-immunopositive neurons *(76,89)*. Following transplantation of mouse ES cells differentiated by being grown on top of the bone marrow stromal cell line MS5 and then exposed to a cocktail of cytokines (Shh, FGF-8, BDNF, and AA) into the striatum of a mouse PD model more than 70% of the TH-positive cells survived. These grafted cells caused a reduction greater than 70% in the drug-induced motor asymmetry in the mice *(78)*.

Hemiparkinsonian rats grafted with cells derived from the Nurr1-overexpressing mouse ES cells showed marked behavioral improvement in several tests of drug-induced and spontaneous motor function *(82)*. Interestingly, the most efficient differentiation into DAergic neurons (80% of all cells exhibiting neuronal markers) was achieved when the mouse ES cells were both engineered to overexpress Nurr1 and were exposed to a combination of Shh and FGF-8 *(82)*. A particularly impressive aspect of this study was that the grafts reversed deficits in spontaneous behaviors (e.g., in the paw-reaching and -stepping tests), as opposed to just drug-induced rotation. In addition, the grafted TH-immunopositive neurons were found to display electrophysiological characteristics consistent with a DAergic phenotype. In another study, mouse ES cells overexpressing Bcl-XL exhibited more extensive fiber outgrowth and supported a more pronounced amelioration of behavioral symptoms than transplanted wild-type ES cells *(84)*. These observations were similar with earlier findings showing that overexpression of Bcl-2, another related antiapoptotic molecule, increases the extent of their fiber outgrowth into the host brain in grafted primary mouse midbrain DAergic neurons *(90,91)*.

As mentioned in the DAergic Differentiation of ES Cells by Soluble Factors section, it is possible to derive DAergic neurons from human ES cells in vitro. However, successful integration and survival of transplanted DAergic differentiated human ES cells have not been reported yet. Thus, an important future step is to examine under what conditions human ES-derived DAergic neurons are able to survive, integrate, and function when transplanted into the rat and the primate models of PD.

ALTERNATIVE NON-NEURAL STEM CELL SOURCES

Somatic stem cells have been claimed to possess a broad differentiation potential, and several studies over the past 5 yr have suggested that stem cells from tissues outside of the brain are also capable of generating neurons *(92)*. Such stem cells may be considered alternatives in PD, and this chapter describes two potential sources: bone marrow–derived cells *(93–95)* and umbilical cord stem cells *(96)*. Other potential sources include adult multipotent progenitor cells derived from bone marrow *(97)* and skin-derived stem cells *(98)*, but only a few studies on these cell sources have been published to date.

Bone Marrow Cells

Adult bone marrow cells have been recently proposed as an alternative source of neural donor tissue. Unlike ES cells, which are derived from the inner cell mass of a blastocyst, or neural stem cells isolated from different brain regions, adult bone marrow cells are divided into two subpopulations: hematopoietic and bone marrow stromal cells. The bone marrow–derived stem cells are interesting to mention in the context of cell therapy for PD, as it has been repeatedly suggested that they can generate DAergic neurons. When differentiated, hematopoietic stem cells normally give rise to myeloid and lymphoid cells *(99)*, whereas bone marrow stromal cells differentiate into fat *(100)*, tendon and cartilage *(101)*, bone *(102)* and muscle *(103)*, as well as constituting a microenvironment that is required for the proliferation of hematopoietic stem cells *(104)*. It has not yet been reported whether hematopoietic stem cells (short- and/or long-term renewal progenitors), characterized by specific cell surface markers (i.e., lineage-negative, c-Kit-positive, and Sca-1-positive), can differentiate into lineages other than blood cells. However, unsorted bone marrow cells may be capable of differentiating into neural cells in vitro *(93,105,106)*.

Furthermore, two landmark studies in 2000 suggested that bone marrow cells could differentiate into neurons in vivo *(107,108)*. In these studies, infusion of bone marrow stem cells into neonatal or irradiated adult mice showed that small fractions of neuronal cells (0.3–2.3%) in the central nervous system contained donor cell markers. However, these findings have been questioned by a series of follow-up studies, because bone marrow stem cells have been shown to adopt the functional features of neural lineages via cell fusion, rather than by signal-mediated differentiation *(109–112)*. Another study based on bone marrow transplantation has highlighted the possible transdifferentiation potential of bone marrow stromal cells *(113)*. Yet, the current consensus is that there is insufficient evidence that bone marrow stem cells can express neuronal markers upon transplantation, only that they can differentiate into microglial cells after transplantation *(109,111,114)*.

A recent study suggested that engineered human bone marrow cells could be suitable for autologous transplantation in PD *(94)*. In this study, human bone marrow cells transfected with the Notch intracellular domain protein and treated with neurotrophic factors differentiated into TH-expressing neurons. Transplantation of these TH-expressing neurons into the striatum of 6-OHDA-lesioned rats resulted in reductions in apomorphine-induced motor asymmetry *(94)*, but effects on spontaneous behavior were not shown, and there was no direct evidence of production and release of DA. Because apomorphine-induced rotation can also decrease in response to nonspecific damage of the striatum, these findings did not unequivocally prove that the grafted cells were DAergic. Nevertheless, this initial study definitely warrants follow-up experiments; if bone marrow cells can generate DAergic neurons, it would have tremendous implications for cell therapy in PD. In a similar manner, rat and human bone marrow cells have been transplanted into the cerebral ventricles of 15-d-old rat embryos *(115)*. These cells migrated throughout the developing brain, and after 1 mo they were found to express the neuronal marker NeuN in the cortex and olfactory bulb. From that study, and contrasting previous reports where bone marrow was grafted into the adult brain, the embryonic brain may contain factors vital to the promotion of neuronal differentiation.

Interestingly, considerable overlap of gene expression profiles between ES, neural stem cells, and hematopoietic stem cells has been found *(116)*. This overlap suggests that common pathways exist between different stem cell classes undergoing differentiation. Thus, it is conceivable that a differentiation protocol, established as effective in generating neurons from embryonic and neural stem cells, may be applied to the other classes of stem cells and may stimulate the formation of the same specific neuronal phenotypes.

An alternative explanation for the presence of neural markers in bone marrow–derived cells is that the bone marrow niche contains several different types of committed tissue-specific stem cells. These cells are merely waiting for a signal indicating that they need to migrate back to the organ where they were derived and differentiate into cells specific for that particular organ *(117)*. Thus, the bone marrow cells prone to neural differentiation may have originated from neural tissue, rather than directly from the bone marrow lineage, and are kept as a reservoir in the bone marrow niche. This could explain why, in response to the mitogens bFGF and epidermal growth factor, some bone marrow cells form neurospheres that give rise to neural cells when differentiated *(118)* or express neuronal markers prior to differentiation *(119)*. There has been little characterization of these subgroups thus far.

Among all the bone marrow studies, the concept that adult bone marrow stromal cells could be used in the future for brain repair seems to be the most significant. However, the potential appears to lie more specifically in certain cell subgroups located within the bone marrow. A greater focus on these subpopulations may yield more encouraging results than those presently obtained.

Umbilical Cord Blood Cells

Blood cells taken from the umbilical cord have been shown to have the capability to differentiate into both neuronal and glial cell types *(96,120,121)*. Blood in the human umbilical cord contains hematopoietic and mesenchymal stem cells; yet, in contrast to the adult bone marrow, the advantage of using cord blood cells is that the tissue can be obtained at birth and cryopreserved for several years. Such cells could be used as routine starting material for the isolation and expansion of cells for autologous transplantation *(120)*. Thus, blood cells of the umbilical cord could be transdifferentiated into other cell types, human cord blood banks could be created for use in autologous transplantation.

THE OPTIMAL CELL FOR PD THERAPY: PROS AND CONS WITH SOURCES CURRENTLY AVAILABLE

Embryonic Neural Stem Cells as Cell Therapy for PD

Although DAergic neurons can be generated in large numbers from neural precursors under special conditions *(25)*, they are still not a realistic option as donor cells for transplantation in PD. Limitations in long-term propagation of the cells and difficulties in achieving stable DAergic differentiation continue to be major problems. As mentioned in the DAergic Differentiation by Epigenetic Factors section, the efficiency of DAergic differentiation decreases with time in midbrain neural precursors after proliferation for extended periods in vitro *(30)*. Furthermore, although expanded precursors can differentiate into DA-producing neurons in vitro, they do not always maintain a DAergic phenotype after intracerebral implantation *(31,36,59)*. Improved understanding of the biology of neural stem cells could reveal more clues about the signals that favor the differentiation of stem cells into dopaminergic neurons and help resolve the puzzling fact that cells that have differentiated in vitro do not always preserve their phenotype.

ES Cells as Cell Source for PD

The time necessary to expand human ES cells into large numbers of human DAergic neurons is relatively long (at least 39 d [*79,80*; see Fig. 3]). The length of the existing differentiation protocol may be a limitation for a routine production of DAergic neurons (i.e., increased risk of contamination).

The risk that undifferentiated cells are included among transplanted human ES-derived DAergic neurons is a major issue. Obviously, this can lead to the formation of lethal teratomas in the host brain. Chromosomal aberrations identified in mid-term cultured human ES cells are a cause for concern *(122)*. This suggests that long-term propagation of human ES cells can lead to abnormalities in the cells intended to be used for grafting purposes. Eventually, existing human ES cell lines could accumulate chromosomal aberrations and mutations that either inhibit differentiation into DAergic neurons or cause differentiation along unpredictable pathways. Both scenarios would render them highly unsuitable for grafting and require that new "mutation-free" human ES cell lines are derived. The most important current shortcoming with human ES cell lines as a potential source of cells for grafting in patients with PD is the absence of convincing transplantation data in experimental animals. Indeed, the survival after grafting, integration in the host brain, and the release of DA, leading to the decline in motor dysfunctions, remains to be shown by the human ES-derived DAergic neurons in animal models of PD.

Goals to Achieve for Cell Therapy in PD

Importantly, grafts of stem cell-derived neurons should be performed in such a manner that dyskinesias are not triggered in patients with PD. Currently, we do not understand why graft-induced dyskinesias can develop in some patients. Nevertheless, it will probably be possible to graft cells to PD patients that reproducibly never cause dyskinesias. It may require a well-characterized and homogenous source of cells. Most likely, it will be necessary to target these cells to specific host brain regions, and possibly only certain types of PD patients (regarding age of onset, duration of disease, and history of medication), will be suitable to avoid the dyskinesias. Even if the problem is resolved, stem cell therapy has its own set of issues that need attention. The optimal differentiation stage of the stem cells at the time of transplantation is likely to be an important factor to understand. Final differentiation occurs once progenitors have exited the cell cycle, and it is often characterized by expression of the enzymes or neurotransmitters required for neural function *(123,124)*. We need to determine whether the differentiated ES cells or neural stem cell culture should be harvested *before* or *after* the cells undergo their genetically determined final cell division. Based on the authors' experience grafting primary DAergic neurons obtained from aborted embryos, the optimal time for harvesting might be when they commence a period of vigorous neurite growth directed toward appropriate targets in the developing brain *(125)*. Although there are reasons to believe that predifferentiated cells may be particularly sensitive to the trauma involved in transplantation, evidence also shows that an undifferentiated DAergic progenitor will not develop into a DAergic neuron in the adult striatum *(125)*.

Graft immune rejection is another limitation of cell-based therapies that could be overcome using stem cells. Obviously, if somatic stem cells could be used, it may be possible to isolate the patient's own cells and perform an autograft, which raises no immune issue. Also, human ES-based therapy offers solutions to the immune problems. Creation of human blastocysts via the transfer of the nucleus of a patient's somatic cell to an enucleated oocyte would allow the derivation of human ES cell lines immunocompatible with patient. This is the basis of therapeutic cloning. The procedure is ethically controversial; aspects of the methodology are similar to those used for reproductive cloning—a practice, if applied to humans, considered unethical in most societies. Despite all the ethical discussions around human cloning, a group in South Korea has already published the derivation of a human ES cell line from a cloned human blastocyst *(126)*. This cell line is claimed to be able to give rise to any tissue that would not be rejected by the donor of the somatic nucleus.

Purification of DAergic Progenitors for Cell Therapy in PD

Thus far, ES cells have not been shown to produce a specific, pure population of cells after differentiation. Moreover, some of the ES cells in cultures derived from EBs do not undergo differentiation. Such cells should be removed or eliminated prior to transplantation; otherwise, they could be the source of unwanted cell proliferation. When developing cell therapy for PD, it would be valuable to purify the DAergic cell population using a methodology like fluorescence-activated cell sorting (FACS). Then, it would be possible to graft predetermined, homogenous populations of cells and define the optimal cell type for grafting. In some studies, another approach has been used in attempts to obtain more homogenous grafts. For example, mouse ES cells have been treated with antimitotic drugs prior to transplantation to avoid the overgrowth of mesenchymal tissue and tumor formation. Because the DAergic neurons derived from the ES cell cultures are postmitotic, they are not affected by the antimitotic agent *(77)*.

Currently, there is a lack of an adequate panel of cell surface markers that can help identify cells at different stages of differentiation. Green fluorescent protein (GFP) expression, under the control of specific promoters related to DAergic differentiation, is one method to identify DAergic neurons. Already, labeling of TH-expressing DAergic neurons has been shown to be a simple and useful system for the purification of DAergic neurons by FACS. Using this approach, GFP-fluorescent DAergic neurons can be used for the functional identification of molecules governing mesencephalic DAergic differentiation and for preclinical research, including pharmaceutical drug screening and transplantation *(127)*. Another study has shown successful FACS of mesencephalic DAergic neurons using the Pitx3-regulated GFP reporter system *(128)*. The *pitx3* gene is expressed in mesencephalic DAergic neurons and is involved in the generation of the mesencephalic DAergic phenotype *(129)*. In this case, ES cells were generated from transgenic mice that express GFP inserted into the transcription factor Pitx3. This demonstrates that the natural expression of specific transcription factors can be used as a selection marker for development along the DAergic lineage. Thus, the use of reporter expression systems can be a powerful tool for the purification of differentiating DAergic neurons for transplantation. In the adult central nervous system, expression of the transmembrane protein Δ-like 1 has been observed in monoaminergic nuclei in the adult brain, including SNpc and the ventral tegmental area *(130)*. This suggests that the Δ-like 1 protein could be used to select dopaminergic neurons. If a panel of cell surface markers is developed, human ES-derived DAergic neurons could be sorted based on markers characteristic of nigral DAergic neurons before transplantation. Thereby, it would be possible to enrich for DAergic neurons and to remove unwanted cell material from the cultured cells.

CONCLUSION

The key to restorative stem cell therapy is to induce the differentiation of stem and progenitor cells into the desired phenotype. The elucidation of regulatory cascades influencing the specification and development of neuronal types is essential in embryonic (neural) stem cell research. Although our current understanding of factors influencing DAergic neuron development is in its infancy, the field is advancing rapidly. Possibly, the future will show that somatic stem cells can be differentiated into DAergic neurons. As evident from data discussed in this chapter, a great deal is known already about mouse and human ES and neural stem cells and their capacity to differentiate into DAergic neurons. These stem cells may someday facilitate the development of neural transplantation as therapy for PD.

ACKNOWLEDGMENTS

Our work described in this chapter was supported by grants from the Danish Academy of Technical Sciences, the National Institute of Health, USA grant number 5 R21 NS043717-02, the United States Army Medical Research Acquisition Activity (USAMRAA) award number. W81XWH-04-1-0366, Swedish Research Council Project number K2002-99SX-14472-01A, the Swedish Parkinson's Disease Foundation, Konung Gustaf V:s and Drottning Victorias Stiftelse, Syskonen Svensson's Foundation, the USAMRMC grant, and the Michael J. Fox Foundation for Parkinson's Research.

REFERENCES

1. Lindvall, O. and Bjorklund, A. (1983) Dopamine and norepinephrine-containing neuron systems; their anatomy in the rat brain. In: *Chemical Neuroanatomy*. Emerson, P. C., ed. Raven, New York, pp. 229–255.
2. Bjorklund, A. and Lindvall, O. (1984) Dopamine-containing systems in the CNS. In: Bjorklund, A. and Hokfelt, T., eds. *Handbook of chemical neuroanatomy*. Elsevier, Amsterdam, pp. 55–123.
3. Jenner, P. (2003) Dopamine agonists, receptor selectivity and dyskinesia induction in Parkinson's disease. *Curr. Opin. Neurol.* **16(Suppl 1),** S3–S7.
4. Lindvall, O., Brundin, P., Widner, H., et al. (1990) Grafts of fetal dopamine neurons survive and improve motor function in Parkinson's disease. *Science* **247,** 574–577.
5. Kordower, J. H., Freeman, T. B., Chen, E. Y., et al. (1998) Fetal nigral grafts survive and mediate clinical benefit in a patient with Parkinson's disease. *Mov. Disord.* **13,** 383–393.
6. Tabbal, S., Fahn, S., and Frucht, S. (1998) Fetal tissue transplantation [correction of transplanation] in Parkinson's disease. *Curr. Opin. Neurol.* **11,** 341–349.
7. Lang, A. E. and Obeso, J. A. (2004) Time to move beyond nigrostriatal dopamine deficiency in Parkinson's disease. *Ann. Neurol.* **55,** 761–765.
8. Lang, A. E. and Obeso, J. A. (2004) Challenges in Parkinson's disease: restoration of the nigrostriatal dopamine system is not enough. *Lancet Neurol.* **3,** 309–316.
9. Lindvall, O. and Hagell, P. (2001) Cell therapy and transplantation in Parkinson's disease. *Clin. Chem. Lab. Med.* **39,** 356–361.

10. Piccini, P., Brooks, D. J., Bjorklund, A., et al. (1999) Dopamine release from nigral transplants visualized in vivo in a Parkinson's patient. *Nat. Neurosci.* **2**, 1137–1140.

11. Freed, C. R., Greene, P. E., Breeze, R. E., et al. (2001) Transplantation of embryonic dopamine neurons for severe Parkinson's disease. *N. Engl. J. Med.* **344**, 710–719.

12. Hagell, P., Piccini, P., Bjorklund, A., et al. (2002) Dyskinesias following neural transplantation in Parkinson's disease. *Nat. Neurosci.* **5**, 627–628.

13. Olanow, C. W., Goetz, C. G., Kordower, J. H., et al. (2003) A double-blind controlled trial of bilateral fetal nigral transplantation in Parkinson's disease. *Ann. Neurol.* **54**, 403–414.

14. Ma, Y., Feigin, A., Dhawan, V., et al. (2002) Dyskinesia after fetal cell transplantation for parkinsonism: a PET study. *Ann. Neurol.* **52**, 628–634.

15. Isacson, O., Bjorklund, L. M., and Schumacher, J. M. (2003) Toward full restoration of synaptic and terminal function of the dopaminergic system in Parkinson's disease by stem cells. *Ann. Neurol.* **53(Suppl 3)**, S135–S146; discussion S46–S48.

16. Sauer, H. and Brundin, P. (1991) Effects of cool storage on survival and function of intrastriatal ventral mesencephalic grafts. *Restor. Neurol. Neurosci.* **2**, 123–135.

17. Brundin, P., Karlsson, J., Emgard, M., et al. (2000) Improving the survival of grafted dopaminergic neurons: a review over current approaches. *Cell Transplant* **9**, 179–195.

18. Hagell, P. and Brundin, P. (2001) Cell survival and clinical outcome following intrastriatal transplantation in Parkinson disease. *J. Neuropathol. Exp. Neurol.* **60**, 741–752.

19. Sortwell, C. E. (2003) Strategies for the augmentation of grafted dopamine neuron survival. *Front Biosci.* **8**, S522–S532.

20. Gage, F. H. (2000) Mammalian neural stem cells. *Science* **287**, 1433–1438.

21. Reynolds, B. A., Tetzlaff, W., and Weiss, S. (1992) A multipotent EGF-responsive striatal embryonic progenitor cell produces neurons and astrocytes. *J. Neurosci.* **12**, 4565–4574.

22. Reynolds, B. A. and Weiss, S. (1992) Generation of neurons and astrocytes from isolated cells of the adult mammalian central nervous system. *Science* **255**, 1707–1710.

23. Svendsen, C. N., ter Borg, M. G., Armstrong, R. J., et al. (1998) A new method for the rapid and long term growth of human neural precursor cells. *J. Neurosci. Methods* **85**, 141–152.

24. Svendsen, C. N. and Rosser, A. E. (1995) Neurones from stem cells? *Trends Neurosci.* **18**, 465–467.

25. Studer, L., Tabar, V., and McKay, R. D. (1998) Transplantation of expanded mesencephalic precursors leads to recovery in parkinsonian rats. *Nat. Neurosci.* **1**, 290–295.

26. Bouvier, M. M. and Mytilineou, C. (1995) Basic fibroblast growth factor increases division and delays differentiation of dopamine precursors in vitro. *J. Neurosci.* **15**, 7141–7149.

27. Nishino, H., Hida, H., Takei, N., et al. (2000) Mesencephalic neural stem (progenitor) cells develop to dopaminergic neurons more strongly in dopamine-depleted striatum than in intact striatum. *Exp. Neurol.* **164,** 209–214.
28. Martinez-Serrano, A., Rubio, F. J., Navarro, B., et al. (2001) Human neural stem and progenitor cells: in vitro and in vivo properties, and potential for gene therapy and cell replacement in the CNS. *Curr. Gene Ther.* **1,** 279–299.
29. Arsenijevic, Y. and Weiss, S. (1998) Insulin-like growth factor-I is a differentiation factor for postmitotic CNS stem cell-derived neuronal precursors: distinct actions from those of brain-derived neurotrophic factor. *J. Neurosci.* **18,** 2118–2128.
30. Yan, J., Studer, L., and McKay, R. D. (2001) Ascorbic acid increases the yield of dopaminergic neurons derived from basic fibroblast growth factor expanded mesencephalic precursors. *J. Neurochem.* **76,** 307–311.
31. Ostenfeld, T., Joly, E., Tai, Y. T., et al. (2002) Regional specification of rodent and human neurospheres. *Brain Res. Dev. Brain Res.* **134,** 43–55.
32. Studer, L., Csete, M., Lee, S. H., et al. (2000) Enhanced proliferation, survival, and dopaminergic differentiation of CNS precursors in lowered oxygen. *J. Neurosci.* **20,** 7377–7383.
33. Storch, A., Paul, G., Csete, M., et al. (2001) Long-term proliferation and dopaminergic differentiation of human mesencephalic neural precursor cells. *Exp. Neurol.* **170,** 317–325.
34. Carvey, P. M., Ling, Z. D., Sortwell, C. E., et al. (2001) A clonal line of mesencephalic progenitor cells converted to dopamine neurons by hematopoietic cytokines: a source of cells for transplantation in Parkinson's disease. *Exp. Neurol.* **171,** 98–108.
35. Brundin, P. and Bjorklund, A. (1998) Survival of expanded dopaminergic precursors is critical for clinical trials. *Nat. Neurosci.* **1,** 537.
36. Kim, J. Y., Koh, H. C., Lee, J. Y., et al. (2003) Dopaminergic neuronal differentiation from rat embryonic neural precursors by Nurr1 overexpression. *J. Neurochem.* **85,** 1443–1454.
37. Sakurada, K., Ohshima-Sakurada, M., Palmer, T. D., and Gage, F. H. (1999) Nurr1, an orphan nuclear receptor, is a transcriptional activator of endogenous tyrosine hydroxylase in neural progenitor cells derived from the adult brain. *Development* **126,** 4017–4026.
38. Zetterstrom, R. H., Solomin, L., Jansson, L., et al. (1997) Dopamine neuron agenesis in Nurr1-deficient mice. *Science* **276,** 248–250.
39. Ostenfeld, T., Caldwell, M. A., Prowse, K. R., et al. (2000) Human neural precursor cells express low levels of telomerase in vitro and show diminishing cell proliferation with extensive axonal outgrowth following transplantation. *Exp. Neurol.* **164,** 215–226.
40. Sommer, L. and Rao, M. (2002) Neural stem cells and regulation of cell number. *Prog. Neurobiol.* **66,** 1–18.
41. Frederiksen, K., Jat, P. S., Valtz, N., et al. (1988) Immortalization of precursor cells from the mammalian CNS. *Neuron* **1,** 439–448.
42. Cepko, C. L. (1989) Immortalization of neural cells via retrovirus-mediated oncogene transduction. *Annu. Rev. Neurosci.* **12,** 47–65.

43. Ryder, E. F., Snyder, E. Y., and Cepko, C. L. (1990) Establishment and characterization of multipotent neural cell lines using retrovirus vector-mediated oncogene transfer. *J. Neurobiol.* **21,** 356–375.
44. Snyder, E. Y., Deitcher, D. L., Walsh, C., et al. (1992) Multipotent neural cell lines can engraft and participate in development of mouse cerebellum. *Cell* **68,** 33–51.
45. Matsuura, N., Lie, D. C., Hoshimaru, M., et al. (2001) Sonic hedgehog facilitates dopamine differentiation in the presence of a mesencephalic glial cell line. *J. Neurosci.* **21,** 4326–4335.
46. Wu, K. J., Grandori, C., Amacker, M., et al. (1999) Direct activation of TERT transcription by c-MYC. *Nat. Genet.* **21,** 220–224.
47. Lundblad, V. and Wright, W. E. (1996) Telomeres and telomerase: a simple picture becomes complex. *Cell* **87,** 369–375.
48. Bodnar, A. G., Ouellette, M., Frolkis, M., et al. (1998) Extension of life-span by introduction of telomerase into normal human cells. *Science* **279,** 349–352.
49. Sharma, H. W., Sokoloski, J. A., Perez, J. R., et al. (1995) Differentiation of immortal cells inhibits telomerase activity. *Proc. Natl. Acad. Sci. USA* **92,** 12343–12346.
50. Bryan, T. M., Englezou, A., Gupta, J., et al. (1995) Telomere elongation in immortal human cells without detectable telomerase activity. *Embo. J.* **14,** 4240–4248.
51. Harley, C. B., Futcher, A. B., and Greider, C. W. (1990) Telomeres shorten during ageing of human fibroblasts. *Nature* **345,** 458–460.
52. Flax, J. D., Aurora, S., Yang, C., et al. (1998) Engraftable human neural stem cells respond to developmental cues, replace neurons, and express foreign genes. *Nat. Biotechnol.* **16,** 1033–1039.
53. Villa, A., Snyder, E. Y., Vescovi, A., and Martinez-Serrano, A. (2000) Establishment and properties of a growth factor-dependent, perpetual neural stem cell line from the human CNS. *Exp. Neurol.* **161,** 67–84.
54. Lotharius, J., Barg, S., Wiekop, P., et al. (2002) Effect of mutant alpha-synuclein on dopamine homeostasis in a new human mesencephalic cell line. *J. Biol. Chem.* **277,** 38884–38894.
55. Kitchens, D. L., Snyder, E. Y., and Gottlieb, D. I. (1994) FGF and EGF are mitogens for immortalized neural progenitors. *J. Neurobiol.* **25,** 797–807.
56. Hoshimaru, M., Ray, J., Sah, D. W., and Gage, F. H. (1996) Differentiation of the immortalized adult neuronal progenitor cell line HC2S2 into neurons by regulatable suppression of the v-myc oncogene. *Proc. Natl. Acad. Sci. USA* **93,** 1518–1523.
57. Sah, D. W., Ray, J., and Gage, F. H. (1997) Bipotent progenitor cell lines from the human CNS. *Nat. Biotechnol.* **15,** 574–580.
58. Wagner, J., Akerud, P., Castro, D. S., et al. (1999) Induction of a midbrain dopaminergic phenotype in Nurr1-overexpressing neural stem cells by type 1 astrocytes. *Nat. Biotechnol.* **17,** 653–659.
59. Kim, T. E., Lee, H. S., Lee, Y. B., et al. (2003) Sonic hedgehog and FGF8 collaborate to induce dopaminergic phenotypes in the Nurr1-overexpressing neural stem cell. *Biochem. Biophys. Res. Commun.* **305,** 1040–1048.

60. Ye, W., Shimamura, K., Rubenstein, J. L., et al. (1998) FGF and Shh signals control dopaminergic and serotonergic cell fate in the anterior neural plate. *Cell* **93,** 755–766.
61. Snyder, E. Y., Yoon, C., Flax, J. D., and Macklis, J. D. (1997) Multipotent neural precursors can differentiate toward replacement of neurons undergoing targeted apoptotic degeneration in adult mouse neocortex. *Proc. Natl. Acad. Sci. USA* **94,** 11663–11668.
62. Rubio, F. J., Bueno, C., Villa, A., et al. (2000) Genetically perpetuated human neural stem cells engraft and differentiate into the adult mammalian brain. *Mol. Cell Neurosci.* **16,** 1–13.
63. Renfranz, P. J., Cunningham, M. G., and McKay, R. D. (1991) Region-specific differentiation of the hippocampal stem cell line HiB5 upon implantation into the developing mammalian brain. *Cell* **66,** 713–729.
64. Villa, A., Navarro, B., and Martinez-Serrano, A. (2002) Genetic perpetuation of in vitro expanded human neural stem cells: cellular properties and therapeutic potential. *Brain Res. Bull.* **57,** 789–794.
65. Westerman, K. A. and Leboulch, P. (1996) Reversible immortalization of mammalian cells mediated by retroviral transfer and site-specific recombination. *Proc. Natl. Acad. Sci. USA* **93,** 8971–8976.
66. Martin, G. R. (1980) Teratocarcinomas and mammalian embryogenesis. *Science* **209,** 768–776.
67. Smith, A. G. (2001) Embryo-derived stem cells: of mice and men. *Annu. Rev. Cell Dev. Biol.* **17,** 435–462.
68. Richards, M., Fong, C. Y., Chan, W. K., et al. (2002) Human feeders support prolonged undifferentiated growth of human inner cell masses and embryonic stem cells. *Nat. Biotechnol.* **20,** 933–936.
69. Hovatta, O., Mikkola, M., Gertow, K., et al. (2003) A culture system using human foreskin fibroblasts as feeder cells allows production of human embryonic stem cells. *Hum. Reprod.* **18,** 1404–1409.
70. Rosler, E. S., Fisk, G. J., Ares, X., et al. (2004) Long-term culture of human embryonic stem cells in feeder-free conditions. *Dev. Dyn.* **229,** 259–274.
71. Thomson, J. A., Itskovitz-Eldor, J., Shapiro, S. S., et al. (1998) Embryonic stem cell lines derived from human blastocysts. *Science* **282,** 1145–1147.
72. Zhang, S. C., Wernig, M., Duncan, I. D., et al. (2001) In vitro differentiation of transplantable neural precursors from human embryonic stem cells. *Nat. Biotechnol.* **19,** 1129–1133.
73. Reubinoff, B. E., Pera, M. F., Fong, C. Y., et al. (2000) Embryonic stem cell lines from human blastocysts: somatic differentiation in vitro. *Nat. Biotechnol.* **18,** 399–404.
74. Lee, S. H., Lumelsky, N., Studer, L., et al. (2000) Efficient generation of midbrain and hindbrain neurons from mouse embryonic stem cells. *Nat. Biotechnol.* **18,** 675–679.
75. Itskovitz-Eldor, J., Schuldiner, M., Karsenti, D., et al. (2000) Differentiation of human embryonic stem cells into embryoid bodies compromising the three embryonic germ layers. *Mol. Med.* **6,** 88–95.

76. Kawasaki, H., Mizuseki, K., Nishikawa, S., et al. (2000) Induction of midbrain dopaminergic neurons from ES cells by stromal cell-derived inducing activity. *Neuron* **28,** 31–40.
77. Kawasaki, H., Suemori, H., Mizuseki, K., et al. (2002) Generation of dopaminergic neurons and pigmented epithelia from primate ES cells by stromal cell-derived inducing activity. *Proc. Natl. Acad. Sci. USA* **99,** 1580–1585.
78. Barberi, T., Klivenyi, P., Calingasan, N. Y., et al. (2003) Neural subtype specification of fertilization and nuclear transfer embryonic stem cells and application in parkinsonian mice. *Nat. Biotechnol.* **21,** 1200–1207.
79. Park, S., Lee, K. S., Lee, Y. J., et al. (2004) Generation of dopaminergic neurons in vitro from human embryonic stem cells treated with neurotrophic factors. *Neurosci. Lett.* **359,** 99–103.
80. Perrier, A. L., Tabar, V., Barberi, T., et al. (2004) Derivation of midbrain dopamine neurons from human embryonic stem cells. *Proc. Natl. Acad. Sci. USA* **101,** 12543–12548.
81. Farkas, L. M., Dunker, N., Roussa, E., et al. (2003) Transforming growth factor-beta(s) are essential for the development of midbrain dopaminergic neurons in vitro and in vivo. *J. Neurosci.* **23,** 5178-5186.
82. Kim, J. H., Auerbach, J. M., Rodriguez-Gomez, J. A., et al. (2002) Dopamine neurons derived from embryonic stem cells function in an animal model of Parkinson's disease. *Nature* **418,** 50–56.
83. Chung, S., Sonntag, K. C., Andersson, T., et al. (2002) Genetic engineering of mouse embryonic stem cells by Nurr1 enhances differentiation and maturation into dopaminergic neurons. *Eur. J. Neurosci.* **16,** 1829–1838.
84. Shim, J. W., Koh, H. C., Chang, M. Y., et al. (2004) Enhanced in vitro midbrain dopamine neuron differentiation, dopaminergic function, neurite outgrowth, and 1-methyl-4-phenylpyridium resistance in mouse embryonic stem cells overexpressing Bcl-XL. *J. Neurosci.* **24,** 843–852.
85. Park, S., Kim, E. Y., Ghil, G. S., et al. (2003) Genetically modified human embryonic stem cells relieve symptomatic motor behavior in a rat model of Parkinson's disease. *Neurosci. Lett.* **353,** 91–94.
86. Pera, M. F. (2001) Human pluripotent stem cells: a progress report. *Curr. Opin. Genet. Dev.* **11,** 595–599.
87. Bjorklund, L. M., Sanchez-Pernaute, R., Chung, S., et al. (2002) Embryonic stem cells develop into functional dopaminergic neurons after transplantation in a Parkinson rat model. *Proc. Natl. Acad. Sci. USA* **99,** 2344–2349.
88. Tropepe, V., Hitoshi, S., Sirard, C., et al. (2001) Direct neural fate specification from embryonic stem cells: a primitive mammalian neural stem cell stage acquired through a default mechanism. *Neuron* **30,** 65–78.
89. Morizane, A., Takahashi, J., Takagi, Y., et al. (2002) Optimal conditions for in vivo induction of dopaminergic neurons from embryonic stem cells through stromal cell-derived inducing activity. *J. Neurosci. Res.* **69,** 934–939.
90. Holm, K. H., Cicchetti, F., Bjorklund, L., et al. (2001) Enhanced axonal growth from fetal human bcl-2 transgenic mouse dopamine neurons transplanted to the adult rat striatum. *Neuroscience* **104,** 397–405.

91. Schierle, G. S., Leist, M., Martinou, J. C., et al. (1999) Differential effects of Bcl-2 overexpression on fibre outgrowth and survival of embryonic dopaminergic neurons in intracerebral transplants. *Eur. J. Neurosci.* **11,** 3073–3081.
92. Corti, S., Locatelli, F., Strazzer, S., et al. (2003) Neuronal generation from somatic stem cells: current knowledge and perspectives on the treatment of acquired and degenerative central nervous system disorders. *Curr. Gene Ther.* **3,** 247–272.
93. Woodbury, D., Schwarz, E. J., Prockop, D. J., and Black, I. B. (2000) Adult rat and human bone marrow stromal cells differentiate into neurons. *J. Neurosci. Res.* **61,** 364–370.
94. Dezawa, M., Kanno, H., Hoshino, M., et al. (2004) Specific induction of neuronal cells from bone marrow stromal cells and application for autologous transplantation. *J. Clin. Invest.* **113,** 1701–1710.
95. Kohyama, J., Abe, H., Shimazaki, T., et al. (2001) Brain from bone: efficient "meta-differentiation" of marrow stroma-derived mature osteoblasts to neurons with Noggin or a demethylating agent. *Differentiation* **68,** 235–244.
96. Buzanska, L., Machaj, E. K., Zablocka, B., et al. (2002) Human cord blood-derived cells attain neuronal and glial features in vitro. *J. Cell. Sci.* **115,** 2131–2138.
97. Jiang, Y., Henderson, D., Blackstad, M., et al. (2003) Neuroectodermal differentiation from mouse multipotent adult progenitor cells. *Proc. Natl. Acad. Sci. USA* **100(Suppl 1),** 11854–11860.
98. Joannides, A., Gaughwin, P., Schwiening, C., et al. (2004) Efficient generation of neural precursors from adult human skin: astrocytes promote neurogenesis from skin-derived stem cells. *Lancet* **364,** 172–178.
99. Weissman, I. L., Anderson, D. J., and Gage, F. (2001) Stem and progenitor cells: origins, phenotypes, lineage commitments, and transdifferentiations. *Annu. Rev. Cell. Dev. Biol.* **17,** 387–403.
100. Harigaya, K., Cronkite, E. P., Miller, M. E., and Shadduck, R. K. (1981) Murine bone marrow cell line producing colony-stimulating factor. *Proc. Natl. Acad. Sci. USA* **78,** 6963–6966.
101. Ashton, B. A., Allen, T. D., Howlett, C. R., et al. (1980) Formation of bone and cartilage by marrow stromal cells in diffusion chambers in vivo. *Clin. Orthop.* **151,** 294–307.
102. Rickard, D. J., Sullivan, T. A., Shenker, B. J., et al. (1994) Induction of rapid osteoblast differentiation in rat bone marrow stromal cell cultures by dexamethasone and BMP-2. *Dev. Biol.* **161,** 218–228.
103. Ferrari, G., Cusella-De Angelis, G., Coletta, M., et al. (1998) Muscle regeneration by bone marrow-derived myogenic progenitors. *Science* **279,** 1528–1530.
104. Dexter, T. M., Moore, M. A., and Sheridan, A. P. (1977) Maintenance of hemopoietic stem cells and production of differentiated progeny in allogeneic and semiallogeneic bone marrow chimeras in vitro. *J. Exp. Med.* **145,** 1612–1616.

105. Jiang, Y., Vaessen, B., Lenvik, T., et al. (2002) Multipotent progenitor cells can be isolated from postnatal murine bone marrow, muscle, and brain. *Exp. Hematol.* **30,** 896–904.

106. Deng, W., Obrocka, M., Fischer, I., and Prockop, D. J. (2001) In vitro differentiation of human marrow stromal cells into early progenitors of neural cells by conditions that increase intracellular cyclic AMP. *Biochem. Biophys. Res. Commun.* **282,** 148–152.

107. Brazelton, T. R., Rossi, F. M., Keshet, G. I., and Blau, H. M. (2000) From marrow to brain: expression of neuronal phenotypes in adult mice. *Science* **290,** 1775–1779.

108. Mezey, E., Chandross, K. J., Harta, G., et al. (2000) Turning blood into brain: cells bearing neuronal antigens generated in vivo from bone marrow. *Science* **290,** 1779–1782.

109. Weimann, J. M., Charlton, C. A., Brazelton, T. R., et al. (2003) Contribution of transplanted bone marrow cells to Purkinje neurons in human adult brains. *Proc. Natl. Acad. Sci. USA* **100,** 2088–2093.

110. Terada, N., Hamazaki, T., Oka, M., et al. (2002) Bone marrow cells adopt the phenotype of other cells by spontaneous cell fusion. *Nature* **416,** 542–545.

111. Alvarez-Dolado, M., Pardal, R., Garcia-Verdugo, J. M., et al. (2003) Fusion of bone-marrow-derived cells with Purkinje neurons, cardiomyocytes and hepatocytes. *Nature* **425,** 968–973.

112. Weimann, J. M., Johansson, C. B., Trejo, A., and Blau, H. M. (2003) Stable reprogrammed heterokaryons form spontaneously in Purkinje neurons after bone marrow transplant. *Nat. Cell. Biol.* **5,** 959–966.

113. Cogle, C. R., Yachnis, A. T., Laywell, E. D., et al. (2004) Bone marrow transdifferentiation in brain after transplantation: a retrospective study. *Lancet* **363,** 1432–1437.

114. Vitry, S., Bertrand, J. Y., Cumano, A., and Dubois-Dalcq, M. (2003) Primordial hematopoietic stem cells generate microglia but not myelin-forming cells in a neural environment. *J. Neurosci.* **23,** 10724–10731.

115. Munoz-Elias, G., Marcus, A. J., Coyne, T. M., et al. (2004) Adult bone marrow stromal cells in the embryonic brain: engraftment, migration, differentiation, and long-term survival. *J. Neurosci.* **24,** 4585–4595.

116. Ramalho-Santos, M., Yoon, S., Matsuzaki, Y., et al. (2002) "Stemness": transcriptional profiling of embryonic and adult stem cells. *Science* **298,** 597–600.

117. Ratajczak, M. Z., Kucia, M., Reca, R., et al. (2004) Stem cell plasticity revisited: CXCR4-positive cells expressing mRNA for early muscle, liver and neural cells 'hide out' in the bone marrow. *Leukemia* **18,** 29–40.

118. Suzuki, H., Taguchi, T., Tanaka, H., et al. (2004) Neurospheres induced from bone marrow stromal cells are multipotent for differentiation into neuron, astrocyte, and oligodendrocyte phenotypes. *Biochem. Biophys. Res. Commun.* **322,** 918–922.

119. Goolsby, J., Marty, M. C., Heletz, D., et al. (2003) Hematopoietic progenitors express neural genes. *Proc. Natl. Acad. Sci. USA* **100,** 14926–14931.

120. Kogler, G., Sensken, S., Airey, J. A., et al. (2004) A new human somatic stem cell from placental cord blood with intrinsic pluripotent differentiation potential. *J. Exp. Med.* **200,** 123–135.

121. Jang, Y. K., Park, J. J., Lee, M. C., et al. (2004) Retinoic acid-mediated induction of neurons and glial cells from human umbilical cord-derived hematopoietic stem cells. *J. Neurosci. Res.* **75,** 573–584.

122. Draper, J. S., Smith, K., Gokhale, P., et al. (2004) Recurrent gain of chromosomes 17q and 12 in cultured human embryonic stem cells. *Nat. Biotechnol.* **22,** 53–54.

123. Edlund, T. and Jessell, T. M. (1999) Progression from extrinsic to intrinsic signaling in cell fate specification: a view from the nervous system. *Cell* **96,** 211–224.

124. Goridis, C. and Brunet, J. F. (1999) Transcriptional control of neurotransmitter phenotype. *Curr. Opin. Neurobiol.* **9,** 47–53.

125. Sinclair, S. R., Fawcett, J. W., and Dunnett, S. B. (1999) Dopamine cells in nigral grafts differentiate prior to implantation. *Eur. J. Neurosci.* **11,** 4341–4348.

126. Hwang, W. S., Ryu, Y. J., Park, J. H., et al. (2004) Evidence of a pluripotent human embryonic stem cell line derived from a cloned blastocyst. *Science* **303,** 1669–1674.

127. Sawamoto, K., Nakao, N., Kobayashi, K., et al. (2001) Visualization, direct isolation, and transplantation of midbrain dopaminergic neurons. *Proc. Natl. Acad. Sci. USA* **98,** 6423–6428.

128. Zhao, S., Maxwell, S., Jimenez-Beristain, A., et al. (2004) Generation of embryonic stem cells and transgenic mice expressing green fluorescence protein in midbrain dopaminergic neurons. *Eur. J. Neurosci.* **19,** 1133–1140.

129. Smidt, M. P., Asbreuk, C. H., Cox, J. J., et al. (2000) A second independent pathway for development of mesencephalic dopaminergic neurons requires Lmx1b. *Nat. Neurosci.* **3,** 337–341.

130. Jensen, C. H., Meyer, M., Schroder, H. D., et al. (2001) Neurons in the monoaminergic nuclei of the rat and human central nervous system express FA1/dlk. *Neuroreport* **12,** 3959–3963.

3

Neural Transplantation in the Nonhuman Primate Model of Parkinson's Disease

Kimberly B. Bjugstad and John R. Sladek Jr.

ABSTRACT

Neural transplantation research for Parkinson's disease has followed a circuitous and, at times, an unpredictable path. Based primarily on successful rodent studies, clinical trials using adrenal medullary tissue or fetal mesencephalic tissue were initiated throughout the world, but highly variable results in both transplant paradigms sent researchers back to the animal models for further study. Based on a then, newly available neurotoxin, MPTP was used in nonhuman primates to better model the neuropathology and behavior of Parkinson's disease. Using the MPTP, dopamine-depleted primate model, researchers have been able to address questions that were unanswered before advancing to clinical trials and which rodent models could not or did not answer. New investigations pursued questions regarding immunorejection, sources of dopamine-producing cells, transplant location, and even characteristics of the host. While this may give the impression that neural transplantation research had taken a step backward, the nonhuman primate model has helped to lay a foundation from which other transplant approaches could be compared before moving into clinical trials. Polymer encapsulated cells and neural progenitor cell lines are two such approaches that will be examined in the nonhuman primate model before being attempted in the Parkinson's patient population. While feasibility has been proven in the primate model, clinical applicability is still dependent on the creation of a stable source of donor cells.

Key Words: Parkinson's disease; MPTP; primate; neural transplantation; dopamine.

From: *Contemporary Neuroscience: Cell Therapy, Stem Cells, and Brain Repair*
Edited by: C. D. Sanberg and P. R. Sanberg © Humana Press Inc., Totowa, NJ

INTRODUCTION

The 1980s were a decade of "Reagan-omics." A Bush was president, and the singer Prince was still named "Prince." It was also during this decade that three important events occurred that would impact research in Parkinson's disease (PD). In 1983, a paper was published about a small group of intravenous drug users who presented (almost overnight) with parkinsonian-like symptoms *(1,2)*. Langston et al. discussed the patients' use of a synthetic meperidine analog, methylphenyl-tetrahydropyridine (MPTP), which seemed to be responsible for the complete destruction of the substantia nigra (SN). Later that same year, an article was published demonstrating that MPTP could induce parkinsonian-like symptoms in nonhuman primates, and the motor impairments could be temporarily reversed with levodopa (L-Dopa) treatment *(3)*. Furthermore, it was confirmed that MPTP destroyed the dopaminergic neurons of the SN and consequently depleted the striatum of dopamine *(3)*.

The second important event was the publication of a clinical study by Backlund et al., in which two patients with PD received tissue transplants from their own dopaminergic adrenal medulla into their striatum *(4)*. The group's decision to proceed with clinical trials was based on successful transplant experiments performed in the unilaterally lesioned, dopamine-depleted hemiparkinsonian rat model. The rationale behind these experiments was simple. The striatum is dopamine-depleted. L-Dopa medications work by replacing the dopamine in the striatum, but there can be complications in how the patients manage the timing of their medication, and after several years, those medications begin to fail. Therefore, if dopamine-producing cells were placed into the striatum, the striatum would have a constant dopamine presence, relieving some of the burden of oral medications. Four years later, Goetz et al. published a second clinical trial on adrenal medulla autografts, presenting data on 19 PD patients *(5)*. The adrenal medulla transplants increased the duration of the medicine's effectiveness (i.e., on time) of patients and decreased the severity of their symptoms between dosing (i.e., off time; *5*). Unfortunately, the posttransplant morbidity rate was unacceptable. Of the original patients, 10% developed severe and permanent medical or behavioral disabilities *(5)*.

The last event was another clinical transplant study conducted for the first time in the United States *(6)*. Freed et al. initially transplanted one PD patient with dopaminergic ventral mesencephalic (VM) human fetal brain tissue into the striatum. Later, Freed and colleagues transplanted a group of 34 PD patients with fetal VM tissue *(7)*. The improvements in this group were highly variable. Some patients did extremely well and even reduced or stopped all their previous PD medications. Some patients worsened initially but returned to their original baseline measures over a couple of years. A few patients did not change, and a small percentage became dyskinetic *(8)*. Overall, Freed's group found that younger patients (<60 yr old) were more responsive to the VM transplants (as indicated by their Unified Parkinson's Disease Rating Scale) than older patients. However, the oldest patient (75 yr) was one of the best responders to the transplant. Again, these clinical trials were initiated based primarily on successful experiments in the hemi-parkinsonian rat model.

The difference between successful rodent studies and marginal human studies might be that the phylogenetic dissimilarity is too great for rodent results to translate directly to humans. For example, the volume of innervation by a single rat neuron is 1/20 the volume that a nonhuman primate neuron innervates *(9)*. Transplanted neurons need the capacity and the time to innervate the host brain. Primates have a longer life span than rats, which is more compatible for studying transplant survival and efficacy. In addition, there are differences between the two species in dopamine metabolism and dopaminergic receptor subtypes and distributions *(10–13)*. Many questions remain simply because they could not be answered using a rodent model. At this time, the available nonhuman primate model shared more behavioral similarities with the rodent PD model than with PD patients. Both the primate and rat models are created by lesioning the dopaminergic fibers using 6-hydroxydopamine (6-OHDA). This technique induces bradykinesia, rotation, and sensorimotor biases—only a partial replication of the impairments seen in PD. With the discovery of MPTP and its effects on the primate nigrostriatal system (but not on the rat nigrostriatal system), a new PD model was available that could answer lingering questions from the earlier clinical trials. The MPTP primate model mimics the behavioral impairments seen in PD (e.g., tremor and freezing) more closely than the 6-OHDA primate model *(14–17)*.

These three events have been at the forefront of PD research for the last 20 yr. Nonhuman primate studies have investigated the failure of adrenal medulla tissue transplants and the variable success of fetal tissue transplants. The cellular replacement experiments have expanded to include engineered cell lines and stem cell implants, with the anticipation that clinical trials will soon be initiated with these new options. A new millennium has begun, another Bush is president, and the "artist formerly known as Prince" is Prince again. Hopefully, what we have learned about treating PD with dopaminergic transplants in primates will allow us to return to clinical research with more confidence and better results. Celebrities, such as Michael J. Fox, have become strong public advocates, and some states (e.g., California, Wisconsin, and New Jersey) have proposed major new initiatives in stem cell research that will help facilitate future clinical trials.

THE MPTP PRIMATE MODEL: A BRIEF REVIEW

MPTP is a protoxin that crosses the blood–brain barrier and is converted into the toxin methyl-phenylpridinium (MPP+) by monoamine oxidase B. It then can be taken up by the dopamine neuron via the dopamine transporter (DAT). MPP+ kills the dopaminergic neurons via its actions on mitochondrial respiration. In neurons that are already under high levels of oxidative stress because of dopamine metabolism, MPP+ inhibits complex I in the electron transport cascade, increasing the availability of unpaired electrons (for review, see ref. *18*). MPP+ selectively kills the dopaminergic neurons; however, the neurons of the SN are especially susceptible for two reasons *(19)*. First, neuromelanin—an iron-based pigment specific to SN dopamine neurons—exacerbates free-radical formation. In addition, neuromelanin can bind MPP+, creating an intracellular reservoir of neurotoxins *(20)*. Second, the dopaminergic neurons of the SN have a lower capacity to safely sequester MPP+ because of a lower concentration of vesicular monoamine transport *(21)*. In primates treated with MPTP, there is a decreased number of dopamine neurons found in the SN, the ventral tegmental area, and the retrorubral area, with the greatest cell loss in the SN *(3,19,22*; Fig. 1A).

As a result of dopamine cell loss in the SN, several changes occur in the primate striatum that mimic dopamine depletion in PD. There is a significant decline in dopamine and homovanillic acid (HVA, a metabolite of dopamine) content in the striatum. In MPTP-treated primates, these changes are relatively equal between the two parts of the striatum—the caudate and the putamen. However, in PD, the putamen appears to be more dopamine-depleted than the caudate *(23)*. MPTP also decreases presynaptic DAT binding in the striatum and increases the density of postsynaptic D1 and D2 dopamine receptors *(24,25)*. An interesting phenomenon of primates (human and nonhuman) is the presence of an endogenous population of dopamine neurons that reside in the striatum (Fig. 1B). When the striatum is dopamine-depleted, as with PD or MPTP treatment, there is an increased number of striatal dopamine neurons *(26,27)*. In both cases, it appears that this may be a compensatory reaction to the lack of dopamine input from the SN *(27)*. Despite a few differences, MPTP mimics PD in the primate brain quite well. The similar neuroanatomical and neurochemical changes also produce similar behavioral deficits.

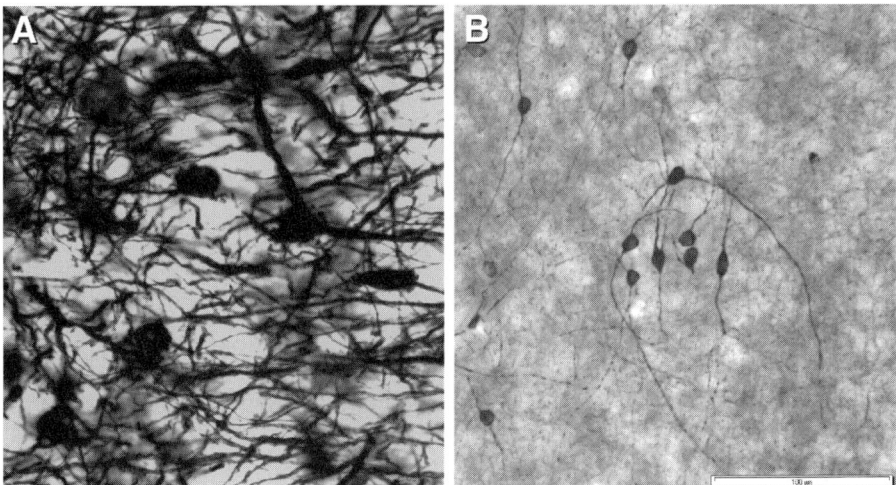

Fig. 1. Dopaminergic neurons of the nigrostriatal pathway in the primate brain.
(**A**) Large dopamine neurons are found in the SN. During PD and after MPTP, most
of the nigral neurons are lost. (**B**) Tyrosine hydroxylase–positive neurons are found
in both the caudate and putamen of primates. After dopamine depletion, as seen in
PD and with MPTP treatment, the number of these endogenous dopamine neurons
can triple. The bar in part **B** is used for both pictures and is 100 μm.

MPTP typically is given either intramuscularly over several days or intravenously and produces bilateral dopamine depletion that is relatively stable
over time *(23,28)*. Yet, the corresponding behavioral effects can be variable.
MPTP-treated monkeys can show severe parkinsonian behaviors, including
akinesia, rigidity, tremor, flexed posture, poverty of movement, delayed initiation of movement, drooling, and difficulty in swallowing *(3,17,29–32)*.
These monkeys often become moribound and require "nursing care." These
motor deficits are permanent; however, they can be reversed temporarily
with L-Dopa therapy *(3,33)*. The severe impairment roughly corresponds to
more than a 90% loss of striatal dopamine content *(30,34)*. Monkeys with an
80–90% loss of dopamine are more functional but still have many clinical
signs of PD *(30)*. The dopamine loss is typically greater in the caudate than
in the putamen of these moderately effected monkeys, and within a year,
many show signs of spontaneous behavioral recovery, even though their
striatal dopamine levels remain reduced *(28,34)*. Several MPTP-treated
monkeys remain asymptomatic, despite a decline in striatal dopamine by as
much as 80% *(34)*.

A unilateral parkinsonian primate model also can be created using MPTP
infused via the carotid artery or through intracranial infusion of 6-OHDA.

Like the 6-OHDA unilaterally lesioned rat, these monkeys are functional in daily activities but when challenged with a dopaminergic agonist, they have rotational biases. This model then is less like PD than the bilaterally, dopamine-depleted MPTP-treated primate, but the unilaterally lesioned monkeys present fewer veterinary requirements.

ADRENAL MEDULLA TISSUE TRANSPLANTS

The adrenal medulla is the inner part of the adrenal gland, which is found just above the kidneys. The tissue contains fibroblasts, endothelial cells, smooth muscle cells, Schwann cells, neurons, and chromaffin cells (35). The chromaffin cells secrete a variety of factors and neurotransmitters, including dopamine. In general, this type of transplant is an autograft, where the host also serves as the tissue donor. This would normally eliminate the need for immunosuppression, but autologous adrenal transplants in the parkinsonian primates indicate that immunorejection does occur.

Early transplant protocols used a two-step method for transplanting adrenal tissue. First, a "cavity" was created in the brain at the transplant site using a small amount of sterile gelatin foam. This cavity was thought to "set the stage" for the transplant, as it was found that the brain releases neurotrophic factors during the first few days following damage (36,37). The adrenal tissue would then be transplanted into a site that was enriched with neurotrophins and blood vessels to increase graft survival. One study found that adrenal grafts placed into the striatum alleviated some motor impairments seen in hemiparkinsonian monkeys approx 3 mo after transplantation (38). Monkeys showed improvement in reaching with their contralateral limb and had decreased apomorphine-induced rotations. As the post-transplant time increased, the improvements began to reverse (38,39). At 6 mo, the transplanted monkeys were sacrificed, and no tyrosine hydroxylase–positive donor cells were found. In a similar study, both adrenal-grafted and cavity–only-treated monkeys had a moderate behavioral improvement, but neither improvement was as profound as that seen with fetal ventral mesencephalic tissue implants (40). (See the Fetal Ventral Mesencephalic Tissue Transplants section for more details on fetal tissue transplants in primates.) Examination of the striatum and graft site showed dopaminergic fiber reinnervation of the striatum originating from the olfactory bulb and nucleus accumbens in both the grafted and cavity-only groups (40). The researchers concluded that adrenal grafts were no better than simple cavitation of the striatum in relieving parkinsonism in primates (40).

The ineffectiveness of adrenal tissue grafts might be explained in the low survival rate of chromaffin cells, regardless of precavitation (38–42).

Most studies report finding a high frequency of macrophages, reactive gliosis, and necrosis around the implant site *(38,39,42–45)*. In one study, adrenal tissue was implanted into normal nonparkinsonian monkeys *(44)*. The graft site had increased macrophage infiltration at 1 wk posttransplant, but this infiltration appeared to decrease over the 1-yr posttransplant time period *(44)*. Peripheral indicators of immunorejection were also analyzed, and no changes in the number of T cells, B cells, natural killer cells, or peripheral monocytes were found, suggesting the immunorejection is limited to the host brain.

In an attempt to improve adrenal graft survival, the sural nerve was cografted with adrenal tissue. The sural nerve is a lower branch of the sciatic nerve, providing sensory information from the leg. Sural nerve cografts introduce a source of nerve growth factor (NGF) from the Schwann cells within the tissue *(46–48)*. Cografts significantly improved chromaffin cell survival *(41,49)*. Furthermore, parkinsonian monkeys with adrenal transplants and sural nerve cografts performed better in skilled reaching tasks than animals with only adrenal grafts *(49)*. Efforts to increase dopamine cell survival are still being investigated. With adrenal medulla grafts, elevating graft survival has been focused more on adding trophic-producing cografts and preventing immunorejection. In fetal mesencephalic tissue transplants, there are many more avenues to investigate to improve dopamine cell survival.

FETAL VM TISSUE TRANSPLANTS

Fetal VM tissue transplanted into the dopamine-depleted caudate of the striatum always had better success than adrenal medulla grafts *(50,51)*. Despite being allografts (the host is not the donor but is the same species as the donor), fetal VM tissue appeared to induce less immunorejection and gliotic scarring than did adrenal grafts *(52–54)*. The immune reaction was not substantially different than that seen in sham surgeries *(44)*. Fetal VM tissue placed into the dopamine-depleted caudate increased striatal dopamine content, normalized the HVA:dopamine ratio, increased the density of DAT sites, and decreased dopamine receptor sensitivity *(25,54–58)*. Additionally, in MPTP-treated monkeys, it was found that VM transplants increased cellular metabolic activity, as indicated by higher levels of mitochondrial cytochrome oxidase *(59)*. All these measures of dopamine and metabolic activity are altered initially by MPTP treatment and subsequent dopamine depletion of the striatum (see The MPTP Primate Model section). In most cases, fetal VM grafts survived, developed large mature neurons, extended neurites into the host-surround, and made synaptic connections

(50,52,54,55,60–70; Fig. 2A,B). The improvement in striatal functioning and dopamine content correlated with behavioral recovery *(25,51,65,71)*.

By 3–6 mo after transplantation, many parkinsonian monkeys were less behaviorally impaired than nontransplanted monkeys *(17,50–52,55, 63,65,68,71–73)*. In unilaterally dopamine-depleted monkeys, the fetal neural tissue transplants had improved voluntary reaching tasks and exhibited less drug-induced rotations *(65,68,72)*. In a bilaterally dopamine-depleted monkey with motor impairments severe enough to require constant monitoring, the fetal VM grafts stimulated such improvement that the animal was considered nearly normal in behavior and regained the ability to feed without assistance *(63)*. However, in that same study, another severely parkinsonian monkey showed no improvement, even at 7-mo post-transplant *(63)*. Upon examination of the VM graft of this animal, few surviving dopamine neurons of graft origin were found, whereas numerous cells survived in the animal that had improved *(63)*. Despite the large numbers of animals that benefited from the VM transplants, some did not improve *(51,56,63,68,71)*. Furthermore, within a single monkey, some motor impairments were improved with the fetal VM transplants, whereas others remained unchanged. For example, in unilaterally lesioned monkeys, the transplants had reduced drug-induced rotations, but contralateral side biases were not affected (i.e., somasensory neglect, head position, or forced reaching tasks; *39,72*).

Although fetal VM transplants appeared to have a greater effect than the adrenal medulla transplants, there was still variability in dopamine measures and behavioral effects. These inconsistencies led to other questions. Do the fetal neural transplants have to be from the VM? Does donor age affect the survival of the grafts? Is the caudate the best area to transplant?

Several studies transplanted fetal nondopaminergic cerebellar tissue or striatal eminence into the striatum of dopamine-depleted monkeys. Not surprisingly, they found no change in motor impairments, and no synaptic connections were made between the host and the atypical tissue grafts *(17,52,66,70,71)*. Another study found that cerebellar tissue, when transplanted with spinal cord cografts, initiated a dopaminergic sprouting into the striatum from the olfactory bulb and nucleus accumbens *(70)*. Most likely, the dopaminergic sprouting resulted from the presence of the spinal cord cografts and was independent of the cerebellar graft. Like the sural nerve cografts transplanted with adrenal tissue *(41)*, the spinal cord would be a source of Schwann cell–derived NGF *(46–49)*. Regardless, the authors suggested that replenishing the striatum with dopamine may not necessarily have to rely on transplanting dopamine-secreting cells, but the striatum could be reinnervated from neighboring dopaminergic sites *(70)*.

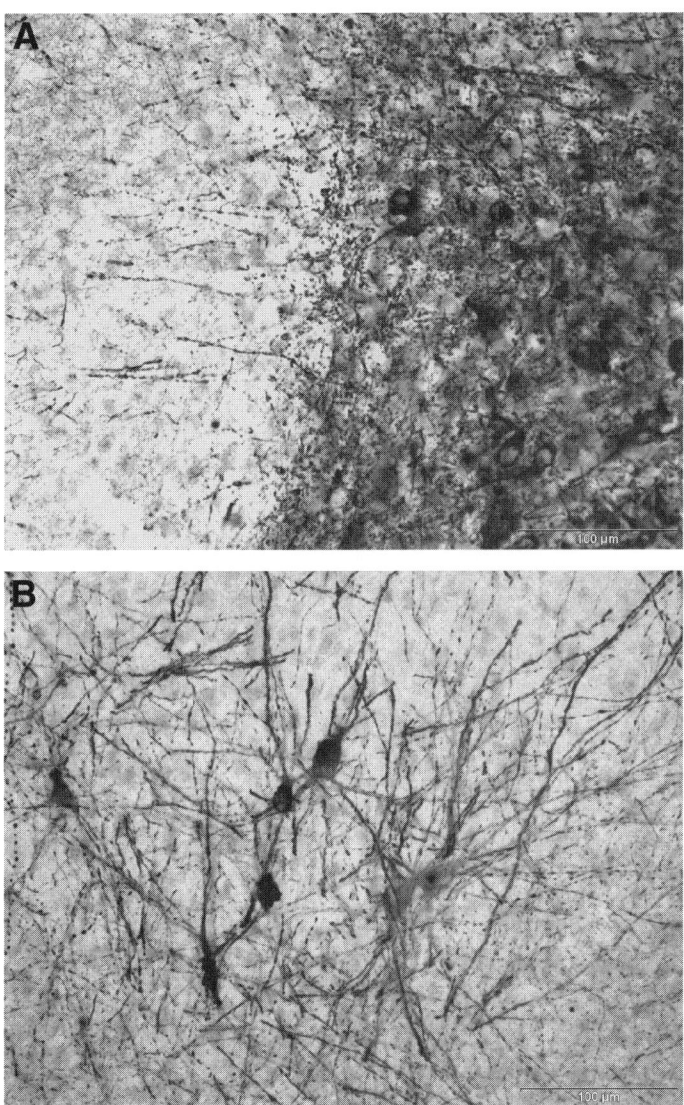

Fig. 2. Fetal VM tissue transplanted into the caudate. This monkey was treated with MPTP and transplanted with embryonic day-44 VM tissue from a St. Kitts African green monkey. Data are presented by Sladek et al. in ref. *60*. (**A**) A photomicrograph of the graft-host border. The graft (left) contains a high density of dopaminergic neurons with neuritic processes extending out into the host caudate. (**B**) Several donor-derived dopamine neurons were found just ventral to the main graph and better demonstrate the extensive neuritic outgrowth and modest migration that can occur.

69

The fetal age of the donor, or more precisely, the stage of VM neurogenesis, was found to be a critical factor influencing graft survival and alleviation of motor impairments in parkinsonian monkeys. *Early-stage tissue* is VM tissue harvested at the time of neurogenesis *(54)*. This is approx 40–45-d postconception in the St. Kitts African green monkeys, 36–40 d in the Rhesus monkey, and approx 74 d in the marmoset *(54,65)*. In humans, this corresponds roughly to Carnegie stage 18–19; in rats, it is equivalent to embryonic day 15. *Moderate-stage* tissue is harvested just after dopamine neurogenesis, and *late-stage* tissue is from any point well after VM neurogenesis *(54)*.

When transplanted into the striatum, early-stage VM tissue has shown the best results. Early VM tissue is more likely to induce changes in motor impairment and has better graft survival than moderate- and late-stage tissue transplants *(42,52,54,55,60,64,68,71)*. One study found that early-stage tissue increased striatal dopamine content to 20% of control levels, and thousands of dopamine neurons survived in the grafts *(54)*. These moderate changes in dopamine content and survival significantly improved the functioning of MPTP-treated monkeys *(71)*. In monkeys transplanted with late-stage VM, less than 1000 donor-derived dopamine neurons survived, and there was no change in motor impairments *(54,71)*. At the beginning of dopamine neurogenesis, a negative correlation exists between donor age and graft survival *(52,55,60,64)*.

Until recently, nearly all the studies have transplanted dopaminergic tissue into the caudate nucleus of the striatum. The caudate nucleus is rather large and easily accessible, compared to the putamen and SN, which are deeper structures; however, it is the SN that is lost in MPTP treatment and in PD, and it is the putamen that is more depleted of dopamine in PD. Studies that examine fetal VM tissue transplanted into the putamen have shown greater variability in graft survival than when tissue is transplanted into the caudate, but there was a general increase in DAT density *(52,56,71,72)*. One study found that the VM grafts placed into the putamen extended neurites across the internal capsule to reinnervate the caudate *(52)*. The neuritic outgrowth was likely goal-directed, as no dopaminergic fibers were seen extending from the graft into nondopaminergic areas *(52)*. Some unique behavioral improvements have also been seen. In unilaterally lesioned monkeys, when tissue was transplanted into the putamen, their contralateral reaching tasks improved, and a decrease was seen in contralateral somatosensory neglect—behaviors that typically remain impaired in monkeys with caudate transplants *(42)*.

Few studies have investigated VM transplants in the SN. Generally, only half as many grafted cells survived in the SN, compared to grafts placed into the caudate *(55,74)*. Variable reinnervation, usually a few millimeters, can be seen extending from the graft into the host-surround, but no significant reinnervation of the caudate was found *(74)*. Similar to transplants placed into the putamen, changes in behavior resulting from nigral transplants are mainly in contralateral reaching tasks, which were not always affected by caudate transplants *(74)*.

Studies conducted in parkinsonian monkeys have determined several points that were not demonstrated (or not done) in rodent studies prior to clinical trials. First, in primates, it was shown that adrenal medulla grafts, despite being autografts, provoked a strong immunoreaction in the brain, which is probably responsible for low-graft survival. Second, primate studies demonstrated that fetal VM tissues, although allografts, were well tolerated by the brain, had relatively good survival and integration with the host, and somewhat relieved motor impairments. Third, the parkinsonian primate model was used to illustrate that VM tissue taken at the point of dopamine neurogenesis and transplanted into the caudate provides better results than when other fetal tissues were used, when tissue was taken at later stages of neurogenesis, or when tissue was transplanted into other sites. Regardless of what seems to be a promising clinical future for fetal VM grafts, not every cell in the VM graft survives, not all behaviors are affected, and not every subject benefits from fetal VM transplants. Fetal VM tissue does not produce a consistent improvement in either humans or monkeys, and with a strong ethical argument against using this tissue, many researchers seek out cell alternatives that they hope will be just as promising, if not more so.

ENCAPSULATED CELL LINES AND NEURAL STEM CELLS

With new technologies, neural transplant studies for PD seem to have returned to where they began. Polymer encapsulation has given new life to adrenal medulla transplants, specifically to the chromaffin cells, which are isolated from the adrenal tissue and inserted into polymer spheres. The polymer spheres are porous to allow the exchange of dopamine or other small proteins. The size of the polymers' pores can be manipulated to prevent large-molecular-weight molecules (e.g., immunoglobulin G) to enter the sphere, thus providing immunoprotection for the encapsulated cells *(75)*. In primate studies, bovine chromaffin cells were encapsulated and transplanted into the striatum of MPTP-treated monkeys. Immunoreactivity to the encapsulated cells was not substantially different from the immunoreactivity created with sham penetration *(76)*. In monkeys with encapsulated chromaffin cells, a significant decrease in drug-induced rotations was seen which was maintained for the 9-mo study period *(77)*. PC12 cells are a cell line derived from a tumor of the rat adrenal medulla that, when differentiated, will produce dopamine. Using encapsulation techniques, PC12 cells were transplanted into the striatum of parkinsonian monkeys. PC12-transplanted animals performed better at contralateral reaching tasks than did animals transplanted with empty capsules or animals transplanted with encapsulated bovine chromaffin cells *(76)*. Two other research groups found that encapsulated PC12 cells produced measurable levels of dopamine, and ^{18}F-DOPA uptake could be seen within the areas of the transplanted capsules. However, there was no clear evidence of striatal reinnervation or sprouting of dopamine nerve terminals, suggesting that the PC12 cells within the capsules merely dispense dopamine and do not make functional connections with the host environment *(78)*. Although the capsules provide immunoprotection for chromaffin and PC12 cells, they only work if they remain intact. Capsules can rupture, and when they do, a strong immunoreaction can be seen in response to the unprotected cells *(76,79)*.

Neural stem cells (NSC) are pluripotent cells derived from the brain and can give rise to any of the three primary cell types of the brain: neurons, astrocytes, or oligodendrocytes *(80)*. The potential for these cells to treat neurodegenerative diseases like PD is significant, but it is a new science, and many questions need answers. Regarding in vivo studies, what is known thus far is this: NSC survive, they migrate, and they interact with the host brain to affect changes in the host, as indicated by studies in rodents *(81–88)*. Significantly fewer studies have investigated the potential of NSC in primate models of neurodegeneration.

One group that has transplanted NSC bilaterally into the caudate and unilaterally into the SN of MPTP-treated monkeys has seen NSC survival for at least 4 mo in both immunosuppressed and nonsuppressed monkeys (Fig. 3). The NSC also appeared to migrate away from the implant sites and align along the nigrostriatal pathway and in the SN *(89–91;* Fig. 4). The presence of NSC in the unimplanted SN suggests that NSC from the

Fig. 3. *(opposite page)* NSC implanted into MPTP-treated monkeys survive in animals that were immunosuppressed and those that were not. However, in both groups, less than 10% of the NSC were identified after 4 mo posttransplant.

Fig. 4. *(opposite page)* Human neural stem cells (hNSC) were implanted into the caudate and SN of MPTP-treated monkeys. This figure illustrates the ventral portion of the SN. Large dopaminergic neurons (open arrows) can be seen juxtaposed with hNSC (closed arrows). Many hNSC are closely associated with the neuritic processes of the neurons, rather than randomly dispersed in the neuropile.

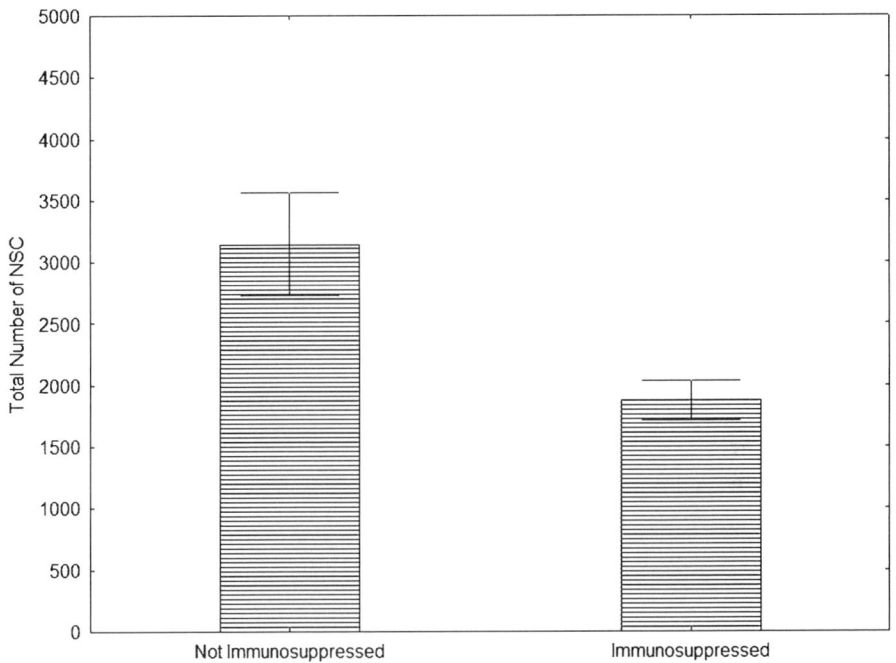

Fig. 3.

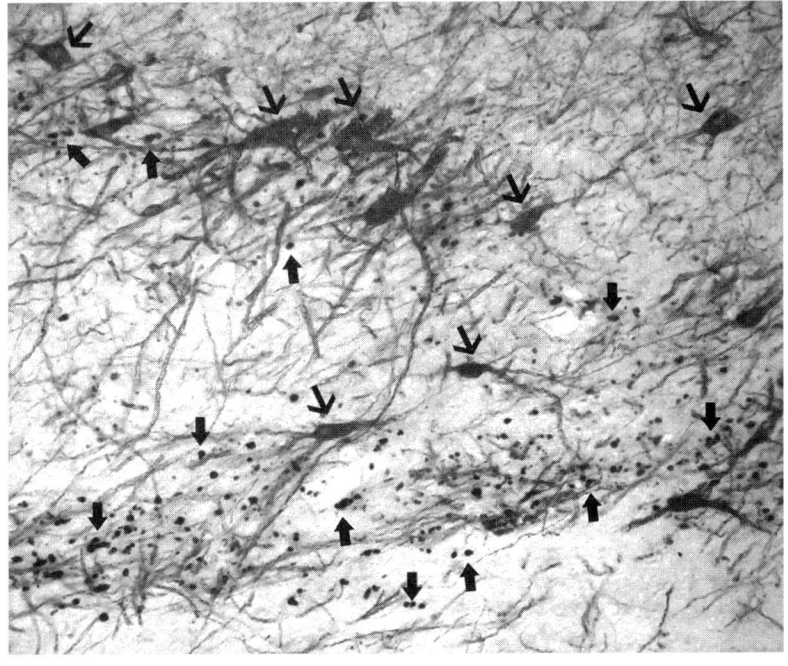

Fig. 4.

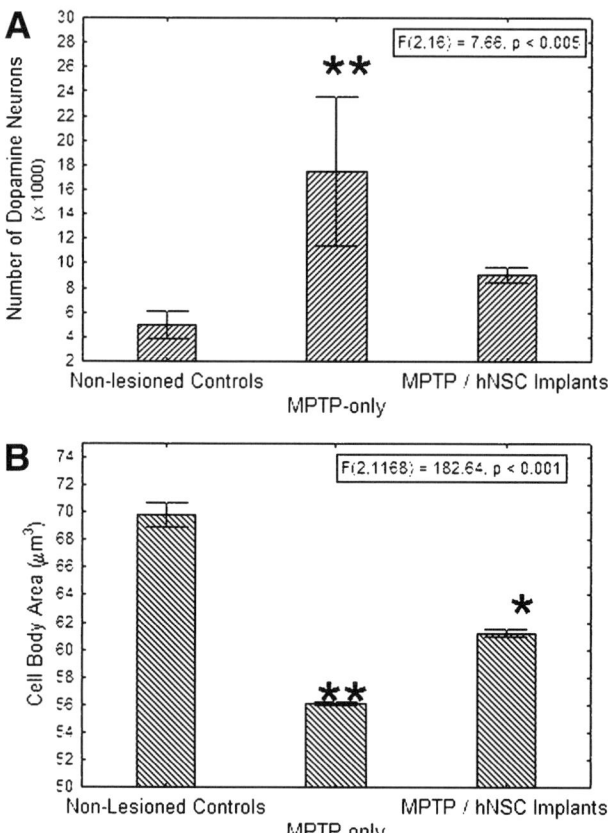

Fig. 5. Trophic effects of NSC can be seen in the restoration of the striatal dopamine neurons. **(A)** MPTP treatment induces an increase in the number of striatal dopamine neurons. In monkeys treated with MPTP and transplanted with NSC, the number of striatal dopamine neurons decreases to levels found in normal controls. **(B)** Although the number of striatal dopamine neurons increases after MPTP treatment, the cells are much smaller than the dopamine neurons found in normal controls. In MPTP-treated primates with NSC transplants, the size of the striatal dopamine neurons reduces and approaches control levels. **Significantly different from both control and MPTP/human NSC implant groups; *significantly different from both control and MPTP-only groups; $p < 0.05$.

implanted side migrated to the unimplanted SN, or (more likely) that the NSC migrated caudally from the implanted caudate via the few surviving nigrostriatal fibers.

The purpose of NSC migration in MPTP-treated, dopamine-depleted monkeys is suggested by the changes observed in the striatal dopamine neurons. MPTP treatment induces a compensatory increase in striatal dopamine neurons, which are typically smaller than normal. In MPTP-treated monkeys also implanted with NSC, striatal dopamine neurons decreased in number and increased in size, which is more representative of neurons found in normal monkeys (*91–95*; Fig. 5). These results indicate that the NSC are not only migrating but are also interacting with the host to reverse MPTP-induced changes in the striatum. This reversal may be caused by the release of neurotrophic factors via NSC (*86,88,91*), enhanced function from surviving SN neurons induced by NSC, or the replacement of lost SN neurons by differentiated NSC. The uniqueness of NSC transplants reflects the dynamic role the host brain has in guiding the survival, migration, and function of transplants.

CONCLUSION

In the last 20 yr, neural transplants in the parkinsonian primate model have provided knowledge about immunorejection in the brain, donor age and neurogenesis, transplant location, optimal tissue choices, and tissue alternatives. At present, fetal VM tissue transplanted into the caudate of a dopamine-depleted monkey can replenish the nigrostriatal dopamine system at least to a level that produces significant motor improvements for most behaviors in a certain number of animals. Fortunately, innovative technologies provoke new questions that lead researchers and patients to find even better treatments, if not cures. NSC may be one of those treatments. In parkinsonian primates, NSC migrate to the site of neuropathology and provide trophic support to partially restore the striatal dopamine system. The effects on motor impairments, and the extent of differentiation to replace lost cells, are currently being investigated. It is an exciting new field, where the parkinsonian primate model will have an important role.

"How'd you like to have your own personal biological repair kit standing by at the hospital? Sound like magic? Welcome to the future of medicine."

Ron Reagan Jr., 2004 Democratic National Convention, Boston, MA (July 27, 2004) regarding the use of stem cells to treat neurodegenerative disorders.

REFERENCES

1. Langston, J. W., Ballard, P., Tetrud, J. W., and Irwin, I. (1983) Chronic parkinsonism in humans due to a product of meperidine-analog synthesis. *Science* **219,** 979–980.
2. Weingarten, H. L. (1988) 1-Methyl-4-Phenyl-1,2,3,6-Tetrahydropyridine (MPTP): One designer drug and serendipity. *J. Forensic Sci.* **23,** 588–595.

3. Burns, R. S., Chiueh, C. C., Markey, S. P., et al. (1983) A primate model of parkinsonism: selective destruction of dopaminergic neurons in the pars compacta of the substantia nigra by N-methyl-4-phenyl-1,2,3,6-tetrahydropyridine. *Proc. Natl. Acad. Sci. USA* **80,** 4546–4550.

4. Backlund, E. O., Granberg, P. O., Hamberger, B., et al. (1985) Transplantation of adrenal medullary tissue to striatum in parkinsonism. First clinical trials. *J. Neurosurg.* **62,** 169–173.

5. Goetz, C. G., Olanow, C. W., Koller, W. C., et al. (1989) Multicenter study of autologous adrenal medullary transplantation to the corpus striatum in patients with advanced Parkinson's disease. *N. Engl. J. Med.* **320,** 337–341.

6. Freed, C. R., Breeze, R. E., Rosenberg, N. L., et al. (1990) Transplantation of human fetal dopamine cells for Parkinson's disease. Results at 1 year. *Arch. Neurol.* **47,** 505–512.

7. Freed, C. R., Greene, P. E., Breeze, R. E., et al. (2001) Transplantation of embryonic dopamine neurons for severe Parkinson's disease. *N. Engl. J. Med.* **344,** 710–719.

8. Ma, Y., Feigin, A., Dhawan, V., et al. (2002) Dyskinesia after fetal cell transplantation for parkinsonism: a PET study. *Ann. Neurol.* **52,** 628–634.

9. Lindvall, O. (1989) Transplantation into the human brain: present status on future possibilities. *J. Neurol. Neurosurg. Psychiatry* **Suppl,** 39–54.

10. Bacopoulos, N. G., Maas, J. W., Hattox, S. E., and Roth, R. H. (1978) Regional distribution of dopamine metabolites in human and primate brain. *Commun. Psychopharmacol.* **2,** 281–286.

11. Elsworth, J. D., Laurence, M. S., Roth, R. H., et al. (1991) D1 and D2 dopamine receptors independently regulate spontaneous blink rate in the vervet monkey. *J. Pharm. Exp. Ther.* **259,** 595–600.

12. Garrick, N. A., Redmond, D. E., Jr., and Murphy, D. L. (1979) *Primate-Rodent Monoamine Oxidase Differences. Monoamine Oxidase: Structure, Function, and Altered Functions.* Academic, New York, pp. 351–359.

13. Pifl, C., Reither, H., and Hornykiewicz, O. (1991) Lower efficacy of the dopamine D1 agonist, SK38393, to stimulate adenylyl cyclase activity in primate than in rodent striatum. *Eur. J. Pharmacol.* **202,** 273–276.

14. Taylor, J. R., Elsworth, J. D., Roth, R. H., et al. (1990) Cognitive and motor deficits in the acquisition of an object retrieval/detour task in MPTP-treated monkeys. *Brain* **113,** 617–637.

15. Taylor, J. R., Elsworth, J. D., Roth, R. H., et al. (1990) Improvements in MPTP-induced object retrieval deficits and behavioral deficits after fetal nigral grafting in monkeys. *Prog. Brain Res.* **82,** 543–559.

16. Taylor, J. R., Elswoth, J. D., Roth, R. H., et al., eds. (1994) *Behavioral Effects of MPTP Administration in the Vervet Monkey: A Primate Model of Parkinson's Disease. Toxin-Induced Models of Neurological Disorders.* Plenum Press, New York, pp. 139–174.

17. Taylor, J. R., Elsworth, J. D., Roth, R. H., et al. (1991) Grafting of fetal substantia nigra to striatum reverses behavioral deficits induced by MPTP in primates: a comparison with other types of grafts as controls. *Exp. Brain Res.* **85,** 335–348.

18. Speciale, S. G. (2002) Insights into parkinsonian neurodegeneration. *Neurotoxicol. Teratol.* **24,** 607–620.
19. German, D. C., Manaye, K. F., Sonsalla, P. K., and Brooks, B. A. (1992) Midbrain dopaminergic cell loss in Parkinson's disease and MPTP-induced parkinsonism: sparing of calbindin-D28k-containing cells. *Ann. NY Acad. Sci.* **648,** 42–62.
20. Snyder, S. H. and D'Amato, R. J. (1986) MPTP: a neurotoxin relevant to the pathophysiology of Parkinson's disease. The 1985 George C. Cotzias lecture. *Neurology* **36,** 250–258.
21. Speciale, S. G., Liang, C.-L., Sonsalla, P. K., et al. (1998) The neurotoxin 1-methyl-4-phenylpyridinum is sequestered within neurons that contain the vesicular monoamine transporter. *Neuroscience* **84,** 1177–1185.
22. Deutch, A. Y., Elsworth, J. D., Goldstein, M., et al. (1986) Preferential vulnerability of A8 dopamine neurons in the primate to the neurotoxin 1-methyl-4-phenyl-1,2,3,6-tetrahydropyridine. *Neurosci. Lett.* **68,** 51–56.
23. Pifl, C., Schingnitz, G., and Hornykiewicz, O. (1988) The neurotoxin MPTP does not reproduce in the rhesus monkey the interregional pattern of striatal dopamine loss typical of human idiopathic Parkinson's disease. *Neurosci. Lett.* **92,** 228–233.
24. Pope-Coleman, A., Tinker, J. P., and Schneider, J. S. (2000) Effects of GM1 ganglioside treatment on pre- and postsynaptic dopaminergic markers in the striatum of parkinsonian monkeys. *Synapse* **36,** 120–128.
25. Elsworth, J. D., al-Tikriti, M. S., Sladek, J. R., Jr., et al. (1994) Novel radioligands for the dopamine transporter demonstrate the presence of intrastriatal nigral grafts in the MPTP-treated monkey: correlation with improved behavioral function. *Exp. Neurol.* **126,** 300–304.
26. Porritt, M. J., Barchelor, P. E., Hughes, A. J., et al. (2000) New dopaminergic neurons in Parkinson's disease striatum. *Lancet* **356,** 44–45.
27. Betarbet, R., Turner, R., Chockkan, V., et al. (1997) Dopaminergic neurons intrinsic to the primate striatum. *J. Neurosci.* **17,** 6761–6768.
28. Eidelberg, E., Brooks, B. A., Morgan, W. W., et al. (1986) Variability and functional recovery in the N-methyl-4-phenyl-1,2,3,6-tetrahydropyridine model of parkinsonism in monkeys. *Neuroscience* **18,** 817–822.
29. Taylor, J. R., Elsworth, J. D., Roth, R. H., et al. (1997) Severe long-term 1-methyl-4-phenyl-1,2,3,6-tetrahydropyridine-induced parkinsonism in the vervet monkey (Cercopithecus aethiops sabaeus). *Neuroscience* **81,** 745–755.
30. Chiueh, C. C., Burns, R. S., Markey, S. P., et al. (1985) Primate model of parkinsonism: selective lesion of nigrostriatal neurons by 1-methyl-4-phenyl-1,2,3,6-tetrahydropyridine produces an extrapyramidal syndrome in rhesus monkeys. *Life Sci.* **36,** 213–218.
31. Matsumura, M. (2001) Experimental parkinsonism in primates. *Stereotact. Funct. Neurosurg.* **77,** 91–97.
32. Jenner, P. (2003) The contribution of the MPTP-treated primate model to the development of new treatment strategies for Parkinson's disease. *Parkinsonism Relat. Disord.* **9,** 131–137.

33. Nomoto, M., Jenner, P., and Marsden, C. D. (1985) The dopamine D2 agonist LY 141865, but not the D1 agonist SKF 38393, reverses parkinsonism induced by 1-methyl-4-phenyl-1,2,3,6-tetrahydropyridine (MPTP) in the common marmoset. *Neurosci. Lett.* **57,** 37–41.

34. Elsworth, J. D., Taylor, J. R., Sladek, J. R., Jr., et al. (2000) Striatal dopaminergic correlates of stable parkinsonism and degree of recovery in old-world primates one year after MPTP treatment. *Neuroscience* **95,** 399–408.

35. Hansen, J. T., Notter, M. F., Okawara, S. H., and Gash, D. M. (1988) Organization, fine structure, and viability of the human adrenal medulla: considerations for neural transplantation. *Ann. Neurol.* **24,** 599–609.

36. Bakay, R. A., Fiandaca, M. S., Sweeney, K. M., et al. (1988) Delayed stereotactic transplantation technique in non-human primates. *Prog. Brain Res.* **78,** 463–471.

37. Nieto-Sampedro, M., Lewis, E. R., Cotman, C. W., et al. (1982) Brain injury causes a time-dependent increase in neuronotrophic activity at the lesion site. *Science* **217,** 860–861.

38. Bankiewicz, K. S., Plunkett, R. J., Kophin, I. J., et al. (1988) Transient behavioral recovery in hemiparkinsonian primates after adrenal medullary allografts. *Prog. Brain Res.* **78,** 543–549.

39. Hansen, J. T., Kordower, J. H., Fiandaca, M. S., et al. (1988) Adrenal medullary autografts into the basal ganglia of Cebus monkeys: graft viability and fine structure. *Exp. Neurol.* **102,** 65–75.

40. Plunkett, R. J., Bankiewicz, K. S., Cummins, A. C., et al. (1990) Long-term evaluation of hemiparkinsonian monkeys after adrenal autografting or cavitation alone. *J. Neurosurg.* **73,** 918–926.

41. Kordower, J. H., Fiandaca, M. S., Notter, M. F., et al. (1990) NGF-like trophic support from peripheral nerve for grafted rhesus adrenal chromaffin cells. *J. Neurosurg.* **73,** 418–428.

42. Morihisa, J. M., Nakamura, R. K., Freed, W. J., et al. (1984) Adrenal medulla grafts survive and exhibit catecholamine-specific fluorescence in the primate brain. *Exp. Neurol.* **84,** 643–653.

43. Fiandaca, M. S., Kordower, J. H., Hansen, J. T., et al. (1988) Adrenal medullary autografts into the basal ganglia of Cebus monkeys: injury-induced regeneration. *Exp. Neurol.* **102,** 76–91.

44. Bakay, R. A., Boyer, K. L., Freed, C. R., and Ansari, A. A. (1998) Immunological responses to injury and grafting in the central nervous system of non-human primates. *Cell Transplant.* **7,** 109–120.

45. Morihisa, J. M., Nakamura, R. K., Freed, W. J., et al. (1987) Transplantation techniques and the survival of adrenal medulla autografts in the primate brain. *Ann. NY Acad. Sci.* **495,** 599–605.

46. Cochran, M. and Black, M. M. (1985) PC12 neurite regeneration and long-term maintenance in the absence of exogenous nerve growth factor in response to contact with Schwann cells. *Brain Res.* **349,** 105–116.

47. Rush, R. A. (1984) Immunohistochemical localization of endogenous nerve growth factor. *Nature* **312,** 364–367.

48. Riopelle, R. J., Boegman, R. J., and Cameron, D. A. (1981) Peripheral nerve contains heterogeneous growth factors that support sensory neurons in vitro. *Neurosci. Lett.* **25,** 311–316.
49. Watts, R. L., Mandir, A. S., and Bakay, R. A. (1995) Intrastriatal cografts of autologous adrenal medulla and sural nerve in MPTP-induced parkinsonian macaques: behavioral and anatomical assessment. *Cell Transplant.* **4,** 27–38.
50. Bakay, R. A., Fiandaca, M. S., Barrow, D. L., et al. (1985) Preliminary report on the use of fetal tissue transplantation to correct MPTP-induced Parkinson-like syndrome in primates. *Appl. Neurophysiol.* **48,** 358–361.
51. Redmond, D. E., Sladek, J. R., Jr., Roth, R. H., et al. (1986) Fetal neuronal grafts in monkeys given methylphenyltetrahydropyridine. *Lancet* **1,** 1125–1127.
52. Fine, A., Hunt, S. P., Oertel, W. H., et al. (1988) Transplantation of embryonic marmoset dopaminergic neurons to the corpus striatum of marmosets rendered parkinsonian by 1-methyl-4-phenyl-1,2,3,6-tetrahydropyridine. *Prog. Brain Res.* **78,** 479–489.
53. Dubach, M., Schmidt, R. H., Martin, R., et al. (1988) Transplant improves hemiparkinsonian syndrome in nonhuman primate: intracerebral injection, rotometry, tyrosine hydroxylase immunohistochemistry. *Prog. Brain Res.* **78,** 491–496.
54. Elsworth, J. D., Sladek, J. R., Jr., Taylor, J. R., et al. (1996) Early gestational mesencephalon grafts, but not later gestational mesencephalon, cerebellum or sham grafts, increase dopamine in caudate nucleus of MPTP-treated monkeys. *Neuroscience* **72,** 477–484.
55. Collier, T. J., Sortwell, C. E., Elsworth, J. D., et al. (2002) Embryonic ventral mesencephalic grafts to the substantia nigra of MPTP-treated monkeys: feasibility relevant to multiple-target grafting as a therapy for Parkinson's disease. *J. Comp. Neurol.* **442,** 320–330.
56. Elsworth, J. D., Brittan, M. S., Taylor, J. R., et al. (1996) Restoration of dopamine transporter density in the striatum of fetal ventral mesencephalon-grafted, but not sham-grafted, MPTP-treated parkinsonian monkeys. *Cell Transplant.* **5,** 315–325.
57. Elsworth, J. D., Brittan, M. S., Taylor, J. R., et al. (1998) Upregulation of striatal D2 receptors in the MPTP-treated vervet monkey is reversed by grafts of fetal ventral mesencephalon: an autoradiographic study. *Brain Res.* **795,** 55–62.
58. Collier, T. J., Elsworth, J. D., Taylor, J. R., et al. (1994) Peripheral nerve-dopamine neuron co-grafts in MPTP-treated monkeys: augmentation of tyrosine hydroxylase-positive fiber staining and dopamine content in host systems. *Neuroscience* **61,** 875–889.
59. Collier, T. J., Redmond, D. E., Jr., Roth, R. H., et al. (1997) Metabolic energy capacity of dopaminergic grafts and the implanted striatum in parkinsonian nonhuman primates as visualized with cytochrome oxidase histochemistry. *Cell Transplant.* **6,** 135–140.
60. Sladek, J. R., Jr., Elsworth, J. D., Roth, R. H., et al. (1993) Fetal dopamine cell survival after transplantation is dramatically improved at a critical donor gestational age in nonhuman primates. *Exp. Neurol.* **122,** 16–27.

61. Sladek, J. R., Jr., Collier, T. J., Elsworth, J. D., et al. (1993) Can graft-derived neurotrophic activity be used to direct axonal outgrowth of grafted dopamine neurons for circuit reconstruction in primates? *Exp. Neurol.* **124,** 134–139.
62. Fiandaca, M. S., Bakay, R. A., Sweeney, K. M., and Chan, W. C. (1988) Immunologic response to intracerebral fetal neural allografts in the rhesus monkey. *Prog. Brain Res.* **78,** 287–296.
63. Sladek, J. R., Jr., Redmond, D. E., Jr., Collier, T. J., et al. (1988) Fetal dopamine neural grafts: extended reversal of methylphenyltetrahydropyridine-induced parkinsonism in monkeys. *Prog. Brain Res.* **78,** 497–506.
64. Collier, T. J., Sladek, C. D., Gallagher, M. J., et al. (1988) Cryopreservation of fetal rat and non-human primate mesencephalic neurons: viability in culture and neural transplantation. *Prog. Brain Res.* **78,** 631–636.
65. Annett, L. E., Martel, F. L., Rogers, D. C., et al. (1994) Behavioral assessment of the effects of embryonic nigral grafts in marmosets with unilateral 6-OHDA lesions of the nigrostriatal pathway. *Exp. Neurol.* **125,** 228–246.
66. Sortwell, C. E., Blanchard, B. C., Collier, T. J., et al. (1998) Pattern of synaptophysin immunoreactivity within mesencephalic grafts following transplantation in a parkinsonian primate model. *Brain Res.* **791,** 117–124.
67. Sladek, J. R., Jr., Collier, T. J., Elsworth, J. D., et al. (1998) Intrastriatal grafts from multiple donors do not result in a proportional increase in survival of dopamine neurons in nonhuman primates. *Cell Transplant.* **7,** 87–96.
68. Annett, L. E., Torres, E. M., Clarke, D. J., et al. (1997) Survival of nigral grafts within the striatum of marmosets with 6-OHDA lesions depends critically on donor embryo age. *Cell Transplant.* **6,** 557–569.
69. Sladek, J. R., Jr., Collier, T. J., Haber, S. N., et al. (1986) Survival and growth of fetal catecholamine neurons transplanted into primate brain. *Brain Res. Bull.* **17,** 809–818.
70. Bankiewicz, K. S., Plunkett, R. J., Jacobowitz, D. M., et al. (1991) Fetal nondopaminergic neural implants in parkinsonian primates. Histochemical and behavioral studies. *J. Neurosurg.* **74,** 97–104.
71. Taylor, J. R., Elsworth, J. D., Sladek, J. R., Jr., et al. (1995) Sham surgery does not ameliorate MPTP-induced behavioral deficits in monkeys. *Cell Transplant.* **4,** 13–26.
72. Annett, L. E., Torres, E. M., Ridley, R. M., et al. (1995) A comparison of the behavioural effects of embryonic nigral grafts in the caudate nucleus and in the putamen of marmosets with unilateral 6-OHDA lesions. *Exp. Brain Res.* **103,** 355–371.
73. Bankiewicz, K. S., Plunkett, R. J., Mefford, I., et al. (1990) Behavioral recovery from MPTP-induced parkinsonism in monkeys after intracerebral tissue implants is not related to CSF concentrations of dopamine metabolites. *Prog. Brain Res.* **82,** 561–571.
74. Starr, P. A., Wichmann, T., van Horne, C., and Bakay, R. A. (1999) Intranigral transplantation of fetal substantia nigra allograft in the hemiparkinsonian rhesus monkey. *Cell Transplant.* **8,** 37–45.
75. Hancock, E. E. and Cima, L. G. (1994) Macromolecular nutrient limitations of encapsulated cells. In *Biomaterials for Drug and Cell Delivery* (Mikos, A. G.,

Murphy, R. M., Bernstein, H., and Peppas, N. A., eds.), Materials Research Society, Pittsburgh, PA, pp. 171–178.

76. Aebischer, P., Goddard, M., Signore, A. P., and Timpson, R. L. (1994) Functional recovery in hemiparkinsonian primates transplanted with polymer-encapsulated PC12 cells. *Exp. Neurol.* **126,** 151–158.

77. Xue, Y. L., Wang, Z. F., Zhong, D. G., et al. (2000) Xenotransplantation of microencapsulated bovine chromaffin cells into hemiparkinsonian monkeys. *Artif. Cells Blood Substit. Immobil. Biotechnol.* **28,** 337–345.

78. Subramanian, T., Emerich, D. F., Bakay, R. A., et al. (1997) Polymer-encapsulated PC-12 cells demonstrate high-affinity uptake of dopamine in vitro and 18F-Dopa uptake and metabolism after intracerebral implantation in nonhuman primates. *Cell Transplant.* **6,** 469–477.

79. Yoshida, H., Date, I., Shingo, T., et al. (2003) Stereotactic transplantation of a dopamine-producing capsule into the striatum for treatment of Parkinson disease: a preclinical primate study. *J. Neurosurg.* **98,** 874–881.

80. Villa, A., Snyder, E. Y., Vescovi, A., and Martinez-Serrano, A. (2000) Establishment and properties of a growth factor-dependent, perpetual neural stem cell line from the human CNS. *Exp. Neurol.* **161,** 67–84.

81. Ourednik, V., Ourednik, J., Flax, J. D., et al. (2001) Segregation of human neural stem cells in the developing primate forebrain. *Science* **293,** 1820–1824.

82. Nishino, H., Hida, H., Takei, N., et al. (2000) Mesencephalic neural stem (progenitor) cells develop to dopaminergic neurons more strongly in dopamine-depleted striatum than in intact striatum. *Exp. Neurol.* **164,** 209–214.

83. Brundin, L., Brismar, H., Danilov, A. I., et al. (2003) Neural stem cells: a potential source for remyelination in neuroinflammatory disease. *Brain Pathol.* **13,** 322–328.

84. Jeong, S. W., Chu, K., Jung, K. H., et al. (2003) Human neural stem cell transplantation promotes functional recovery in rats with experimental intracerebral hemorrhage. *Stroke* **34,** 2258–2263.

85. Tang, Y., Shah, K., Messerli, S. M., et al. (2003) In vivo tracking of neural progenitor cell migration to glioblastomas. *Hum. Gene Ther.* **14,** 1247–1254.

86. Lu, P., Jones, L. L., Snyder, E. Y., and Tuszynski, M. H. (2003) Neural stem cells constitutively secrete neurotrophic factors and promote extensive host axonal growth after spinal cord injury. *Exp. Neurol.* **181,** 115–129.

87. Ourednik, J., Ourednik, V., Lynch, W. P., et al. (2002) Neural stem cells display an inherent mechanism for rescuing dysfunctional neurons. *Nat. Biotechnol.* **20,** 1103–1110.

88. Aboody, K. S., Brown, A., Rainov, N. G., et al. (2000) Neural stem cells display extensive tropism for pathology in adult brain: evidence from intracranial gliomas. *Proc. Natl. Acad. Sci. USA* **97,** 12846–12851.

89. Sladek, J. R., Bjugstad, K. B., Teng, T. D., et al. (2003) The migration of neural stem cells grafted into dopamine-depleted primates follows the nigrostriatal pathway. Society for Neuroscience Abstracts, Abstract 300. 12.

90. Sladek, J. R., Bjugstad, K. B., Teng, T. D., et al. (2004) Implanted neural stem cells migrate along the nigrostriatal pathway and affect the dopamine cells in a

Parkinson monkey model. Experimental Biology meeting abstracts. Accessed at select.biosis.org/faseb. *FASEB J.* **18,** Abstract 8949.

91. Bjugstad, K. B., Snyder, E. Y., Redmond, D. E., et al. (2004) hNSC migrate along the nigrostriatal pathway and restore striatal dopaminergic neurons to control levels. American Society of Neural Transplantation and Repair Abstracts.

92. Bjugstad, K. B., Redmond, D. E., Ourednik, V., et al. (2003) Identification of multiple populations of tyrosine hydroxylase positive cells in striatum of MPTP treated monkeys implanted with human neural stem cells. American Society of Neural Transplantation and Repair Abstracts.

93. Sladek, J. R., Bjugstad, K. B., Teng, T. D., et al. (2003) Three tyrosine hydroxylase positive cell types emerge in the striatum of MPTP-treated monkeys after human neural stem cell implantation. Experimental Biology meeting abstracts. Accessed at select.biosis.org/faseb. *FASEB J.* **17,** Abstract 83635.

94. Bjugstad, K. B., Redmond, D. E., Teng, T. D., et al. (2003) Neural stems cells transplanted into primate substantia nigra (SN) and striatum influence endogenous tyrosine hydroxylase (TH) neurons. Society for Neuroscience Abstracts, Abstract 300. 11.

95. Bjugstad, K. B., Redmond, D. E., Jr., Teng, Y. D., et al. (2005) Neural stem cells implanted into MPTP-treated monkeys increase the size of endogenous dopamine neurons found in the striatum: a return to control measures. *Cell Transplant.* **14,** 183–192.

4

Cell-Based Therapy for Huntington's Disease

Claire M. Kelly, Stephen B. Dunnett, and Anne E. Rosser

ABSTRACT

Huntington's disease (HD) is an inherited neurodegenerative disorder resulting from an expansion of the CAG (cytosine-adenine-guanine) repeat sequence that encodes for glutamine residues. The disease manifests in the third and fourth decades of life, and death is imminent usually within 20 yr. Its pathology is marked by cell loss in the striatal nuclei. Following the marked cell loss, enlargement of the ventricles is observed, as well as cortical degeneration. The aim of neural transplantation is to replace the cells lost in the striatum as a result of the disease and to reform the damaged circuitry. Clinical trials using human fetal tissue have shown a proof of principle for neural transplantation as therapy for the disease. However, there are logistical and ethical issues associated with the current regime using human fetal tissue, and an alternative cell source is therefore required. This chapter outlines the various cell sources that are being explored as possible alternatives for use in neural transplantation for HD.

Key Words: Huntington's disease; neural transplantation; cell therapy.

INTRODUCTION

George Huntington described the disease that took his name in his now well-known paper written in 1872 *(1)*. Huntington's disease (HD) is a progressive and devastating neurodegenerative disorder that affects approx 5–10 per 100,000 in the caucasian community *(173)*. It is inherited in an autosomal, dominant manner and can be reliably diagnosed via an accurate DNA test. The disease belongs to a family of trinucleotide repeat disorders, in which there is expansion of the CAG (cytosine-adenine-guanine) repeat sequence that encodes for glutamine residues. To date, eight triplet repeat disorders have been described: HD, dentatorubral-pallidoluysian atrophy, spinobulbar muscular atrophy, and some spinocerebellar ataxias *(2)*.

From: *Contemporary Neuroscience: Cell Therapy, Stem Cells, and Brain Repair*
Edited by: C. D. Sanberg and P. R. Sanberg © Humana Press Inc., Totowa, NJ

The clinical symptoms of HD include movement disorders (predominantly chorea, bradykinesia, rigidity, and dystonia), as well as a dysphasia and dysarthria, intellectual impairment (initially frontal in nature), and psychiatric disturbance. Although most patients have some symptoms across all categories by the middle to late stages of the disease, the predominant symptoms in the early stage vary from patient to patient. Furthermore, despite that disease onset is usually defined by the appearance of a motor disorder, subtle cognitive decline may precede the onset of motor symptoms by up to 10 yr (3). The disease most commonly manifests in the third and fourth decades of life and is relentlessly progressive; death occurs approx 20 yr after diagnosis (4,5). Presently, symptomatic treatments are limited, and there are no proven disease-modifying therapies available.

The gene for HD, now known as *huntingtin* (*htt*), comprises 67 exons (6) and is found on the end of chromosome 4 (4p16.3) (7). In exon 1, the gene contains a repeated triplet CAG sequence that encodes for the amino acid glutamine, which is the region that is expanded in the mutant gene. The *htt* gene codes for the 350-Kda protein Huntingtin (Htt), the normal function of which is still unclear. In the normal gene, there are fewer than 36 triplet repeats; however, those with repeats of 35–38 triplets are considered to be intermediate, and the extent to which any phenotype results from repeats in this range remains to be determined (8). A correlation exists between repeat length and age of disease onset, but this is largely due to the early onset in patients with very large repeat numbers. Yet, for the majority with repeat numbers in the 40s and low 50s, the correlation is poor.

There have been considerable advances in the knowledge of HD's cellular pathology over the last 5–10 yr, but the precise mechanism by which the gene induces cell death remains elusive. It appears that the mutant Htt produces its effect in a "gain of function" manner; however, loss of function may have a more minor role (9). Evidence supporting gain of function has been derived from the observation that exogenous mutant Htt can cause degeneration in a variety of cell culture systems, despite the presence of endogenous wild-type Htt, and small fragments have been shown to be more toxic than larger ones. Accumulating data show that aggregation of the mutant protein has a significant impact, producing microaggregates as well as larger cellular aggregates, which are now thought of as pathological hallmarks of the disease (10). In vivo, postmortem antibody staining suggests that intraneuronal aggregates are predominantly comprised of truncated N-terminal fragments, rather than full-length mutant Htt (11). The aggregates are believed to be the result of misfolded, expanded polyglutamine repeat sequences, leading to impaired cellular metabolism and cell death

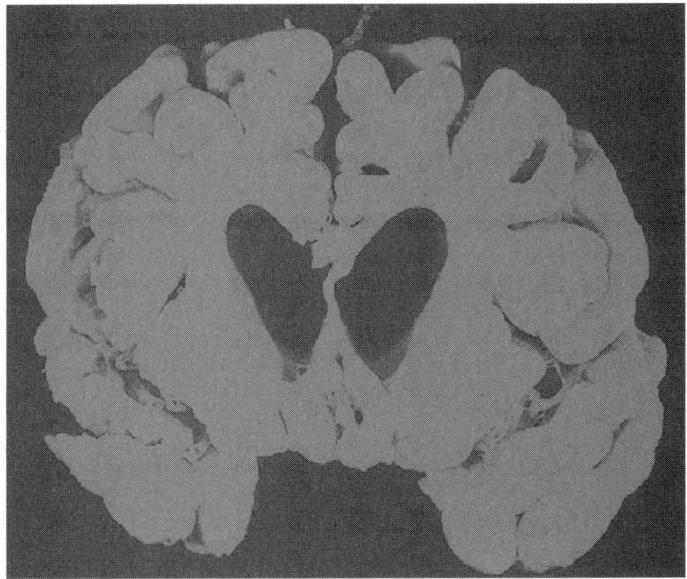

Fig. 1. Coronal section through HD brain. Note the degree of atrophy of the caudate nuclei and the cerebral cortex. Courtesy of Prof. P. Harper.

(12). Whether intracellular aggregates are part of the pathogenic process by interfering with cellular trafficking *(13)*, are beneficial to the cell *(14)*, or simply represent a disease marker, with no interfering or beneficial effects, is still controversial. Wild-type huntingtin may be important in a wide range of cellular processes; a number of potential effects of the mutant protein have been postulated as potentially relevant to the disease state. For example, the cyclic adenosine monophosphate response element–mediated pathway shows the early disruption and is significantly downregulated, compared to the retinoic acid response element and nuclear factor-κB pathways *(15)*. This suggests that reduced CRE-dependent transcription may contribute to disease pathogenesis, polyglutamine expansion may also trigger apoptosis via the activation of caspases 1 and 8 *(16)* pathways, and mutant Htt seems to interfere with proteasome function, thus preventing toxic protein fragments from being removed from the cell *(17,18)*. A rational approach to the development of a disease-modifying therapy will require a more complete understanding of these processes.

At a macroscopic level, the pathology of HD is characterized by neuronal loss in the head of the caudate and putamen of the striatum (Fig. 1) with the medium spiny projection neurones are more affected than striatal interneurones *(5)*. Eventually, there is significant atrophy of striatal structures

because of the neuronal loss, with a compensatory expansion of lateral ventricles. Positron emission tomography scans have shown a progressive loss
of D2 binding in the HD brain *(19)*. The remaining striatum is hypo-
metabolic, and energy production and oxidative metabolism are significantly
reduced *(20)*. As the disease progresses, the pathology becomes more widespread, including wide areas of the neocortex, where overall brain weight may
decrease by 25–30%. Gliosis is seen with marked neuronal loss. Neuronal
loss in the cortex is found to be layer-specific, with the greatest loss seen in
layer VI, and significantly further loss is seen in layers III and V *(2,5)*.

NEURAL TRANSPLANTATION
OF PRIMARY FETAL STRIATAL TISSUE

The relatively focal loss of medium spiny GABAergic projection neurons
in the striatum presents an opportunity to explore neural transplantation as a
strategy for cell replacement and circuit reconstruction. The medium spiny
neurons of the caudate nucleus and putamen form part of a complex circuitry of parallel feedback loops involving discrete areas of cortex and subcortical structures. The medium spiny neurons receive major inputs from
the cerebral cortex, thalamus, and substantia nigra pars compacta, and their
primary outputs occur via GABAergic projections to the globus pallidus
and the substantia nigra pars reticulata. Experimental studies conducted in
animals over the past two decades have established that striatal neurons that
are lost through a lesion can be functionally replaced by transplantation of
the homologous population of fetal neurons. To achieve this, the developing
fetal striatum is dissected, dissociated using enzymatic digestion of the tissue or diced into small tissue pieces less than 1 mm^3, and transplanted stereotaxically into the striatum (see Fig. 2). Following transplantation, these
cells continue developing, innervate the surrounding neuropil, and repair
the circuitry that has been damaged from disease.

Transplantation into animal models of HD has demonstrated the ability
of embryonic striatal neurons to survive, mature, make synaptic connections, and ameliorate functional deficits in both rodent and primate models
of HD.

The success of neural transplantation depends on harvesting the fetal tissue from the appropriate part of the developing central nervous system
(CNS) at the appropriate gestational age, and the preparation be optimized
to maximize cell viability. During development, the striatum forms within
two ridges in the floor of the embryonic lateral ventricles: the lateral and
medial ganglionic eminences (LGE and MGE; see Fig. 3). The greatest concentration of DARPP-32-positive medium, spiny neurons is derived from

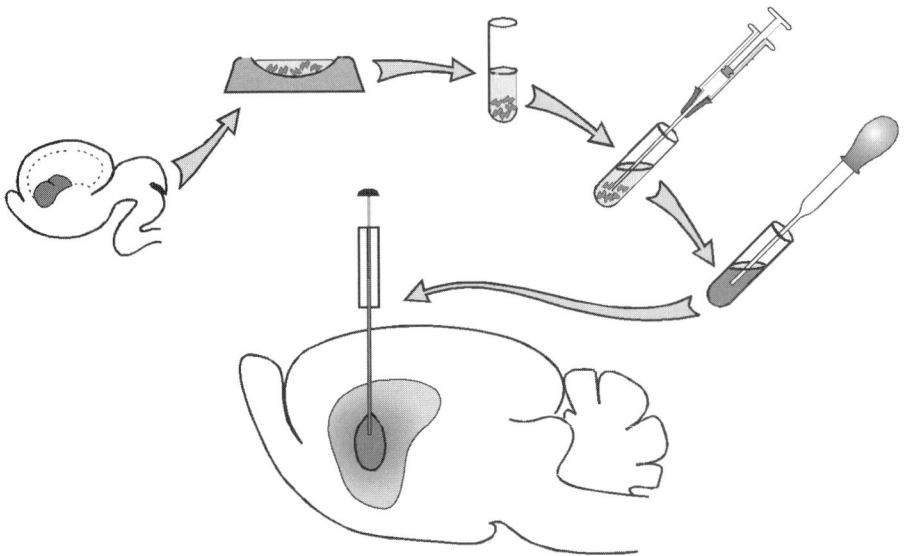

Fig. 2. Schematic illustration of the dissection, preparation, and implantation of neural fetal striatal cells into the lesioned striatum of adult rats.

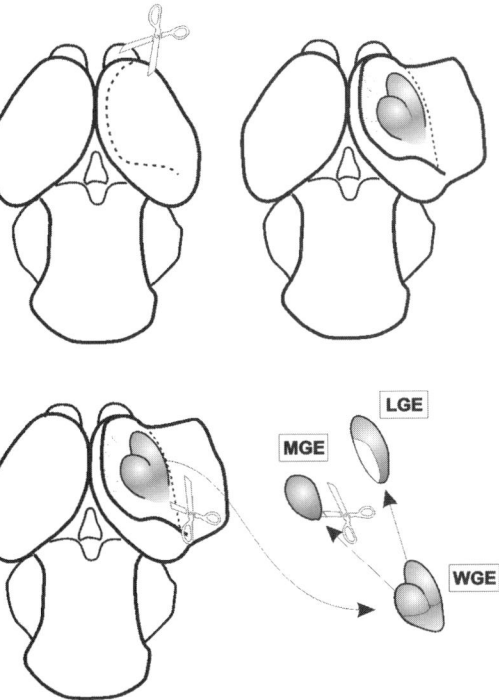

Fig. 3. Schematic dissection of the whole, medial, and lateral ganglionic eminences (WGE, MGE, and LGE) from the developing fetal brain *(165)*.

the LGE *(21)*, and the striatal interneurons are predominantly derived from the MGE *(22)*, but it is still unclear whether maximal functional benefits will be achieved by transplanting cells from selective dissection of the LGE *(23)* or by implantation of cells derived from both parts combined (the so-called *whole ganglionic eminence* [WGE]) *(24,25)*. More empirical studies are required to make this determination.

The gestational age of the tissue is an important factor in establishing appropriate neuronal differentiation and optimal survival of the grafts. Thus, Fricker et al. has shown that grafts derived from younger E14 fetal donors (corresponding to the stage of peak neurogenesis in the developing ganglionic eminence) that are transplanted into adult rat hosts result in larger grafts, better differentiation of the DARPP-32 phenotype, and improved recovery on tests of skilled paw use, compared to grafts from older fetuses. In consideration of these findings, recent pilot clinical trials of cell transplantation in HD conducted in France, the United States, and United Kingdom all use tissue of embryos from fetuses at approx 6–9 wk postconception *(26–28)*. To date, staging has been determined by the analogy of developmental stages between species *(29)*. Thus, the E14 fetuses used for optimal functional effect by Fricker et al. correspond to stage 18 in the Carneige series of developmental stages *(29)*. A similar pattern of graft survival and functional recovery has been reported in the marmoset *(30)* using donor tissue of 73–75-d gestation, which corresponds to a similar Carnegie stage (18–21) in this species *(31)*. In humans, the corresponding stage is reached between 44 and 53 d postconception (when fetal size is approx 13–24 mm crown-rump length [CRL]) *(29)*. Because the striatal eminence in humans is difficult to dissect below 20 mm CRL, tissue for transplantation may most readily be harvested toward the end of this time frame. Although validated by exhibiting tissue survival at this stage of development when xenografted into immunosuppressed rat hosts, the optimal gestational age for human fetal striatal tissue has not yet been systematically determined experimentally. Animal xenotransplantation studies of humans to rats have found surviving grafts with tissue derived from human fetal donors as large as CRL 110 mm. This may suggest that human fetal striatal tissues have significant plasticity over an extended gestational period; however, older tissues have not produced functional benefit in xenograft experiments thus far. At this early stage of progress for the field optimization of tissue preparation procedures, comparing parameters of survival, differentiation, and functional effect is urgently required to address these ambiguities.

Preparing tissue as a cell suspension after dissection involves mechanical dissociation of the tissue. Fricker et al. *(32)* found that trypsinization of rat striatal tissue prior to dissociation resulted in larger surviving grafts that also showed improvements in rotational behavior. Moreover, these grafts contained more striatal tissue and more DARPP-32-positive medium, spiny neurons than did grafts implanted as tissue fragments. However, whether cell suspension or tissue pieces provide optimum grafts remains unclear. Only one study examines this issue directly, and no histological differences were reported. A modest improvement in functional recovery was seen on one test in animals receiving tissue fragment grafts, compared to suspension grafts prepared from the same rat striatal donor tissue *(33,34)*.

Clinical trials of human fetal neurotransplants are ongoing for PD, and early trials are now underway for HD based on extensive experimental work in animal models. The PD trials that began in the late 1980s used primary human fetal mesencephalic tissue as the host tissue and transplanted it into the host striatum, which is the target area of these cells. Placing a graft into the substantia nigra is not viable, as the cells are unable to reliably project the distance from the substantia nigra to their targets in the striatum. This mesencephalic tissue contains fate-committed dopaminergic neuroblasts, which have the capacity to differentiate into fully mature dopaminergic neurons following transplantation, provided that the biological principles arising from animal work are followed. These principles include harvesting tissue between specific gestational ages and optimizing tissue preparation methodologies, as described above. Considering the PD trials, which took these principles into account and used effective longitudinal assessment, results have demonstrated improvements in a range of motor skills, and many (but not all) of the patients have been able to reduce or even eliminate their daily intake of levodopa (L-dopa) *(35,36)*. However, there is variable success with this approach, which may be a direct result of variations in transplant methodology, as well as differences in patient selection criteria *(37–39)*. Recent trials have also highlighted the possibility of dyskinetic side effects in some patients *(37)*; the reasons for these effects are currently being investigated *(40)*.

Parallel clinical trials of neural transplantation in HD are at a much earlier stage than the PD trials and are currently underway in a small number of centers around the world. The French trial, based in Créteil, was the first to provide efficacy data according to a systematic long-term evaluation of five patients. Three of these patients received bilateral striatal implants and were reported to show substantial improvement over several years *(41)*. More recently, the French trial was expanded to include other French-speaking regions in Europe. A total of 40 patients have now received transplants and are currently undergoing follow-up; efficacy data are not available yet.

In another study in Florida, six of seven patients appeared to show improvement, but one declined significantly, so that the overall group changes were not significant *(42)*. One patient died after 18 mo, owing to cardiovascular disease. Postmortem analysis of this patient's brain showed surviving graft tissue that was not affected by the underlying disease progression, at least at that time point *(38)*. The graft tissue was positive for striatal markers, such as acetylcholinesterase, calbindin, calretinin, dopamine and tyrosine hydroxylase. Moreover, there was no sign of immune rejection in the graft region *(38)*. In the same study, three patients developed subdural hemorrhages, and two required surgical drainage *(28)*. These events may have been related to the stage of disease, which was more advanced than the disease stages in the French or UK studies. More advanced cases of HD tend to have more cerebral atrophy, with an increased risk of intracranial bleeding perioperatively. A small number of patients have received grafts in several other centers with reports of safety *(174,175)*, and although efficacy studies are underway in these centers, systematic reports have not yet been published.

Initial studies of cell transplantation in HD provide accumulating evidence of the safety conditions and preliminary efficacy. However, the limited availability of fetal tissue, along with the difficulty in ensuring a high degree of standardization and quality control when a continuous source of fresh donor tissue is required from elective abortion, suppresses the widespread use of neural transplantation as a practical therapy. Ethical and legislative concerns regarding abortion, as well as the large number of donors required to support each operation, already hinder the quantity of patients who can receive grafts to a few specialist centers in a restricted amount of countries. Moreover, the shifting preference for medical, rather than surgical, abortions may further impede the availability of tissues to supply the few programs already in progress. These issues have stimulated the search for alternative sources of donor cells or tissues that circumvent the problems associated with primary fetal tissue collection.

ALTERNATIVE TISSUE SOURCES

A desired characteristic of an alternative cell source is the generation of large, stable cell populations to circumvent the supply issue and to allow regular characterization to ensure the quality and character of the tissue, without the need for separate characterization of every collection. Second, tissue storage methods should be refined and validated to allow the cells to be delivered on demand to advance optimal clinical management of the recipient, rather than constraining both the surgeon and patient to surgery around an erratic schedule of tissue availability. The trials using primary fetal tissue thus provide a proof of concept of the cell transplantation strategy as the basis for developing a practical therapy using a standardized, quality-controlled source of cells available to any appropriately equipped neurosurgical facility on demand. Several options are now being investigated as potential sources of donor tissue.

Stem Cells

The stem cell is a potential donor source that has attracted significant attention recently. The diverse range of cell types that constitute mature animals arises from a single totipotential stem cell: the zygote. From this totipotential state, germinal populations are established, consisting of proliferate multipotential stem cells at the neurula stage of embryogenesis in the primordia of organs and tissues *(176)*. A multipotential cell has the ability to give rise to all the cell phenotypes specific to a particular tissue or organ. Stem cells undergo self-renewal by symmetric division and can also undergo asymmetric division to produce another stem cell and a more differentiated progeny *(43)*. Some multipotential cells may persist into adulthood, either by remaining quiescent in specific regions of the CNS parenchyma or by continued self-renewal *(43)*. Such cells are now referred to as *tissue-specific stem cells (44,45)*. The presence of these cells in the adult may have an important role in maintaining tissue homeostasis via a transitory amplifying cell population that is multipotential and proliferates rapidly in response to signals associated with plasticity, such as those following injury *(45,46)*.

Many definitions of a stem cell involve the attributes described above and also include multipotentialty. However, the exact definition remains a matter of dispute, and at least some ambiguities in this field result from differences in usage of the term *stem cell*. Therefore, it may be better to use the term *precursors*, rather than stem cells *(47)*, which has the advantage of being inclusive, but the disadvantage is that it specifies few necessary or sufficient properties of the cell, other than that it is not in its final differentiated form. For simplicity, stem cell will be used here to refer to a wide variety of proliferating precursors, but we recognize the limitation of this terminology.

Stem cells from a range of sources are potential donor cells for neural transplantation. However, regardless of the source, therapeutic application will require cells to be directed to differentiate into the precise phenotype needed to replace the cells lost to the disease process and (specifically for this chapter) medium spiny neurons for HD. Stem cell sources considered possible donor cells are now described, along with the extent to which directed differentiation has been achieved. This list is not exhaustive but includes at least the main categories of stem cells currently being explored as alternative cell sources for neural transplantation in HD, as well as a number of other neurodegenerative disorders.

Embryonic Stem Cells

Embryonic stem (ES) cells are isolated from the inner cell mass of the embryo at the blastocyst stage. They are pluripotent and can be propagated in culture for long periods of time in an undifferentiated state *(48–50)*. ES cells have the potential for extensive expansion and may differentiate into all cell types of the body. Significant ethical disputes have been associated with the derivation and use of ES cells, including concerns over the use of human embryos and fears related to human cloning *(51,52)*. As a result of these ethical issues, many countries have restricted or banned ES cell research. Nevertheless, other countries have actively supported the development of ES cell research because of the perceived therapeutic benefit in various diseases. Some countries, such as the United Kingdom, allow cloning of human embryos for therapeutic purposes but impose tight regulations to preclude their use for reproductive cloning.

Culture conditions can be manipulated in such a way that cells clump together to form embryoid bodies (EBs). EBs contain precursors that can generate cells pertaining to any of the three germ layers. Controlling the differentiation of ES cells is important, both to derive the target cell populations and to ensure the absence of cells with a continued proliferative potential. Progress has been made in directing the differentiation of ES cells down a neuronal lineage, e.g., by the addition of retinoic acid and nerve growth factor to the medium *(50,53)*.

A more difficult issue is identifying methods for driving ES-derived neurons down the specific phenotypic lineages required for different applications. Some progress has been achieved for the specific dopaminergic differentiation of ES-derived neurons, with reports of 16–35% tyrosine hydroxylase (TH)–positive neurons being generated by the addition of specific factors to the culture medium *(54–57)*. Expression of the transcription factor Nurr1 enhances the differentiation of ES cells into dopaminergic neurons; 80% TH-positive neurons were generated *(58–62)*. However, many of these studies have based their results on the expression of one marker: TH. Although this is present in dopaminergic neurons, it does not differentiate between the catecholamines dopamine, noradrenaline, and adrenaline and does not indicate that the cells are functional. A detailed analysis of these cells for appropriate receptors, as well as dopamine-synthesizing enzymes' storage and uptake molecules *(58,59)*, must be performed to more fully characterize these cells.

Less is known about the ability of these cells to generate striatal-like cells. Differentiating the cells with chemically defined media resulted in a cell population that expressed neural fate characteristics typical of the forebrain, such as *Dlx5, Dlx1, Lhx5, Tbr1, Pax6, Dbx1, Gsh2,* and *Gsh1*. However, differentiating them to alternative fates was temporally restricted because of a loss of responsiveness to positional cues *(63)*. In the presence of fibroblast growth factor 2 (FGF-2) during the first 8 d in culture, these cells maintain a largely neuronal fate. Yet, with successive passaging, an ontogenic drift toward gliogenesis is evident *(64)*.

Microarray analysis of murine striatal eminences during embryonic development has allowed the identification of genes that are striatal-specific, such as *Gsh2 (178)*, *Dlx*, and the *FoxP* genes *(170)*. Such studies will hopefully identify genes that may be used in vitro and in vivo to direct the differentiation of these cells, as well as ES- and EG-derived neurons down a striatal-specific neuronal lineage for subsequent transplantation.

Another important issue is the potential of ES cells to form tetracarcinomas, because remaining undifferentiated ES cells in grafted cell suspension can continue to divide, forming tumors. For example, Bjorklund et al. *(65)* grafted a mouse ES cell line into a rat model of PD and reported that five out of 25 grafts formed teratoma-like tumors, with consequential death of the animals. One method for eliminating undifferentiated cells is introducing suicide genes, such as the *E. coli gpt* and herpes thymidine kinase (*HSVtk*), into the cells prior to transplantation. Differentiated ES cells are resistant to the effects of ganciclovir. Therefore, the presence of a neomycin resistance gene in the plasmid vector allows selection of the undifferentiated ES cells out of the cell suspension. Undifferentiated HSVtk-positive cells that continue to proliferate can then be destroyed by the conversion of prodrug nucleoside ganciclovir to its phosphorylated form, which is then incorporated into the DNA of replicating cells, resulting in apoptosis of the cells *(66,67)*. The functionality and efficacy of the differentiated cells will also have to be addressed, as well as the possibility of rejection, before they can be considered for clinical trials.

Embryonic Germ Cells

Embryonic germ (EG) cells are diploid primordial cells that migrate from the posterior endoderm of the yolk sac via the gut mesentery during development, thus populating the gonads *(68)*. Once in the gonads, these cells proliferate and finally undergo meiosis to yield spermatozoa or ova.

The population of EG cells taken for culture are obtained from the premeiotic fetal gonads during the first trimester *(69)*.

Mouse EG cell lines vary in their properties depending on the day they were derived, and gene imprinting occurs during migration of the cells to the gonadal ridge. This appears to have effects on the differentiation of these cell lines in vivo and in vitro *(167–169)*. The mouse-derived EG cells can proliferate for prolonged periods of time in culture in the presence of specific growth factors and are pluripotent and chromosomally stable, thus resembling ES cells.

Less is known about human-derived EG cell lines, but early reports suggest that they are not as easy to maintain in culture as mouse EG cells and may spontaneously differentiate in vitro *(70)*. Not much is known of the potential of these cells relating to neural transplantation.

Fetal Neural Stem Cells (NSCs)

ES and EG cells are totipotential or multipotential cell sources and thus require manipulation in vitro to be directed first to a neuronal fate, then to a striatal-specific phenotype. An alternative approach is to identify stem cells that are already committed to a neural lineage (i.e., tissue-specific) and from an even more restricted lineage—striatal precursors, where it may be easier to drive an explicitly striatal phenotype.

Fetal neural stem cells (NSCs), also called embryonic neural precursors, are proliferative cells isolated from discrete regions of the developing brain and are already committed to becoming the major cell types of the CNS. It is becoming increasingly clear that proliferating fetal NSCs are a heterogeneous population of cells *(71)*. They proliferate in response to growth factors, such as epidermal growth factor (EGF) and FGF-2, as spheres of cells known as *neurospheres* *(72–74*; see Fig. 4). In addition to stimulating cell proliferation in culture, FGF-2 and EGF act sequentially on the regulation of differentiation *(75)*. FGF-2 is known to enhance the neuronal differentiation of the cells, whereas EGF promotes astroglial differentiation *(76)*. Several other growth factors may be able to enhance the neuronal differentiation of these cells down particular lineages, including nerve growth factor, insulin-like growth factor, and tumor necrosis factor (TNFα) *(77–80)*. Identifying an appropriate growth factor cocktail that is appropriate to the phenotype associated with each particular application may be a necessary prelude to using these cells for transplantation. However, this is not an easy task.

Cross-species variation exists in the proliferative potential of NSCs *(81)*. Rat cells are found to enter a state of senescence after a relatively short period of time (30–40 d) in contrast to mouse and human cells that have the

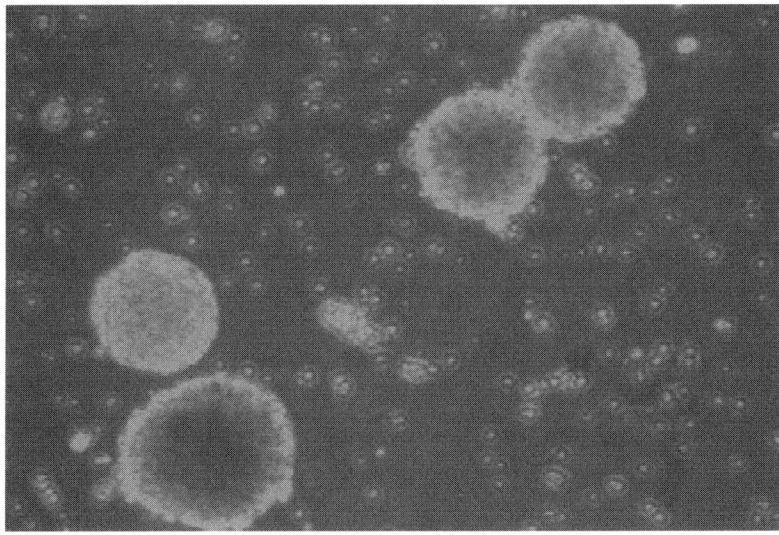

Fig. 4. Fetal NSCs can expand in culture to form free-floating spheres of cells (neurospheres), each of which contains several thousands of cells.

potential to proliferate for much longer time periods in culture *(81)*. This may be related to differences in the tissue culture requirements of the two species *(82)*. Murine NSC cultures contain neural precursor cells that continue to divide in the presence of FGF-2 and EGF, based on a BrdU incorporation assay that showed that β-III tubulin–positive neurons were also BrdU-positive after five passages in culture (Kelly et al., unpublished observations). Human fetal NSCs maintained in culture for 10 d prior to transplantation without passaging resulted in 41% neurons, of which 70% were DARPP-32–positive (Kelly et al., unpublished observations). However, for all species, under presently defined conditions, the proportion of cells that differentiate into neurons is seen to decrease with each passage in culture *(73,82–84)*.

Molecular characterization of fetal NSCs in vitro has shown that they retain a degree of their site-specific identity when environmental cues are absent; however, when cocultured with cells of different origin, they can adopt a new fate *(85,86)*. Expression of site-specific genes (e.g., *Islet1* and *Er81*) is maintained in these precursor cells over time, but with neuronal differentiation, expression of striatal-specific neuronal markers (e.g., DARPP-32 and Islet1) is lost. Yet, they do express homeobox transcription factors, DLX and MEIS2 *(86,87)*. Thus, it appears that NSC expansion in culture limits the differentiation potential of the cells. Further evidence of this derives from transplantation of NSCs into disease models. NSCs expanded for longer periods of time prior to transplantation into the adult rat striatum resulted in lower numbers of neurons in the graft with many grafts not surviving *(88,89)*. One interpretation of these findings is that positional information is lost with continued expansion. Therefore, when long-term expanded cells are placed in an environment, such as the adult CNS, they are not exposed to the developmental signals that they would see in the developing brain. Thus, they are unable to differentiate into neurons that are appropriate to the site from which they were derived (e.g., medium spiny neurons from striatally derived NSCs). However, when grafted to the neonatal brain, similar cells appear to respond to developmental signals and regional determinants by differentiating in a site-specific manner *(90–92)*, suggesting that they retain the capacity to respond to developmental signals if they are present. In comparison to primary tissue grafts, where there is little migration of the grafted cells, fetal NSCs in vivo have been reported to migrate away from the graft *(76,88,90,91)*. Their potential to repair circuitry lost because of disease is still unknown, but animal experiments have shown that posttransplantation differentiated NSCs do send out projections to distant sites *(76,91,93,94)*. The specificity of these projections has not yet been established in detail.

Adult NSCs

Adult NSCs (ANSCs) are an example of an adult tissue-specific stem cell and (as their name suggests) are derived from the mature brain. Using ^{3}H-thymidine autoradiography, Altman and colleagues provided the first clear evidence that a low level of neurogenesis is ongoing in the dentate gyrus of adult rats *(95)*. ANSCs have since been confirmed in two main regions of the CNS: the subgranular layers of the dentate gyrus, from where the newly formed neurons repopulate the dentate gyrus *(96)*, and the subventricular zone of the lateral ventricles *(97)*, from where the newly formed neurons migrate via the rostral migratory stream to the olfactory bulb *(98; see Fig. 5)*. More recently, it has been reported that NSCs may also reside in other regions of the brain, albeit at an even lower concentration, including the cortex *(99,100)* and the medial-rostral part of the substantia nigra pars compacta in the lining of the cerebroventricular system of the midbrain *(101)*. However, these reports remain controversial *(102)*.

The appeal of ANSCs as a donor supply for neural transplantation is the possibility of autologous transplants, thus bypassing the immunological issues of graft rejection, which is severe in the case of xenografts and not entirely benign even for allografts. Furthermore, it may eventually be pos-

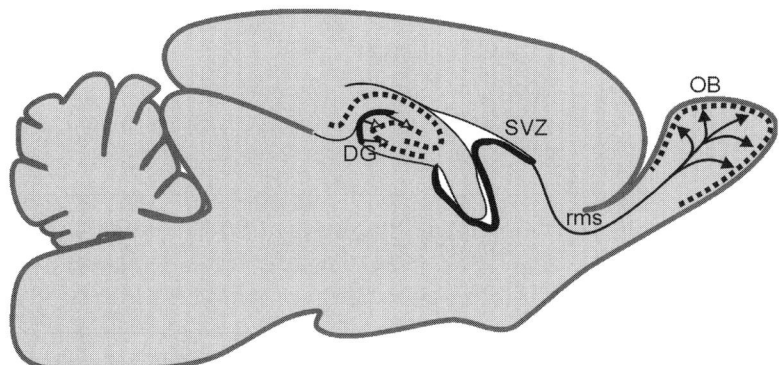

Fig. 5. ANSCs have been identified in the subventricular zone (SVZ) and the dentate gyrus (DG) of the hippocampus. The SVZ cells migrate along the rostral migratory stream (rms) to the olfactory bulb (OB), whereas newly formed neurons in the DG reside within the granule cell layer *(166)*.

sible to recruit such cells for endogenous repair without requiring their isolation and reimplantation. It might be possible to stimulate the resident population of ANSCs to migrate to the site of degeneration, but adult stem cells remain difficult to isolate and grow in culture. In addition, the factors that are required to enhance the proliferation of these cells and their differentiation into the particular phenotypes relevant to the site of degeneration remain unknown.

Trans-*Differentiation* *of Other Tissue-Specific Stem Cell Populations*

Another approach is to attain *trans*-differentiation of non-neural tissue-specific stem cell populations. The classic population—bone marrow–derived stem cells—are more easily harvested than either fetal stem cells or ANSCs, but their disadvantage is that they do not (by default) produce neurally differentiated cells.

During development, mesenchymal stem cells (MSCs) give rise to all mesodermal cell types of the body, including osteoblasts, chondrocytes, adipocytes, and muscle cells. Cells with similar characteristics are found in the adult bone marrow, along with more lineage-restricted cells that contribute to differentiated hematopoietic cells. Controversy currently surrounds the precise categorization of these marrow components *(103)*.

Some evidence shows that *trans*-differentiation can be achieved, but this remains an area of dispute. MSCs have been reported to *trans*-differentiate to ectodermal and endodermal cell fates *(104)*. In vitro, MSCs have differentiated to form neurons and astrocytes. MSCs transplanted into the rat brain survive and express markers of neuroectodermal cells as well as having a functional effect *(104)*. MSCs are not the only cells able to *trans*-differentiate; NSCs have also been shown to have this ability and were seen to differentiate into muscle *(105)*. However, recent evidence suggests that this plasticity may be a result of cell fusion, based on studies that have examined the potential of MSCs to differentiate into hepatocytes *(106,107)*. This issue needs to be clarified for these cells to be considered for neural transplantation.

Hematopoietic stem cells (HSCs), also derived from the bone marrow, continually reconstitute the blood and are the best characterized of the tissue-specific stem cells. Two classes of HSCs have been identified in the mouse: those that survive for approx 2 mo (short term) and those that survive more than 6 mo (long term) *(48)*. Fluorescence-activated cell sorting has been used to positively select cells based on the expression of specific cell surface markers. HSCs can be highly enriched (up to 10,000-fold) and then transplanted into the bone marrow of patients *(108)* for the treatment of oncogenic blood diseases. In an animal model of spinal cord injury, HSCs have been shown to survive for 5 wk after transplantation, differentiate into astrocytes, oligodendrocytes, and neuronal precursors, and show improvement in functional behavior using hindlimb motor function *(109)*, but no mature neurons were identified.

Blood in the human umbilical cord is easily retrieved following labor without risk of harm to the mother or child, and it has been reported to contain multipotential progenitor cells that apparently have the ability to *trans*-differentiate into neuronal and glial cells *(110)*. Transplantation of these cells into the neonatal and adult brain may survive and differentiate into neurons and glia *(111–113)*. Intravenous delivery, rather than neural transplantation, may be a more advantageous method to administer these cells for therapeutic benefit, based on a study by Willing et al. *(112)*, where there was significant improvement in certain behavioral tasks, compared to animals that received neural transplants of cells directly to the striatum. However, further studies are necessary to validate the potential of these cells, and as mentioned previously, the issue of cell fusion needs to be addressed in this context.

XENOGENIC TISSUE

Xenotransplantation offers the opportunity of breeding animals for fetal striatal tissue donation under conditions where the supply can be regulated according to demand. The breeding stock is inbred, well-characterized, and controlled for pathogens, and tissue collection and preparation can be undertaken under standardized sterile GMP conditions. The most likely donor candidate is porcine tissue. The advantages are (1) the extensive experience of animal husbandry within this farm species; (2) the reliability of breeding; (3) the large size of the litters; (4) the possibility of sterile collection under standardized conditions; (5) the comparable size and time course of development of the pig and human brain; (6) and the potential application of transgenic technology to porcine tissue. This would enable the possibility of genetic manipulation, e.g., to modify the immunogenicity of transplanted tissue.

Transplantation of xenogeneic tissues into the immunosuppressed host CNS has been performed using a number of species (e.g., human to rat, pig to rat, rat to mouse, and vice versa) *(81,114–118)*. Both primary and expanded tissue graft experiments have been reported using xenogenic tissue. The grafted tissue has been found to survive transplantation and axonal and glial fiber projections from the grafts as well as make synapses with the host brain.

Studies of the ability of porcine tissue to achieve brain repair have been conducted predominantly in PD models. Primary tissue grafts have been shown to integrate into the host environment, but even with cyclosporine, a slow rejection process ensues, and finding ways to overcome the issue of graft rejection is the subject of much investigation. Primary porcine ventral mesencephalic (VM) tissue grafted into cortically lesioned neonate brains survived and integrated with the host brain and sent out long axonal projections, thus showing porcine tissue's capable response to rat axonal guidance factors *(119)*. Similar findings have been reported after placing porcine VM tissue into the lesioned cortex of adult rats *(120)* and into the striatum in animal models of HD and PD *(172)*. There has been good functional improvement with grafts placed in the striatum of a 6-hydroxydopamine-lesioned rat model, according to compensation on the amphetamine rotation test *(121)*. In these studies, cyclosporine was used as an immunosuppressant to protect the grafted tissues from host rejection. Further improvement of graft survival was achieved by the addition of caspase inhibitors, which increased the number of TH-positive cells in the grafts by 2.5-fold *(122)*. Again, donor age is an important issue. Comparison of donor VM tissue, ranging in ages of E24 to E35, indicated that E26–E27 embryos gave the highest yield based on TH differentiation of the tissue *(123–125)*.

Clinical studies of CNS xenotransplantation are limited. Primary porcine embryonic striatal tissue has been transplanted into the caudate and putamen of 12 immunosuppressed PD patients, with some clinical improvements reported but with little convincing evidence of graft survival *(118)*. The immune response from these grafts was more vigorous than that seen in human-to-rodent models. One patient died 7 mo postoperatively for reasons unrelated to the graft and was found to have a very small number of surviving neurons in the graft region, suggesting that the majority may have been rejected. In the same series, 12 HD patients received porcine striatal grafts, but again, there was little evidence of graft survival or functional effect. In the 12-mo postoperative analysis of these patients, no change was seen in the mean total functional capacity score *(126)*.

Striatal allografts of primary fetal tissue in the HD lesion model have been shown to send out projections in the host brain, as demonstrated by anterograde and retrograde tracing methods *(127,128*; see Fig. 6). Similarly, xenotransplantation of striatal tissue has also been shown to send out projections in the host brain *(76,129–131)*. Neurofilament staining using species-specific antibodies for xeotransplanted tissue has been successful in revealing the extent and richness of short- and long-distance outgrowth from striatal grafts *(76,130)*. Yet, because the tools are not available for tracing allografted tissue, the extent of projections from grafted tissue is more difficult to assess.

Two key issues must be resolved for xenografts to progress as practical therapeutic trials. The first issue relates to the fact (as illustrated by the first pilot clinical trial reported above) that xenografted tissue is largely rejected in the absence of effective immune protection. Two alternatives strategies were adopted in the Diacrin trial: daily treatment with cyclosporine A or treatment with an antibody against major histocompatability complex 1 to block the host T-cell response *(126)*. No clear evidence indicates that either strategy proved effective for yielding good cell survival in patients, and it is surprising that the study had progressed on the basis that preliminary reports of the same strategies in primates were equally ineffective. Combination immunoprotection strategies to promote xenograft survival are an area of active research *(132,133)*, but a clear optimal protocol that will allow reliable long-term survival of xenografted neural tissue in the adult brain in a majority of subjects has not yet been defined.

The second key issue that requires resolution relates to the safety of xenografted tissues. Considering the recent spread of bovine spongiform encephalopathy to humans in the form of new variant Creutzfeld-Jacob disease (CJD), and the difficulty in controlling the spread of animal pathogens, as exemplified by the recent UK foot-and-mouth disease (FMD), there is worldwide concern about eliminating the risk of transmitting animal diseases to humans. This may be particularly dangerous in the context of tissue

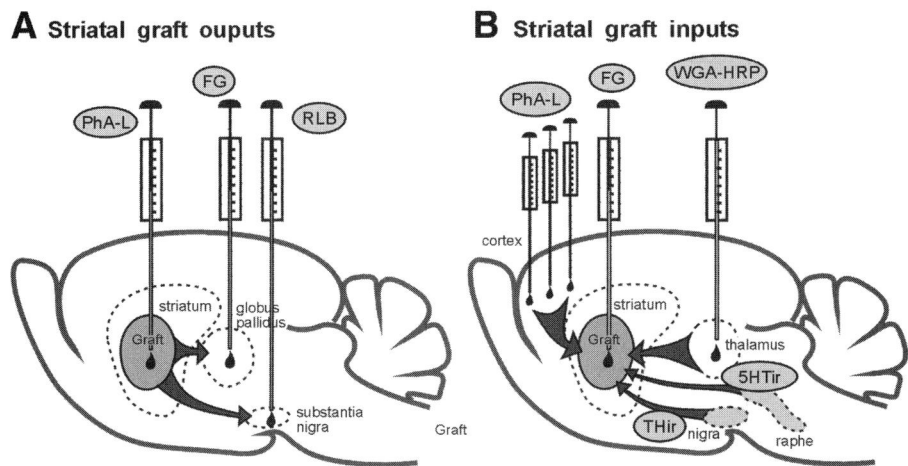

Fig. 6. Schematic of afferent and efferent connections of striatal grafts using anterograde and retrograde pathway tracing and immunohistochemistry. All major striatal inputs and outputs are established in striatal grafts. (Based on ref. *128*.)

transplantation directly into the immunosuppressed CNS. The concern is not just for the recipient, but in the case of porcine endogenous retroviruses, it is whether direct transfer into the brain might provide a route of transmission that allows virus mutation into new forms of viruses that catalyze unpredicted new diseases in humans, even leading to *de novo* epidemics. Although the chances of such mutation are recognized as very low, the cost of occurrence could be devastatingly high. Moreover, the risk of generating a new disease by an unknown mechanism is impossible to absolutely exclude by any known safety procedure. Consequently, the regulatory climate is such that any novel xenograft approach is unlikely to gain approval for trial in the foreseeable future, at least in Europe. In the absence of having suffered the same major bovine spongiform encephalopathy, CJD, and FMD epidemics, US regulations, although strict, are somewhat more permissive, with the result that most academic and commercial research of developing xeno-transplantation as a therapeutic strategy for the CNS has moved westward across the Atlantic Ocean over the last 5 yr.

The capacity for embryonic porcine neural tissue to be expanded in culture depends on gestational age of the tissue, with E22 tissue expanding and surviving transplantation more robustly than that of tissue from an E27 embryo *(134)*. Following transplantation, expanded neural porcine tissue differentiated into neurons and sent out projections to the appropriate target areas. Moreover, expanded porcine cells survived for longer periods of time in the nonimmunosuppressed rat host than did primary porcine CNS tissue *(131)*. Thus, it may be that expansion of the tissue confers a benefit in terms of reduced immunogenicity, and there is some supporting in vitro evidence that the expression of surface histocompatibility antigens is reduced in porcine neural stem cells *(132)*. However, similar to other stem cell sources (see below), a significant problem remains in the expanded cells that retain a specific striatal or dopaminergic phenotype, limiting their functional effects until this problem can be overcome.

GENETICALLY ENGINEERED CELLS

A variety of cells may be engineered in vitro, either to produce molecules of potential importance for CNS release (e.g., in the form of polymer encapsulated cells, as below) or to alter the properties of a cell to render it useful for circuit reconstruction. Of course, these strategies are not necessarily mutually exclusive; trophic factor support may be crucial for transplanted cells to survive and integrate in the host brain, and genetically engineering cells to release trophic factors in the graft region is one possible method for optimizing graft survival.

The herpes simplex viral (HSV) vector was the first virus to be tested to introduce genes into the adult CNS *(135–137)*. More recently, other viral vectors have been introduced, including adenovirus, the recombinant adeno-associated virus (rAAV), lentivirus, and pseudotyped vectors. The rAAV vector is more efficient than the HSV because it achieves much higher levels of expression. The use of such vectors has allowed genes to be transferred to a specific group of cells in the CNS *(138)* and has provided support for the efficacy of such factors as glial cell line–derived neurotrophic factor (GDNF) for PD *(139–142)* and ciliary neurotrophic factor (CNTF) for HD *(143–146)*.

Polymer capsules have been considered as a system for trophic factor delivery to the CNS, because they have the advantages of being relatively cheap to produce and can also be removed from the CNS as required. However, the major drawback is that the effect does not last long *(147)*. With a limited amount of protein required for relatively short periods of time, polymer microspheres are an appealing alternative, considering they are biodegradable, and subsequent surgical procedures are not required for retrieval *(148)*. Yet, improvements in the duration of release have been obtained via the use of encapsulated cells engineered to produce the desired molecules *(149,150)*. Here, cells engineered to secrete specific substances, e.g., neurotrophic factors, are protected from the host immune system by a semipermeable selective biocompatible outer membrane *(145,147,151–153)*.

The outer membrane allows the entry of nutrients to the cells and allows the exit of neuroactive molecules. The advantage of this strategy is that it enables the implantation of xenogeneic cells, which may be much easier to obtain or engineer than human cells. This approach has been used for the delivery of GDNF in animal models of PD *(148,177)* and CNTF in animal models of HD *(149)*.

Regarding HD, several studies have examined the use of polymer encapsulated cells for the delivery of CNTF. Baby hamster fibroblasts have been genetically modified to produce human CNTF and have been incorporated into polymer capsules *(151,154)*. Both rodent and primate studies have been carried out incorporating this method *(145,151,152,154–156)*. These animal studies suggested that CNTF can protect striatal neurons against subsequent damage from an excitotoxic lesion. In addition to protecting specific populations of striatal neurons from lesion-induced cell death, behavioral improvement was observed in skilled motor and cognitive tasks, compared to control animals. Encapsulated CNTF released by BHK cells is now being studied in clinical trials *(157)*. Nevertheless, the use of encapsulated cells for the delivery of growth factors and neurotrophic factors is an attractive alternative and may be required in combination with neural transplantation to provide trophic support to the grafted cells.

Other potential cell sources are immortalized cell lines, the neurally committed lines, such as the Ntera2 cell line, RN33B, and Hib5. Functional benefit has been reported using these cells in various animal models *(158–161)*. The Ntera2 cell line has been the most widely used. These cells are derived from human embryonal carcinomas and are terminally differentiated in vitro with retinoic acid. They have responded to environmental cues when transplanted into the excitotoxically lesioned striatum *(159,160)*, sending out target-specific projections, as well as expressing a site-specific phenotype. Grafting Ntera2 cells into the excitotoxic-lesioned striatum resulted in neuronal differentiation; a preliminary study reported dramatic functional effects *(171)*. However, in a more detailed analysis, the cells did not express any striatal-specific markers, and there was no sustained improvement on skilled paw reaching and cylinder placing *(162)*. Transplantation of the RN33B cell line to the lesioned and nonlesioned striatum of rats has demonstrated their ability to differentiate into neurons in a site-specific way and form connections with target areas, such as the globus pallidus *(163)*. However, only a proportion of the cells showed this differentiation potential. A major disadvantage in using such cell lines is the genotypic variability that arises from the immortalization process *(164)* and the risk that cells continue to proliferate to form tumors after transplantation.

CONCLUSIONS

The replacement and repair of striatal neurons by transplantation in HD may achieve circuit reconstruction and alleviate some of the devastating symptoms associated with the disease. Transplants using primary fetal striatum as the donor tissue are crucial for proof of principle, but ethically and logistically, this donor source will not be suitable for widespread therapeutic application. Alternative cell sources, including xenografts, engineered cell lines, and stem cells, may eventually replace primary fetal tissue and thus provide a cell source that would be widely available to patients. Neural transplantation may not be a suitable therapy for all patients, and the degree of degeneration may be a limiting factor. The extent to which the disease has progressed, and its relationship to transplant survival, are issues that are yet unanswered. Promising results from initial clinical trials lead this to be an exciting field worthy of active and focused investigation.

REFERENCES

1. Huntington, G. (1872) On chorea. *Medical and Surgical Reporter* **26,** 320–321.
2. Reddy, P. H., Williams, M., and Tagle, D. A. (1999) Recent advances in understanding the pathogenesis of Huntington's Disease. *Trends Neurosci.* **22,** 248–254.
3. Lawrence, A. D., Hodges, J. R., Rosser, A. E., et al. (1998) Evidence for specific cognitive deficits in preclinical Huntington's disease. *Brain* **121,** 1329–1341.
4. Naarding, P., Kremer, N. P. H., and Zitman, F. G. (2001) Huntington's disease: a review of the literature on prevalence and treatment of neuropsychiatric phenomena. *Eur. Psychiatry* **16,** 439–445.
5. Ross, C. A. and Margolis, R. L. (2001) Huntington's Disease. *Clin. Neurosci. Res.* **1,** 142–152.
6. Ho, L. W., Carmichael, J., Swatz, J., et al. (2001) The molecular biology of Huntington's disease. *Psychol. Med.* **31,** 3–14.
7. Huntington's Disease Collaborative Research Group (1993) A novel gene containing a trinucleotide repeat that is expanded and unstable on Huntington's disease chromosomes. *Cell* **72,** 971–983.
8. Georgiou-Karistianis, N., Smith, E., Bradshaw, J. L., et al. (2003) Future directions in research with presymptomatic individuals carrying the gene for Huntington's disease. *Brain Res. Bull.* **59,** 331–338.
9. Cattaneo, E. and Calabresi, P. (2002) Mutant huntingtin goes straight to the heart. *Nat. Neurosci.* **5,** 711–712.
10. Wanker, E. E. (2000) Protein aggregation and pathogenesis of Huntington's disease: mechanisms and correlations. *Biol. Chem.* **381,** 937–942.
11. Furtado, S., Suchowersky, O., Rewcastle, B., et al. (1996) Relationship between trinucleotide repeats and neuropathological changes in Huntington's disease. *Ann. Neurol.* **39,** 132–136.

12. Li, S. H. and Li, X. J. (2004) Huntingtin and its role in neuronal degeneration. *Neuroscientist* **10,** 467–475.
13. Saudou, F., Finkbeiner, S., Devys, D., and Greenberg, M. E. (1998) Huntingtin acts in the nucleus to induce apoptosis but death does not correlate with the formation of intranuclear inclusions. *Cell* **95,** 55–66.
14. Arrasate, M., Mitra, S., Schweitzer, E. S., et al. (2004) Inclusion body formation reduces levels of mutant huntingtin and the risk of neuronal death. *Nature* **431,** 805–810.
15. Sugars, K. L., Brown, R., Cook, J., et al. (2004) Decreased cAMP Response element-mediated transcription: an early event in exon 1 and full-length cell models of Huntington's disease that contributes to polyglutamine pathogenesis. *J. Biol. Chem.* **279,** 4988–4999.
16. Li, S. H., Lam, S., Cheng, A. L., Li, X. J. (2000) Intranuclear huntingtin increases the expression of caspase-1 and induces apoptosis. *Hum. Mol. Genet.* **9,** 2859–2867.
17. Venkatraman, P., Wetzel, R., Tanaka, M., et al. (2004) Eukaryotic proteasomes cannot digest polyglutamine sequences and release them during degradation of polyglutamine-containing proteins. *Mol. Cell* **14,** 95–104.
18. Bence, N. F., Sampat, R. M., and Kopito, R. R. (2001) Impairment of the ubiquitin-proteasome system by protein aggregation. *Science* **292,** 1552–1555.
19. Andrews, T. C., Weeks, R. A., Turjanski, N., et al. (1999) Huntington's disease progression. PET and clinical observations. *Brain* **122,** 2353–2363.
20. Kopyov, O. V., Jacques, S., and Eagle, K. S. (1998) Fetal transplantation for the treatment of neurodegenerative diseases- Current status and future potential. *CNS Drugs* **9,** 77–83.
21. Deacon, T. W., Pakzaban, P., and Isacson, O. (1994) The lateral ganglionic eminence is the origin of cells committed to striatal phenotypes: neural transplantation and developmental evidence. *Brain Res.* **668,** 211–219.
22. Olsson, M., Bjorklund, A., and Campbell, K. (1998) Early specification of striatal projection neurons and interneuronal subtypes in the lateral and medial ganglionic eminence. *Neuroscience* **84,** 867–876.
23. Nakao, N., Grasbon-Frodl, E. M., Widner, H., and Brundin, P. (1996) DARPP-32-rich zones in grafts of lateral ganglionic eminence govern the extent of functional recovery in skilled paw reaching in an animal model of Huntington's disease. *Neuroscience* **74,** 959–970.
24. Fricker, R. A., Torres, E. M., and Dunnett, S. B. (1997) The effects of donor stage on the survival and function of embryonic striatal grafts in the adult rat brain. I. Morphological characteristics. *Neuroscience* **79,** 695–710.
25. Watts, C., Brasted, P. J., and Dunnett, S. B. (2000b) Embryonic donor age and dissection influences striatal graft development and functional integration in a rodent model of Huntington's disease. *Exp. Neurol.* **163,** 85–97.
26. Rosser, A. E., Barker, R. A., Armstrong, R. J., et al. (2003) Staging and preparation of human fetal striatal tissue for neural transplantation in Huntington's disease. *Cell Transplant.* **12,** 679–686.

27. Bachoud-Levi, A.-C., Hantraye, P., and Peschanski, M. (2002a) Fetal neural grafts for Huntington's Disease: A prospective View. *Mov. Disord.* **17,** 439–444.
28. Hauser, R. A., Sandberg, P. R., Freeman, T. B., and Stoessl, A. J. (2002b) Bilateral human fetal striatal transplantation in Huntington's disease. *Neurology* **58,** 1704.
29. Butler, H. and Juurlink, B. H. J., eds. (1987) *An Atlas for Staging Mammalian and Chick Embryos.* CRC Press, Inc., Boca Raton, FL.
30. Kendall, A. L., Rayment, F. D., Torres, E. M., et al. (1998) Functional integration of striatal allografts in a primate model of Huntington's disease. *Nat. Med.* **4,** 727–729.
31. Annett, L. E., Torres, E. M., Clarke, D. J., et al. (1997) Survival of nigral grafts within the striatum of marmosets with 6-OHDA lesions depends critically on donor embryo age. *Cell Transplant* **6,** 557–569.
32. Fricker, R. A., Barker, R. A., Fawcett, J. W., and Dunnett, S. B. (1996) A comparative study of preparation techniques for improving the viability of striatal grafts using vital stains, in vitro cultures, and in vivo grafts. *Cell Transplant.* **5,** 599–611.
33. Watts, C. and Dunnett, S. B. (2000) Towards a protocol for the preparation and delivery of striatal tissue for clinical trials of transplantation in Huntington's disease. *Cell Transplant.* **9,** 223–234.
34. Watts, C., Brasted, P. J., and Dunnett, S. B. (2000a) The morphology, integration, and functional efficacy of striatal grafts differ between cell suspensions and tissue pieces. *Cell Transplant.* **9,** 395–407.
35. Hagell, P., Piccini, P., Bjorklund, A., et al. (2002a) Dyskinesias following neural transplantation in Parkinson's disease. *Nat. Neurosci.* **5,** 627–628.
36. Olanow, C. W., Kordower, J. H., and Freeman, T. B. (1996) Fetal nigral transplantation as a therapy for Parkinson's disease. *Trends Neurosci.* **19,** 102–109.
37. Freed, C. R., Greene, P. E., Breeze, R. E., et al. (2001) Transplantation of embryonic dopamine neurons for severe Parkinson's disease. *N. Engl. J. Med.* **344,** 710–719.
38. Freeman, T. B., Cicchetti, F., Hauser, R. A., et al. (2000) Transplanted fetal striatum in Huntington's Disease: Phenotypic development and lack of pathology. *Proc. Natl. Acad. Sci. USA* **97,** 13877–13882.
39. Lindvall, O., Brundin, P., Widner, H., et al. (1990) Grafts of fetla dopamine neurons survive and improve motor function in Parkinson's Disease. *Science* **247,** 574–577.
40. Hagell, P., Piccini, P., Bjorklund, A., et al. (2002b) Dyskinesias following neural transplantation in Parkinson's disease. *Nat. Neurosci.* **5,** 627–628.
41. Bachoud-Levi, A.-C., Remy, P., Nguyen, J.-P., et al. (2002b) Motor and cognitive improvements in patients with Huntington's disease after neural transplantation. *Lancet* **356,** 1975–1979.
42. Hauser, R. A., Furtado, S., Cimino, C. R., et al. (2002a) Bilateral human fetal striatal transplantation in Huntington's disease. *Neurology* **58,** 687–695.
43. Morrison, S. J., Ahah, N. M., and Anderson, D. J. (1997) Regulatory mechanisms in stem cell biology. *Cell* **88,** 287–298.

44. Watt, F. M. and Hogan, B. L. (2000) Out of Eden: stem cells and their niches. *Science* **287,** 1427–1430.
45. Fuchs, E. and Segre, J. A. (2000) Stem cells: a new lease on life. *Cell* **100,** 143–155.
46. Weissman, I. L. (2000) Translating stem and progenitor cell biology to the clinic: barriers and opportunities. *Science* **287,** 1442–1446.
47. van der Kooy, D. and Weiss, S. (2000) Why stem cells? *Science* **287,** 1439–1441.
48. Blau, H. M., Brazelton, T. R., and Weimann, J. M. (2001) The evolving concept of a stem cell: Entity or function? *Cell* **105,** 829–841.
49. Odorico, J. S., Kaufman, D. S., and Thomson, J. A. (2001) Multilineage differentiation from human embryonic stem cell lines. *Stem Cells* **19,** 193–204.
50. Schuldiner, M., Eiges, R., Eden, A., et al. (2001) Induced neuronal differentiation of human embryonic stem cells. *Brain Res.* **913,** 201–205.
51. McHugh, P. R. (2004) Zygote and "clonote"—the ethical use of embryonic stem cells. *N. Engl. J. Med.* **351,** 209–211.
52. Sandel, M. J. (2004) Embryo ethics—the moral logic of stem-cell research. *N. Engl. J. Med.* **351,** 207–209.
53. Reubinoff, B. E., Itsykson, P., Turetsky, T., et al. (2001) Neural progenitors from human embryonic stem cell. *Nat. Biotechnol.* **19,** 1134–1147.
54. Kawasaki, H., Mizuseki, K., Nishikawa, S., et al. (2000) Induction of midbrain dopaminergic neurons from ES cells by stromal cell-derived inducing activity. *Neuron* **28,** 31–40.
55. Lee, S.-H., Lumelsky, N., Studer, L., et al. (2000) Efficient generation of midbrain and hindbrain neurons from mouse embryonic stem cells. *Nat. Biotechnol.* **18,** 675–679.
56. Okabe, S., Forsberg-Nilsson, K., Spiro, A. C., et al. (1996) Development of neuronal precursor cells and functional postmitotic neurons from embryonic stem cells in vitro. *Mech. Dev.* **59,** 89–102.
57. Rolletschek, A., Chang, H., Guan, K. M., et al. (2001) Differentiation of embryonic stem cell-derived dopaminergic neurons is enhanced by survival-promoting factors. *Mech. Dev.* **105,** 93–104.
58. Chung, S., Sonntag, K. C., Andersson, T., et al. (2002) Genetic engineering of mouse embryonic stem cells by Nurr1 enhances differentiation and maturation into dopaminergic neurons. *Eur. J. Neurosci.* **16,** 1829–1838.
59. Kim, J. H., Auerbach, J. M., Rodriguez-Gomez, J. A., et al. (2002) Dopamine neurons derived from embryonic stem cells function in an animal model of Parkinson's disease. *Nature* **418,** 50–56.
60. Kim, T. E., Lee, H. S., Lee, Y. B., et al. (2003) Sonic hedgehog and FGF8 collaborate to induce dopaminergic phenotypes in the Nurr1-overexpressing neural stem cell. *Biochem. Biophys. Res. Commun.* **305,** 1040–1048.
61. Grothe, C., Timmer, M., Scholz, T., et al. (2004) Fibroblast growth factor-20 promotes the differentiation of Nurr1-overexpressing neural stem cells into tyrosine hydroxylase-positive neurons. *Neurobiol. Dis.* **17,** 163–170.

62. Wagner, J., Akerud, P., Castro, D. S., et al. (1999) Induction of a midbrain dopaminergic phenotype in Nurr1-overexpressing neural stem cells by type 1 astrocytes. *Nat. Biotechnol.* **17,** 653–659.
63. Bouhon, I. A., Kato, H., Chandran, S., and Allen, N. D. (2005) Neuronal differentiation of mouse embryonic stem cells in chemically defined medium. *Brain Ref. Bull.* **68,** 62–75.
64. Kato, H., Bouhon, I. A., Chandran, S., and Allen, N. D. (2006) Critical factors influencing fate determination and developmental plasticity of embryonic stem cells derived neural precursor cells. *Stem Cells*, in press.
65. Bjorklund, L. M., Sánchez-Pernaute, R., Chung, S., et al. (2002) Embryonic stem cells develop into functional dopaminergic neurons after transplantation in a Parkinson rat model. *Proc. Natl. Acad. Sci. USA* **99,** 2344–2349.
66. Schuldiner, M., Itskovitz-Eldor, J., and Benvenisty, N. (2003) Selective ablation of human embryonic stem cells expressing a "suicide" gene. *Stem Cells* **21,** 257–265.
67. Fareed, M. U. and Moolten, F. L. (2002) Suicide gene transduction sensitizes murine embryonic and human mesenchymal stem cells to ablation on demand—a fail-safe protection against cellular misbehavior. *Gene Ther.* **9,** 955–962.
68. Molyneaux, K. A., Schaible, K., and Wylie, C. (2001) Quantitative analysis of germ cell movements in tissue explants. *Dev. Biol.* **235,** 63.
69. Shamblott, M. J., Axelman, J., Wang, S. P., et al. (1998) Derivation of pluripotent stem cells horn cultured human primordial germ cells. *Proc. Natl. Acad. Sci. USA* **95,** 13726–13731.
70. Turnpenny, L., Brickwood, S., Spalluto, C. M., et al. (2003) Derivation of human embryonic germ cells: an alternative source of pluripotent stem cells. *Stem Cells* **21,** 598–609.
71. Armstrong, R. J. E. and Svendsen, C. N. (2000) Neural stem cells: from cell biology to cell replacement. *Cell Transplant.* **9,** 139–152.
72. Ciccolini, F. and Svendsen, C. N. (1998) Fibroblast growth factor 2 (FGF-2) promotes acquisition of epidermal growth factor (EGF) responsiveness in mouse striatal precursor cells: identification of neural precursors responding to both EGF and FGF-2. *J. Neurosci.* **18,** 7869–7880.
73. Kelly, C. M., Zietlow, R., and Rosser, A. E. (2003) The effects of various concentrations of FGF 2 on the proliferation and neuronal differentiation of murine embryonic neural precursor cells in vitro. *Cell Transplant.* **12,** 215–223.
74. Tropepe, V., Sibilia, M., Ciruna, B. G., et al. (1999) Distinct neural stem cells proliferate in response to EGF and FGF in the developing mouse telencephalon. *Dev. Biol.* **208,** 166–188.
75. Vescovi, A. L., Reynolds, B. A., Fraser, D. D., and Weiss, S. (1993) bFGF regulates the proliferative fate of unipotent (neuronal) and bipotent (neuronal/astroglial) EGF-generated CNS progenitor cells. *Neuron* **11,** 951–966.
76. Armstrong, R. J. E., Watts, C., Svendsen, C. N., et al. (2000) Survival, neuronal differentiation and fiber outgrowth of propagated human neural precursor grafts in an animal model of Huntington's disease. *Cell Transplant.* **9,** 1–10.

77. Arsenijevic, Y., Weiss, S., Schneider, B., and Aebischer, P. (2001) Insulin-like growth factor-1 is necessary for neural stem cell proliferation and demonstrates distinct actions of epidermal growth factor and fibroblast growth factor-2. *J. Neurosci.* **27,** 7194–7202.

78. Cattaneo, E. and McKay, R. (1990) Proliferation and differentiation of neuronal stem cells regulated by nerve growth factor. *Nature* **347,** 762–765.

79. Santa-Olla, J. and Covarrubias, L. (1995) Epidermal growth factor (EGF), transforming growth factor-a (TGF-a), and basic fibroblast growth factor (bFGF) differentially influence neural precursor cells of mouse embryonic mesencephalon. *J. Neurosci. Res.* **42,** 172–183.

80. Tropepe, V., Craig, C. G., Morshead, C. M., and van der Kooy, D. (1997) Transforming growth factor-a null and senescent mice show decreased neural progenitor cell proliferation in the forebrain subependyma. *J. Neurosci.* **17,** 7850–7859.

81. Svendsen, C. N., Skepper, J., Rosser, A. E., et al. (1997) Restricted growth potential of rat neural precursors as compared to mouse. *Dev. Brain Res.* **99,** 253–258.

82. Smith, R., Bagga, V., and Fricker-Gates, R. A. (2003) Embryonic neural progenitor cells: the effects of species, region, and culture conditions on long-term proliferation and neuronal differentiation. *J. Hematother. Stem Cell. Res.* **12,** 713–725.

83. Jori, F. P., Galderisi, U., Piegari, E., et al. (2003) EGF-responsive rat neural stem cells: molecular follow-up of neuron and astrocyte differentiation in vitro. *J. Cell. Physiol.* **195,** 220–233.

84. Jain, M., Armstrong, R. J. E., Tyers, P., et al. (2003a) GABAergic immunoreactivity is predominant in neurons derived from expanded human neural precursor cells in vitro. *Exp. Neurol.* **182,** 113–123.

85. Fricker, R. A., Carpenter, M. K., Winkler, C., et al. (1999) Site-specific migration and neuronal differentiation of human neural progenitor cells after transplantation in the adult rat brain. *J. Neurosci.* **19,** 5990–6005.

86. Parmar, M., Skogh, C., Bjorklund, A., and Campbell, K. (2002) Regional specification of neurosphere cultures derived from subregions of the embryonic telencephalon. *Mol. Cell. Neurosci.* **21,** 645–656.

87. Skogh, C., Parmar, M., and Campbell, K. (2003) The differentiation potential of precursor cells from the mouse lateral ganglionic eminence is restricted by in vitro expansion. *Neuroscience* **120,** 379–385.

88. Jain, M., Armstrong, R. J. E., Elneil, S., et al. (2003b) Migration and differentiation of transplanted human neural precursor cells. *Mol. Neurosci.* **14,** 1257–1262.

89. Zietlow, R., Pekarik, V., Armstrong, R. J., et al. (2005) The survival of neural precursor cell grafts is influenced by in vitro expansion. *J. Anat.* **207,** 227–240.

90. Englund, U., Fricker-Gates, R. A., Lundberg, C., et al. (2002a) Transplantation of human neural progenitor cells into the neonatal rat brain: Extensive migration and differentiation with long-distance axonal projections. *Exp. Neurol.* **173,** 1–21.

91. Englund, U., Bjorklund, A., and Wictorin, K. (2002b) Migration patterns and phenotypic differentiation of long-term expanded human neural progenitor cells after transplantation into the adult rat brain. *Dev. Brain Res.* **134,** 123–141.

92. Rosser, A. E., Tyres, P., Dunnett, S. B. (2000) The morphological development of neurons from EGF and FGF-2-driven human CNS precursors depends on their site of integration in the neonatal rat brain. *Eur. J. Neurosci.* **12,** 2405–2413.

93. Eriksson, C., Bjorklund, A., and Wictorin, K. (2003) Neuronal differentiation following transplantation of expanded mouse neurosphere cultures derived from different embryonic forebrain regions. *Exp. Neurol.* **184,** 615–635.

94. Ostenfeld, T., Caldwell, M. A., Prowse, K. R., et al. (2000) Human neural precursor cells express low levels of telomerase in vitro and show diminishing cell proliferation with extensive axonal outgrowth following transplantation. *Exp. Neurol.* **164,** 215–226.

95. Altman, J. and Das, G. D. (1965) Autoradiographic and histological evidence of postnatal hippocampal neurogenesis in rats. *J. Comp. Neurol.* **124,** 319–335.

96. Gage, F. H., Ray, J., and Fisher, L. J. (1995) Isolation, characterization, and use of stem cells from the CNS. *Ann. Rev. Neurosci.* **18,** 159–192.

97. Alvarez-Buylla, A., Seri, B., and Doetsch, F. (2002) Identification of neural stem cells in the adult vertebrate brain. *Brain Res. Bull.* **57,** 751–758.

98. Lois, C. and Alvarez-Buylla, A. (1994) Long-distance neuronal migration in the adult mammalian brain. *Science* **264,** 1145–1148.

99. Gould, E., Reeves, A. J., Graziano, M. S. A., and Gross, C. G. (1999) Neurogenesis in the neocortex of adult primates. *Science* **286,** 548–552.

100. Rietze, R., Poulin, P., and Weiss, S. (2000) Mitotically active cells that generate neurons and astrocytes are present in multiple regions of the adult mouse hippocampus. *J. Comp. Neurol.* **424,** 397–408.

101. Zhao, M., Momma, S., Delfani, K., et al. (2003) Evidence for neurogenesis in the adult mammalian substantia nigra. *PNAS* **100,** 7925.

102. Frielingsdorf, H., Schwarz, K., Brundin, P., and Mohapel, P. (2004) No evidence for new dopaminergic neurons in the adult mammalian substantia nigra. *Proc. Natl. Acad. Sci. USA* **101,** 10177–10182.

103. Ratajczak, M. Z., Kucia, M., Majka, M., et al. (2004) Heterogeneous populations of bone marrow stem cells—are we spotting on the same cells from the different angles? *Folia Histochem. Cytobiol.* **42,** 139–146.

104. Zhao, L.-R., Duan, W.-M., Reyes, M., et al. (2002) Human bone marrow stem cells exhibit neural phenotypes and ameliorate neurological deficits after grafting into the ischemic brain of rats. *Exp. Neurol.* **174,** 11–20.

105. Galli, R, Borello, U., Gritti, A., et al. (2000) Skeletal myogenic potential of human and mouse neural stem cells. *Nature* **3,** 986–991.

106. Wang, X., Willenbring, H., Akkari, Y., et al. (2003) Cell fusion is the principal source of bone-marrow-derived hepatocytes. *Nature* **422,** 897–901.

107. Vassilopoulos, G., Wang, P. R., and Russell, D. W. (2003) Transplanted bone marrow regenerates liver by cell fusion. *Nature* **422,** 901–904.

108. Lagasse, E., Shizuru, J. A., Uchida, N., et al. (2001) Toward regenerative medicine. *Immunity* **14**, 425–436.
109. Koshizuka, S., Okada, S., Okawa, A., et al. (2004) Transplanted hematopoietic stem cells from bone marrow differentiate into neural lineage cells and promote functional recovery after spinal cord injury in mice. *J. Neuropathol. Exp. Neurol.* **63**, 64–72.
110. Sanchez-Ramos, J., Song, S., Kamath, S. G., et al. (2001) Expression of neural markers in human umbilical cord blood. *Exp. Neurol.* **171**, 109–115.
111. Li, H. J., Liu, H. Y., Zhao, Z. M., et al. (2004) Transplantation of human umbilical cord stem cells improves neurological function recovery after spinal cord injury in rats. *Zhongguo Yi Xue Ke Xue Yuan Xue Bao* **26**, 38–42.
112. Willing, A. E., Lixian, J., Milliken, M., et al. (2003) Intravenous versus intrastriatal cord blood administration in a rodent model of stroke. *J. Neurosci. Res.* **73**, 296–307.
113. Zigova, T., Song, S., Willing, A. E., et al. (2002) Human umbilical cord blood cells express neural antigens after transplantation into the developing rat brain. *Cell Transplant.* **11**, 265–274.
114. Armstrong, R. J. E., Huelbrink, C. B., Tyres, P., et al. (2002) The potential for circuit reconstruction by expanded neural precursor cells explored through porcine xenografts in a rat model of Parkinson's disease. *Exp. Neurol.* **175**, 98–111.
115. Deacon, T., Whatley, B., LeBlanc, C., et al. (1999) Pig fetal septal neurons implanted into the hippocampus of aged or cholinergic deafferented rats grow axons and form cross-species synapses in appropriate target regions. *Cell Transplant.* **8**, 111–129.
116. Galpern, W. R., Burns, L. H., Deacon, T. W., et al. (1996) Xenotransplantation of porcine fetal ventral mesencephalon in a rat model of Parkinson's disease: functional recovery and graft morphology. *Exp. Neurol.* **140**, 1–13.
117. Garcia, A. R., Deacon, T. W., Dinsmore, J., and Isacson, O. (1995a) Extensive axonal and glial fiber growth from fetal porcine cortical xenografts in the adult rat cortex. *Cell Transplant.* **4**, 515–527.
118. Isacson, O., Costantini, J. M., Cicchetti, F., et al. (2001) Cell implantation therapies for Parkinson's disease using neural stem, transgenic or xenogenic donor, cells. *Parkinsonism Relat. Disord.* **7**, 205–212.
119. Castro, A. J., Meyer, M., Moller-Dall, A., and Zimmer, J. (2003) Transplantation of embryonic porcine neocortical tissue into newborn rats. *Cell Transplant.* **12**, 733–741.
120. Garcia, A. R., Deacon, T. W., Dinsmore, J., and Isacson, O. (1995b) Extensive axonal and glial fiber growth from fetal porcine cortical xenografts in the adult rat cortex. *Cell Transplant.* **4**, 515–527.
121. Larsson, L. C., Czech, K. A., Brundin, P., and Widner, H. (2000) Intrastriatal ventral mesencephalic xenografts of porcine tissue in rats: immune responses and functional effects. *Cell Transplant.* **9**, 261–272.
122. Cicchetti, F., Costantini, L., Belizaire, R., et al. (2002) Combined inhibition of apoptosis and complement improves neural graft survival of embryonic rat and porcine mesencephalon in the rat brain. *Exp. Neurol.* **177**, 376–384.

123. Barker, R. A., Ratcliffe, E., Richards, A., and Dunnett, S. B. (1999) Fetal porcine dopaminergic cell survival in vitro and its relationship to embryonic age. *Cell Transplant.* **8,** 593–599.

124. Larsson, L. C., Frielingsdorf, H., Mirza, B., et al. (2001) Porcine neural xenografts in rats and mice: donor tissue development and characteristics of rejection. *Exp. Neurol.* **172,** 100–114.

125. Molenaar, G. J., Hogenesch, R. I., Sprengers, M. E., and Staal, M. J. (1997) Ontogenesis of embryonic porcine ventral mesencephalon in the perspective of its potential use as a xenograft in Parkinson's disease. *J. Comp. Neurol.* **382,** 19–28.

126. Fink, J. S., Schumacher, J. M., Ellias, S. L., et al. (2000) Porcine xenografts in Parkinson's disease and Huntington's disease patients: preliminary results. *Cell Transplant.* **9,** 273–278.

127. Wictorin, K., Ouimet, C. C., and Bjorklund, A. (1989) Intrinsic organization and connectivity of intrastriatal striatal transplants in rats as revealed by Darpp-32 immunohistochemistry—specificity of connections with the lesioned host brain. *Eur. J. Neurosci.* **1,** 690–701.

128. Wictorin, K. (1992) Anatomy and connectivity of intrastriatal striatal transplants. *Prog. Neurobiol.* **38,** 611–639.

129. Olsson, M., Bentlage, C., Wictorin, K., et al. (1997) Extensive migration and target innervation by striatal precursors after grafting into the neonatal striatum. *Neuroscience* **79,** 57–78.

130. Wictorin, K., Lagenaur, C. F., Lund, R. D., and Bjorklund, A. (1991) Efferent projections to the host brain from intrastriatal striatal mouse-to-rat grafts: time course and tissue-type specificity as revealed by a mouse specific neuronal marker. *Eur. J. Neurosci.* **3,** 86–101.

131. Armstrong, R. J., Harrower, T. P., Hurelbrink, C. B., et al. (2001b) Porcine neural xenografts in the immunocompetent rat: immune response following grafting of expanded neural precursor cells. *Neuroscience* **106,** 201–216.

132. Harrower, T. P., Richards, A., Cruz, G., et al. (2004) Complement regulatory proteins are expressed at low levels in embryonic human, wild type and transgenic porcine neural tissue. *Xenotransplantation* **11,** 60–71.

133. Armstrong, R. J., Harrower, T. P., Hurelbrink, C. B., et al. (2001a) Porcine neural xenografts in the immunocompetent rat: immune response following grafting of expanded neural precursor cells. *Neuroscience* **106,** 201–216.

134. Armstrong, R. J., Tyers, P., Jain, M., et al. (2003) Transplantation of expanded neural precursor cells from the developing pig ventral mesencephalon in a rat model of Parkinson's disease. *Exp. Brain Res.* **151,** 204–217.

135. During, M. J., Naegele, J. R., O'Malley, K. L., and Geller, A. I. (1994) Long-term behavioral recovery in parkinsonian rats by an HSV vector expressing tyrosine hydroxylase. *Science* **266,** 1399–1403.

136. Song, S., Wang, Y., Bak, S. Y., et al. (1997) An HSV-1 vector containing the rat tyrosine hydroxylase promoter enhances both long-term and cell type-specific expression in the midbrain. *J. Neurochem.* **68,** 1792–1803.

137. Fraefel, C., Song, S., Lim, F., et al. (1996) Helper virus-free transfer of herpes simplex virus type 1 plasmid vectors into neural cells. *J. Virol.* **70,** 7190–7197.

138. Janson, C. G., McPhee, S. W. J., Leone, P., et al. (2001) Viral-based gene transfer to the mammalian CNS for functional genome studies. *Trends Neurosci.* **24,** 706–712.

139. Eslamboli, A., Cummings, R. M., Ridley, R. M., et al. (2003) Recombinant adeno-associated viral vector (rAAV) delivery of GDNF provides protection against 6-OHDA lesion in the common marmoset monkey (Callithrix jacchus). *Exp. Neurol.* **184,** 536–548.

140. Kirik, D., Rosenblad, C., Bjorklund, A., and Mandel, R. J. (2000) Long-term rAAV-mediated gene transfer of GDNF in the rat Parkinson's model: intrastriatal but not intranigral transduction promotes functional regeneration in the lesioned nigrostriatal system. *J. Neurosci.* **20,** 4686–4700.

141. Mandel, R. J., Snyder, R. O., and Leff, S. E. (1999) Recombinant adeno-associated viral vector-mediated glial cell line-derived neurotrophic factor gene transfer protects nigral dopamine neurons after onset of progressive degeneration in a rat model of Parkinson's disease. *Exp. Neurol.* **160,** 205–214.

142. Mandel, R. J., Spratt, S. K., Snyder, R. O., and Leff, S. E. (1997) Midbrain injection of recombinant adeno-associated virus encoding rat glial cell line-derived neurotrophic factor protects nigral neurons in a progressive 6-hydroxy-dopamine-induced degeneration model of Parkinson's disease in rats. *Proc. Natl. Acad. Sci. USA* **94,** 14083–14088.

143. Regulier, E., Pereira, D. A., Sommer, B., et al. (2002) Dose-dependent neuroprotective effect of ciliary neurotrophic factor delivered via tetracycline-regulated lentiviral vectors in the quinolinic acid rat model of Huntington's disease. *Hum. Gene. Ther.* **13,** 1981–1990.

144. Kahn, A., Haase, G., Akli, S., and Guidotti, J. E. (1996) Gene therapy of neurological diseases. *CR Seances Soc. Biol. Fil.* **190,** 9–11.

145. Emerich, D. F. and Winn, S. R. (2004) Neuroprotective effects of encapsulated CNTF-producing cells in a rodent model of Huntington's disease are dependent on the proximity of the implant to the lesioned striatum. *Cell Transplant.* **13,** 253–259.

146. Mittoux, V., Ouary, S., Monville, C., et al. (2002) Corticostriatopallidal neuroprotection by adenovirus-mediated ciliary neurotrophic factor gene transfer in a rat model of progressive striatal degeneration. *J. Neurosci.* **22,** 4478–4486.

147. Emerich, D. F., Hammang, J. P., Baetge, E. E., and Winn, S. R. (1994) Implantation of polymer-encapsulated human nerve growth factor-secreting fibroblasts attenuates the behavioral and neuropathological consequences of quinolinic acid injections into rodent striatum. *Exp. Neurol.* **130,** 141–150.

148. Date, I., Shingo, T., Yoshida, H., et al. (2001) Grafting of encapsulated genetically modified cells secreting GDNF into the striatum of parkinsonian model rats. *Cell Transplant.* **10,** 397–401.

149. Emerich, D. F., Cain, C. K., Greco, C., et al. (1997a) Cellular delivery of human CNTF prevents motor and cognitive dysfunction in a rodent model of Huntington's disease. *Cell Transplant.* **6,** 249–266.

150. Emerich, D. F. (1999) Encapsulated CNTF-producing cells for Huntington's disease. *Cell Transplant.* **8,** 581–582.

151. Emerich, D. F., Lindner, M. D., Winn, S. R., et al. (1996) Implants of encapsulated human CNTF-producing fibroblasts prevent behavioral deficits and striatal degeneration in a rodent model of Huntington's disease. *J. Neurosci.* **16,** 5168–5181.

152. Emerich, D. F., Winn, S. R., Hantraye, P. M., et al. (1997b) Protective effect of encapsulated cells producing neurotrophic factor CNTF in a monkey model of Huntington's disease. *Nature* **386,** 395–399.

153. Emerich, D. F., Bruhn, S., Chu, Y., and Kordower, J. H. (1998) Cellular delivery of CNTF but not NT-4/5 prevents degeneration of striatal neurons in a rodent model of Huntington's disease. *Cell Transplant.* **7,** 213–225.

154. Anderson, K. D., Panayotatos, N., Corcoran, T. L., et al. (1996) Ciliary neurotrophic factor protects striatal output neurons in an animal model of Huntington's disease. *PNAS* **93,** 7346–7351.

155. Kordower, J. H., Isacson, O., Leventhal, L., and Emerich, D. F. (2000) Cellular delivery of trophic factors for the treatment of Huntington's disease: is neuroprotection possible? *Prog. Brain Res.* **127,** 414–430.

156. Mittoux, V., Joseph, J. M., Conde, F., et al. (2000) Restoration of cognitive and motor functions by ciliary neurotrophic factor in a primate model of Huntington's disease. *Hum. Gene Ther.* **11,** 1177–1187.

157. Bachoud-Levi, A. C., Deglon, N., Nguyen, J. P., et al. (2000) Neuroprotective gene therapy for Huntington's disease using a polymer encapsulated BHK cell line engineered to secrete human CNTF. *Hum. Gene. Ther.* **11,** 1723–1729.

158. Lundberg, C., Winkler, C., Whittemore, S. R., and Bjorklund, A. (1996a) Conditionally immortalized neural progenitor cells grafted to the striatum exhibit site-specific neuronal differentiation and establish connections with the host globus pallidus. *Neurobiol. Dis.* **3,** 33–50.

159. Saporta, S., Willing, A. E., Zigova, T., et al. (2001) Comparison of calcium-binding proteins expressed in cultured hNT neurons and hNT neurons transplanted into the rat striatum. *Exp. Neurol.* **167,** 252–259.

160. Miyazono, M., Lee, V. M., and Trojanowski, J. Q. (1995) Proliferation, cell death, and neuronal differentiation in transplanted human embryonal carcinoma (NTera2) cells depend on the graft site in nude and severe combined immunodeficient mice. *Lab. Invest.* **73,** 273–283.

161. Catapano, L. A., Sheen, V. L., Leavitt, B. R., and Macklis, J. D. (1999) Differentiation of transplanted neural precursors varies regionally in adults striatum. *NeuroReport* **10,** 3971–3977.

162. Fricker-Gates, R. A., Muir, J. A., and Dunnett, S. B. (2004) Transplanted hNT cells ("LBS neurons") in a rat model of huntington's disease: good survival, incomplete differentiation, and limited functional recovery. *Cell Transplant.* **13,** 123–136.

163. Lundberg, C., Winkler, C., Whittemore, S. R., and Bjorklund, A. (1996b) Conditionally immortalized neural progenitor cells grafted to the striatum exhibit site-specific neuronal differentiation and establish connections with the host globus pallidus. *Neurobiol. Dis.* **3,** 33–50.

164. Renfranz, P. J., Cunningham, M. G., and McKay, R. D. (1991) Region-specific differentiation of the hippocampal stem cell line HiB5 upon implantation into the developing mammalian brain. *Cell* **66,** 713–729.

165. Rosser, A. E. and Dunnett, S. B. (2001) Neural transplantation for the treatment of Huntington's disease. In *Movement Disorders Surgery* (Krauss, J. K., Jankovic, J., and Grossman, R. G., editors), Lippincott, Williams and Wilkins, Philadelphia, PA, pp. 353–373.

166. Dunnett, S. B. (2001) Stem and precursor cells for cell therapy in the CNS. *Alzheimer's Reports* **4,** 93–101.

167. Durcova-Hills, G., Ainscough, J. F. X., and McLaren, A. (2001) Pluripotential stem cells derived from migrating primordial germ cells. *Differentiation* **68,** 220–226.

168. Durcova-Hills, G., Wianny, F., Merriman, J., et al. (2003) Developmental fate of embryonic germ cells (EGCs), in vivo and in vitro. *Differentiation* **71,** 135–141.

169. McLaren, A. (2001) Mammalian germ cells: Birth, sex, and immortality. *Cell Struct. Func.* **26,** 119–122.

170. Ferland, R. J., Cherry, T. J., Preware, P. O., et al. (2003) Characterisation of Foxp2 and Foxp1 mRNA and protein in the developing and mature brain. *J. Comp. Neurol.* **460,** 266–279.

171. Hurlbert, M. S., Gianani, R. I., Hutt, C., et al. (1999) Neural transplantation of hNT neurons for Huntington's disease. *Cell Transplant.* **8,** 143–151.

172. Isacson, O., Deacon, T. W., Pakzaban, P., et al. (1995) Transplanted xenogeneic neural cells in neurodegenerative disease models exhibit remarkable axonal target specificity and distinct growth patterns of glial and axonal fibres. *Nat. Med.* **1,** 1189–1194.

173. Harper, P. S., ed. (1996) *Huntington's Disease*. W. B. Saunders Company Ltd., Philadelphia, PA.

174. Rosser, A. E., Barker, R. A., Harrower, T., et al. (2002) Unilateral transplantation of human primary fetal tissue in four patients with Huntington's Disease: NEST-UK safety report ISRCTN no 36485475. *J. Neurol. Neurosurg. Psychiatry* **73,** 678–685.

175. Kopyov, O. V., Jacques, S., Lieberman, A., Duma, C. M., and Eagle, K. S. (1998) Safety of intrastriatal neurotransplantation for Huntington's disease patients. *Exp. Neurol.* **149,** 97–108.

176. Larsen, W. J. (1998) *Essentials of Human Embryology*. Elsevier.

177. Sautter, J., Tseng, J. L., Braguglia, D., et al. (1998) Implants of polymer-encapsulated genetically modified cells releasing glial cell line derived neurotrophic factor improve survival, growth and function of fetal dopaminergic grafts. *Exp. Neurol.* **149,** 230–236.

178. Jensen, J. B., Bjorklund, A., and Parmar, M. (2004) Striatal neuron differentiation from neurosphere-expanded progenitors depends on Gsh2 expression. *J. Neurosci.* **24,** 6958–6967.

5

Use of Bone Marrow Stem Cells as Therapy for Behavioral Deficits in Rodent Models of Huntington's Disease

Gary L. Dunbar, Justin D. Oh-Lee, and Laurent Lescaudron

ABSTRACT

This chapter focuses on three new studies that used autologous (i.e., when the donor and recipient are the same individual) and heterologous (i.e., when the donor and recipient are different individuals) transplants of adult, bone-marrow–derived stem cells to counteract the behavioral deficits produced by intrastriatal injections of quinolinic acid (QA) in rats, a rodent model for Huntington's disease (HD). These studies support previous findings that have demonstrated that transplants of adult bone-marrow–derived stem cells can survive and integrate into the host tissue and can counteract functional deficits caused by CNS damage. In the initial study using this HD model, nestin-positive cells were observed 14 d after the transplant and GAD-positive cells were observed at 21 d posttransplant, a finding that supports the contention that adult bone-marrow stem cells possess the plasticity to transdifferentiate into neurons. However, because less than 1% of the cells expressed neuronal phenotypes, and because the functional recovery observed in all three studies occurred within 1 mo after transplantation, it is hypothesized that the ameliorative effects of the transplants were due to the production of neurotrophic factors that facilitated functioning of spared neurons, rather than through the replacement of lost neurons. When compared with transplants of heterologous stem cells, the transplants of autologous stem cells were more efficacious, at least for reducing QA-induced behavioral deficits. Because the use of autologous transplants in HD may prove unfeasible, if the mutant gene becomes expressed in the transplanted autologous stem cells, it may become critically important to find ways to reduce any adverse immunological response or other factors that may be limiting the efficacy of heterologous transplants, if this form of therapy is to become clinically viable for HD.

From: *Contemporary Neuroscience: Cell Therapy, Stem Cells, and Brain Repair*
Edited by: C. D. Sanberg and P. R. Sanberg © Humana Press Inc., Totowa, NJ

In summary, the findings reported in this chapter support the contention that transplantation of adult, bone-marrow–derived stem cells has significant therapeutic potential as a treatment for HD, but research into how to prevent a possible immune reaction of the heterologous transplant may prove to be critical before this approach has clinical utility.

Key Words: Autologous transplant; heterologous transplant; bone-marrow stem cells; Huntington's disease; cognitive and motor deficits.

INTRODUCTION

Pluripotency indicates the ability of a single cell to give rise to all types of differentiated cells in the body. After harvesting embryonic stem (ES) cells during the blastula stage, the three germ layers—ectoderm, mesoderm, and endoderm—are formed. The daughter cells of the ES cells then segregate into groups of organ-specific progenitor cells, which subsequently mature into adult somatic cells. Many of these pluripotent and multipotent stem cells have been identified in almost all adult tissues, including the bone marrow, blood, skin, fat, muscle, and brain.

The use of adult stem cells offer several advantages over ES cells as a source for therapeutic transplants. For instance, adult stem cells are readily accessible and can be harvested directly from the patient, reducing the host/donor immune response. In addition, fewer ethical issues surround the use of adult stem cells. Recent studies showing that transplanted adult stem cells exhibit considerably more plasticity and can transit across germ layers more readily than was previously thought have underscored their potential utility as source for therapeutic transplants *(1,2)*.

The important role that stroma tissue of adult mammalian bone marrow has in providing an adequate environment for cells of the hematopoietic lineage is well documented. This tissue regulates proliferation, differentiation, and maturation of hematopoietic stem cells. In addition, bone marrow stroma contains multipotent stem cells *(3)*, known as *mesenchymal stem cells* (MSCs), which are also referred to as *marrow stromal cells*. The MSCs were first isolated in the mid-1970s *(4)* and were shown to give rise to the differentiated cell lineages found in the stroma, bone, and cartilage. However, recent in vivo evidence in humans and animals indicates that some bone marrow stem cells, such as MSCs, can also differentiate into endothelial cells, cardiac myocytes *(5)*, and skeletal myocytes *(6,7)*. Animal and human studies also have shown that these cells differentiate into a variety of other cells, including: hepatocytes *(8)*, pancreatic islet cells *(9)*, epithelial and nonepithelial kidney cells *(10,11)*, gastrointestinal epithelial cells *(12)*, lung pneumocytes *(13)*, skin keratinocytes *(14)*, brain micro- and macroglia *(15)*, oligodendrocytes *(16)*, astrocytes *(17–19)*, and neuronal cells *(19–25)*. In addition, numerous in vitro studies provide evidence that MSCs give rise to neuronal cells *(2,26,27)*.

MSCs are known to synthesize extracellular matrix proteins, e.g., fibronectin and collagen-type-I *(20,24)*—molecules that stimulate nerve fiber growth in vitro *(28)* and exert neuroprotective effects after brain ischemia in vivo *(29)*. They also produce a number of interleukins (IL-6, IL-7, IL-8, IL-11, IL-12, IL-14, and IL-15), the macrophage colony-stimulating factor, fms-like tyrosine kinase-3 ligand, and stem cell factor *(30,31)*. A molecule of stem cell origin capable of inducing ES cells into becoming dopaminergic cells has also been identified *(32,33)*.

These observations, together with the evidence that MSCs may be able to differentiate into neuronal phenotypes, have led to several studies evaluating the viability of MSC transplants and their potential as therapy for brain disorders (see Table 1). As seen in Table 1, most of these studies have indicated that bone-derived stem cell transplants can survive in vivo and show considerable promise in reversing or preventing functional deficits in animal models of stroke, trauma, and disease. There is also considerable data that MSCs can differentiate into neuronal phenotypes and in some cases, even migrate to the site of brain damage. Given the potential efficacy, and the relative efficiency by which they can be isolated from patients and be altered genetically *(34,35)*, MSCs offer a promising and reliable source for cell and gene therapy, as well as for therapeutic neural transplants.

Although these observations indicate that MSCs can be a promising source for therapeutic transplants, the functional significance of possible immunological responses to these transplants has not been fully explored. In an early investigation of immunological responses, no evidence was found for an inflammatory response or rejection of either rat or human MSCs (hMSCs) transplanted into the rat striatum *(20)*. However, the absence of an immunological response in that study may have been because of the special properties of the MSCs that were transplanted. For example, if hMSCs lack human leucocyte antigen (HLA) class II antigens, as has been reported *(20)*, this could prevent the antigenic presentation to rat CD4[+] T lymphocytes. Furthermore, some investigators have found that hMSCs do not express HLA class I antigens *(24)*, but data from our group indicate that rat MSCs are only negative for class II major histocompatibility complex (MHC) molecules (J. Rossignol, personal communication).

Some evidence shows that MSCs may confer some immunosuppressive effects. In studies using the baboon, intravenous injections of MSCs were able to alter lymphocyte reactivity to allogeneic target cells by reducing

Table 1
Use of Bone Marrow–Derived Stem Cells to Treat Disorders of the CNS

Donor/recipient	Injury	Treatment	Differentiation	Behavioral outcomes	Study
Rat/rat embryo	None	IV injection of MSCs	Neurone, glia	NA	Munoz-Elias et al., 2004 (56)
Rat/rat	Spinal cord contusion	Injection of MSCs in CSF	None, reduced volume cavity	NA	Ohta et al., 2004 (60)
Human/human	Hematologic malignancy	Injection of WBM	Neurons	NA	Mezey et al., 2003 (22)
Rat/rat	QA injection into striatum	TP of WBM autologous in striatum	Neurons	Better in RAWM	Lescaudron et al., 2003 (25)
Human/rat	Stroke, CX	TP of MSCs	Neurons, astrocytes, oligodendrocytes	Better in dot test	Zhao et al., 2002 (24)
Mouse/mouse	None	IV injection of multipotent adult progenitor cells	None	NA	Jiang et al., 2002 (61)
Mouse/mouse	None	TP of MSCs striatum	None	NA	Wehner et al., 2003 (62)
Human/human	Hematologic malignancy	Injection of WBM	Neurons	NA	Weimann et al., 2003 (63)
Mouse/mouse	ASM deficiency	TP of ASM-transfected MSCs in hippocampus	None	Increased neuronal survival, better in rotorod Increase live span	Jin et al., 2002 (23)
Rat/rat	Spinal cord contusion	TP of MSCs in spinal cord	Immature astrocytes	Better in open field	Hofstetter at al., 2002 (64)
Rat/rat	Cortical contusion	TP of MSCs in CX	Neurons, astrocytes	Better in rotorod	Mahmood et al., 2002 (65)
Rat/rat	Sciatic nerve transection	Injection of MSCs into distal stump	Schwann cells nerve function	Improvement of sciatic	Cuevas et al., 2002 (66)
Human/rat	Stroke, CX	IV injection of MSCs	Neurons, astrocytes	Better in dot test	Li et al., 2002 (67)
Rat/rat	Stroke, CX	TP of MSCS in striatum	Neurons, astrocytes	NA	Li et al., 2001a (68)

Species	Model	Treatment	Cell fate	Functional outcome	Reference
Mouse/mouse	MPTP	IV injection of glial cell line–derived neurotrophic factor-transfected MSCs	None, increased dopamine neuron survival	NA	Park et al., 2001 (35)
Mouse/mouse	MPTP	TP of MSCs in striatum	Neurons	Better in rotorod	Li et al., 2001 (69)
Rat/rat	Cortical contusion	IV injection of MSCs	Neurons, astrocytes	Better in rotorod and in neurological tests	Lu et al., 2001b (70)
Rat/rat	Cortical contusion	TP of WBM in striatum and CX	Neurons, astrocytes	Better in rotorod	Mahmood et al., 2001 (71)
Mouse/mouse	Irradiation	IV injection of WBM	Astrocytes, no neurons	NA	Nakano et al., 2001 (72)
Rat/rat	Stroke, CX	TP of WBM in striatum	NA	None in rotorod, in dot test	Chen et al., 2001a (73)
Rat/rat	Stroke, CX	IV injection of MSCs	Neurons, astrocytes	Better in rotorod, in dot test	Chen et al., 2001b (74)
Rat/rat	Spinal cord contusion	TP of MSCs in spinal cord	Neurons	Better locomotor rating score	Chopp et al., 2000 (75)
Rat/rat	Stroke, CX	IV injection of WBM	Neurons, astrocytes	NA	Li et al., 2000 (76)
Mouse/mouse	Irradiation	IV injection of WBM	Neurons, no astrocytes	NA	Brazelton et al., 2000 (21)
Mouse/mouse	PU.1 null	Intraperitoneal injection of WBM	Neurons	NA	Mezey et al., 2000 (22)
Rat/rat	Irradiation, stroke, CX	IV injection of WBM	Astrocytes	NA	Eglitis et al., 1999 (15)
Mouse/mouse	None	TP of MSCs in cerebellum	Neurons, astrocytes	NA	Kopen et al., 1999 (18)
Rat/rat	6-OHDA	TP of tyrosine hydroxylase-transfected MSCs in striatum	Neurons	Reduction in Apo-induced rotation	Schwarz et al., 1999 (34)
Human/rat	None	TP of MSCs in striatum	None	NA	Azizi et al., 1998 (20)
Mouse/mouse	None	TP of WBM	Astrocytes, microglia	NA	Eglitis and Mezey 1997 (15)

Abbreviations: IV, intravenous; NA, not available; MPTP, 1-methyl-4-phenyl-1,2,3,6-tetrahydropyridine; MSC, mesenchymal stem cells; CSF, cerebral spinal fluid; WBM, whole bone marrow; QA, quinolinic acid; TP, transplant; RAWM, radial arm water maze; CX, cortex; ASM, acid sphingomylinase; 6-OHDA, 6-hydroxydopamine.

their proliferation, thereby prolonging skin graft survival *(36)*. In addition, a recent human study further illustrates the potent immunosuppressive effect of MSCs *(37)*. Similarly, Krampera and colleagues *(38)* showed that MSCs have a profound inhibitory effect on the activation of T lymphocytes (both naive and memory T cells) by their cognate peptides. This research suggests that MSCs physically hinder T cells from contacting the antigen-presenting cell in a noncognate fashion. Others observed that this inhibition was not restricted by MHC but was mediated by a soluble factor and through the generation of CD8$^+$ regulatory T cells *(39)*.

Although the role of regulatory T cells in controlling the immune response has been clearly identified in the prevention of allograft rejection, the issues surrounding graft-versus-host disease, chronic inflammation and auto-immunity *(40–42)*, and the effect of T cells and B lymphocytes in the immune response of the brain remain obscure. This is particularly true regarding neurodegenerative diseases. For example, in multiple sclerosis (MS), there is an increase in the number of CD8$^+$ cells near the area of the lesion that can be as much as 10-fold greater than the number of CD4$^+$ cells *(43)*. Although B lymphocytes appear to differentiate into plasmocytes when contacting T cells in the central nervous system (CNS), their presence is limited *(44)*, except in the case of massive T-cell infiltration, which occurs in MS *(45)*.

Furthermore, the functional significance of potential immunological responses in MSC transplants has not been systematically studied in models of neurodegenerative diseases. Previous studies suggest that the immuno-suppressive properties exerted by MSC transplants could mitigate the use of immunosuppressive drugs, conferring an additional advantage when using MSCs for neural transplantation. Recent studies in our laboratories have explored the use of both autologous and heterologous bone marrow–derived stem cell transplants on the behavioral outcomes of rats given intrastriatal injections of the N-methyl-D-aspartate agonist quinolinic acid (QA), which mimics the neuropathological profile and produces analogous behavioral phenotypes similar to those observed in Huntington's disease (HD).

This chapter summarizes the findings of three recent studies conducted to explore the use of autologous and heterologous MSC transplants to counter-act cognitive and motor deficits in a rodent model of HD.

STUDY 1: USE OF AUTOLOGOUS
WHOLE BONE MARROW TRANSPLANTS IN AN HD MODEL

HD is a severe autosomal dominant disorder characterized by progressive locomotor (i.e., choreiform movements), psychological, and cognitive impairments *(46)*. The disease is caused by an expansion of trinucleotide (cytosine-adenine-guanine [CAG]) repeats on the IT15 locus of chromosome 4 *(47)* that codes for the huntingtin protein. Although the gene encoding huntingtin is expressed ubiquitously, selective degeneration occurs, causing GABAergic neuronal loss in the corpus striatum (caudate nucleus, putamen, and globus pallidus). To date, neither curative nor neuroprotective therapies are available to stop disease progression. However, positive results from fetal neural transplants in patients with Parkinson's disease have demonstrated the potential clinical value for this type of therapy *(48,49)*. Recently, intracerebral transplantation of fetal neuroblasts from the striata primordia has been conducted in some HD patients, and beneficial results of this procedure have also been reported *(50,51)*. Nevertheless, the use of primary immature neurons or precursors in transplantation has some limitations owing to the heterogeneity and restricted availability of embryonic tissue. Therefore, other cell sources have been investigated; among them are adult bone marrow–derived stem cells. The first study involved the use of autologous transplantation of adult whole bone marrow (WBM) cells to counteract behavioral deficits in the QA model of HD.

Male Sprague-Dawley rats (450 g) were randomly assigned to one of three experimental conditions: sham-operated rats not receiving any transplants (Sham group); those receiving bilateral intrastriatal injections of QA without any transplants (150 nmol; QA group); and bilateral intrastriatal QA injections, followed by autologous bone marrow transplantation (TP) (QA + TP group). Thirteen days following bilateral surgical injections of either QA or phosphate-buffered saline (PBS) vehicle, WBM cells were aspirated from the left tibia of each rat and immediately washed in D-minimal essential medium (DMEM) and counted. Injections of 7-µL suspension of viable cells (500,000 cells/µL) or the DMEM vehicle were made bilaterally into the striatum of the same rat.

Behavioral testing commenced 10 d after transplant surgery, and the rats were assessed for sensorimotor functioning and spatial learning ability, as measured in a radial arm water maze (RAWM). The RAWM consists of an eight-arm radial maze submerged inside a water tank, with hidden platforms located at the ends of four of the eight arms *(52)*. Each rat was tested in the RAWM twice daily for 22 consecutive days. Rats were allowed to rest 10 s on each platform that they found, at which time the platform was lowered and the animal swam in search of another platform. Rats unable to find all four platforms within 3 min were manually guided to each missed platform, where they were allowed to sit for 10 s. Visuospatial abilities were assessed using a paw-placement test *(53)*, and swim speed measurements were taken

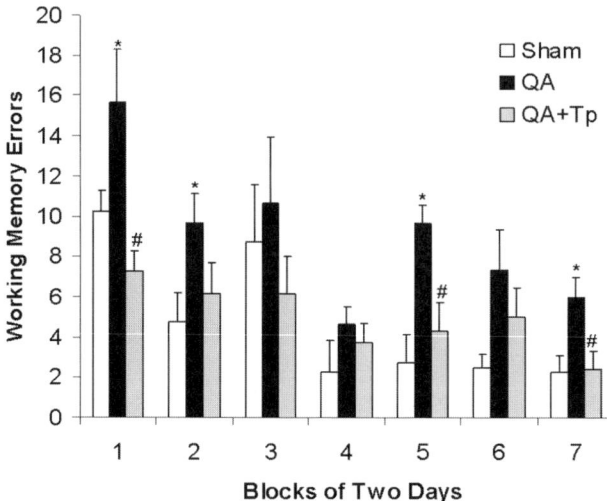

Fig. 1. Bilateral injections of QA caused significant working memory impairment (*p < 0.05; relative to sham-operated control rats) during the first 2 wk of testing in the RAWM spatial learning task. However, QA-treated rats that received transplants of suspended WBM cells were protected from these QA-induced deficits and had significantly fewer working memory errors (#p < 0.05; relative to rats receiving QA without transplants).

prior to the first trial on the first day of testing and following the last trial on the last day of testing. Dependent measures for the RAWM task were the number of working memory errors (i.e., entries into arms that had been already visited) and reference memory errors (i.e., entries into arms that never contained platforms).

Rats with transplants performed at control levels and made significantly fewer working memory errors than QA-treated animals during the first 2 wk of RAWM testing (Fig. 1). Because there were no QA-induced motor deficits on the paw-placement and swim speed tests, these results indicate that the transplants produced a decrease in QA-induced cognitive deficits and did not affect motor performance *per se*.

A histological analysis was conducted on another group of rats that underwent identical surgical procedures but were given intrastriatal injections of PKH26-labeled WBM cells. Results from this analysis revealed that the transplants integrated into the host tissue by the end of the first week (Fig. 2A). Labeled cells were clearly visible at 2-wk posttransplant (Fig. 2B), and the first nestin-positive cells were observed within the striatum (Fig. 2C). However, less than 1% of the transplanted cells ever expressed neuronal

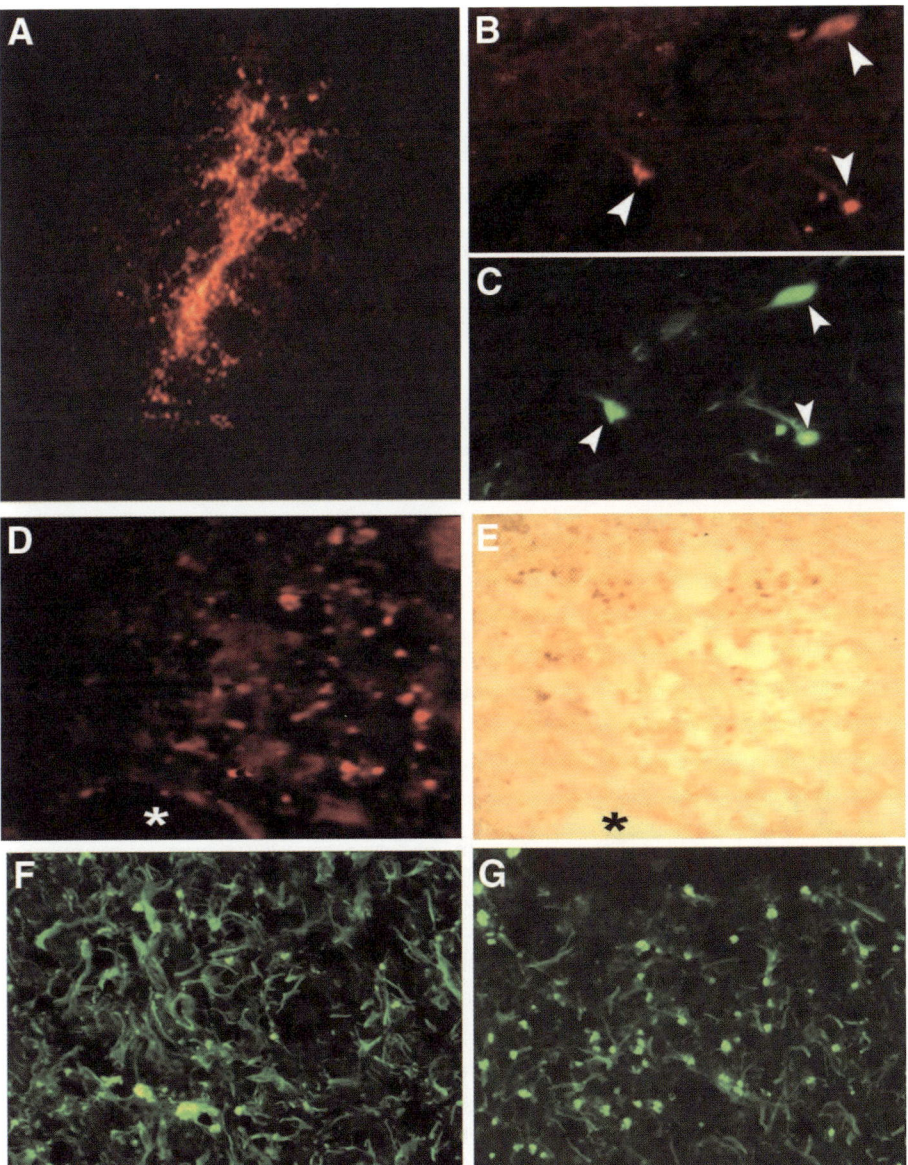

Fig. 2. Microscopic analysis at 1 wk after the transplant revealed numerous viable PKH26-labeled transplanted cells (**A**; ×10). At 2-wk posttransplant, a few (<1%) of these transplanted cells (**B**; ×40) were also labeled positive for nestin (**C**; ×40). At 37-d posttransplant, the transplanted cells remained viable (**D**; ×20; *blood vessel) and were metabolically active, as indicated by robust cytochrome oxidase labeling (**E**; ×20). Although both heterologous (**F**; ×20) and autologous (**G**; ×20) transplants appeared to produce elevated labeling for GFAP in the host tissue near the transplant site, the preliminary analysis suggested that heterologous transplants may have produced slightly more GFAP labeling, possibly indicating an immunological response to the transplants.

markers, and most of the other cells appeared quite primitive and were spherical in shape. Three weeks after the transplantation, a few β-tubulin-III-positive and even fewer $GAD_{65/67}$-positive cells (<0.3%) were observed. Interestingly, those cells were located close to each other, leaving large areas of the striatum devoid of any cells expressing neuronal phenotypes. Labeled cells were clearly visible at 37-d posttransplantation (Fig. 2D), and most of these cells expressed high levels of cytochrome oxidase activity (Fig. 2E), suggesting that they were still metabolically active for over 1 mo after transplantation.

Results of our first study indicated that autologous transplants of WBM can reduce QA-induced learning deficits in a rodent model of HD. Furthermore, this study demonstrated that these transplants are viable for at least 37 d after transplantation and a few (<1%) expressed neuronal phenotypes. However, because the behavioral effects occurred soon after the transplants, and because so few of the cells expressed neuronal phenotypes, the underlying mechanism of recovery is likely the result of causes other than replacement of lost cells and probably involves an increased production of neurotrophic factors, which may protect or accelerate the function of partially affected cells near the site of the lesion. Regardless of the underlying mechanism, these findings suggest that transplantation of adult WBM cells offer significant promise as a potential therapy for HD.

STUDY 2: TRANSPLANTATION OF MSCS IN A UNILATERAL MODEL OF HD

In our second study, we assessed whether heterologous transplants of adult MSCs could reduce motor deficits in a unilateral QA rodent model of HD. The surgical procedures were similar to those used in our first study, but the MSCs were separated from the bone marrow that was aspirated from the femurs of six donor rats using magnetic cell-sorting procedures (VarioMACs; Mitenyi Biotec) with CD90 (Thy-1)-tagged MSCs. The MSCs were plated and transferred through 20 passes during a 3-mo period before transplantation into the striatum of rats that had received unilateral intrastriatal injections of either QA or PBS 1 wk prior to transplantation (TP). Two other groups received DMEM instead of TPs 1 wk after receiving unilateral injections of either QA or PBS into the striatum.

At 1 wk and 4 wk following the initial surgery, all rats were tested on a battery of motor tasks, including the ability to cross a suspended balance beam and stepping ability on a modified "wheelbarrow" or "drag" test *(54)*. In the former task, the dependent measure was the latency to traverse a 2-cm wide by 100-cm long beam. In the latter task, the rat was suspended by holding the base of the tail at an angle, which allowed only the forepaws to come in contact with the surface of a 50-cm long cushion. The rat was then slowly pulled backward across the cushion, and the number of contralateral paw placements, or "hops," were recorded and used as the dependent measures. Rats with unilateral QA injections showed fewer steps or hops with the paw contralateral to the side of the QA injection and tended to let this paw drag across the cushion. For both motor tasks, all rats were given three trials at both the post-QA (1 wk after surgery) and post-TP (3 wk after TP surgery) testing sessions.

Results indicated that heterologous MSC TPs were only modestly successful in counteracting the QA-induced deficits in these motor tasks (Fig. 3). In the balance beam task, only rats receiving the TPs showed significant improvement in latency to traverse the beam following QA-induced deficits on this task (Fig. 3A). However, the MSC TPs were unable to reduce the QA-induced deficits in the drag test (Fig. 3B). The failure of the MSC TPs to exert robust effects in this model may be because of several factors. Relative to our first study, the sorting and plating of MSCs may have reduced the efficacy of these cells to produce neurotrophic factors or may have screened out potentially beneficial hematopoietic stem cells. It was our hope that a greater percentage of MSC cells in the TPs used in this study, relative to those used in our first study, would accentuate the ameliorative effects. Normally, only about 0.1% of the WBM cells are MSCs, and if the MSCs have prominent role in mediating the therapeutic aspects of the bone marrow–derived TPs, increasing the number of MSCs by sorting them from the WBM and growing them in culture should have maximized the efficacy of these TP procedures. However, it is possible that other cells that were also present within the WBM, such as hematopoietic stem cells that produce potentially beneficial cytokines *(55)*, could interact or somehow potentiate the efficacy of the MSCs.

Histological analysis revealed TP profiles that were similar to those observed in our first WBM TP study. This is worthy of note, because despite the 20 passes in culture, the transplanted MSCs appeared viable and metabolically active at 3 wk post-TP. Nonetheless, there was some indication of increased glial fibrillary acidic protein (GFAP) labeling near the implantation site, suggesting a possible host immunological response. Although more controlled studies are needed to directly compare the two approaches, the results of our second study suggest that heterologous MSC TPs may be less efficacious than autologous WBM TPs in reducing behavioral deficits in the QA model of HD.

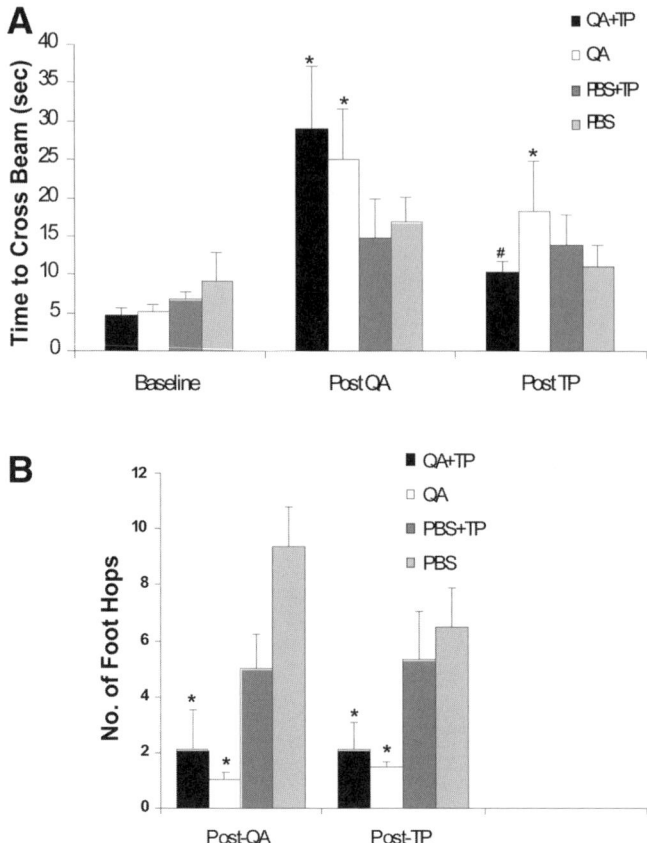

Fig. 3. Rats receiving unilateral intrastriatal injections of QA took longer to cross a 100-cm long suspended balance beam than prior to QA administration (*$p < 0.05$; relative to baseline performance) at 1 wk after receiving the QA. However, QA-treated rats that were given transplants of cultured MSCs showed significant improvements at 3-wk posttransplant (#$p < 0.05$; relative to their performance at 1 wk post-QA injection **(A)**. Interestingly, these same transplants were unable to reduce the QA-induced deficits (*$p < 0.05$; relative to sham-operated controls) on the drag test—a modified gait assessment task **(B)**.

STUDY 3: COMPARISON OF AUTOLOGOUS AND HETEROLOGOUS MARROW STEM CELL TRANSPLANTS

Results of our second study prompted the direct comparison of the efficacy of autologous WBM TPs with heterologous TPs in the same unilateral QA model of HD. The WBM cells in our third study were extracted from the femurs of either the same rat (autologous) or from a different donor (heterologous). The QA surgical procedures, the harvesting of WBM cells, and the transplantation procedures were identical to those used in our first study, with the exception that the QA or vehicle was injected unilaterally and a group of rats were given heterologous TPs. The suspension of WBM cells (approx 500,000 cells in 2 μL in DMEM) or vehicle (DMEM) was injected into the same striatum that received a unilateral injection of QA 1 wk earlier. At 1 and 4 wk after the WBM TP, the motor ability of the rats was assessed using the drag test and accelerating rotorod task. Three trials were given for both tasks on each testing day. The dependent variables were the number of contralateral foot placements (hops) in the drag test and the latency to fall from the rotorod.

All rats receiving QA injections had a significant reduction in the total number of foot hops made by the contralateral forelimb at 7 d postsurgery (Fig. 4A). At 1 wk post-TP, neither the autologous nor the heterologous group showed any change in the QA-induced reduction of foot hops on the drag test. However, at 28-d post-TP, rats that received autologous, but not those receiving heterologous, TPs had a significant increase in the mean number of contralateral foot hops, compared to those made just after QA administration and prior to transplantation. Interestingly, by post-TP day 28, the number of foot hops made by the autologous TP group (but not those made by the heterologous TP group) was still significantly less than their baseline performance before QA administration.

Similar to the results on the drag test, a significant reduction occurred in latency to fall-off of the rotorod following QA administration (Fig. 4B). However, at 28-d post-TP, both the autologous and heterologous TPs groups performed at normal levels on the rotorod test (Fig. 4B). In the autologous TP group, this effect was observed as early as 7 d post-TP and was sustained for the duration of the study. Furthermore, at 28-d post-TP, animals with autologous transplants stayed on the rotorod significantly longer than animals with heterologous transplants. Control animals given DMEM vehicle instead of WBM TPs had no significant recovery in the rotorod test performance (24.5 ± 3.8, 12.8 ± 3.0, and 13.6 ± 5.9 s on the rotorod at baseline, 7-d postlesion, and 28-d posttransplant, respectively.

These results indicate that autologous transplants can counteract QA-induced motor deficits on the drag and rotorod tests more effectively than heterologous transplants. Similar to the previous two studies, the histological analysis indicated very few markers for neuronal phenotypes; thus, the transplant-induced production of trophic factors may underlie functional recovery. Also, because the preliminary histological analysis suggests that GFAP labeling tended to be slightly elevated in the brain specimens con-

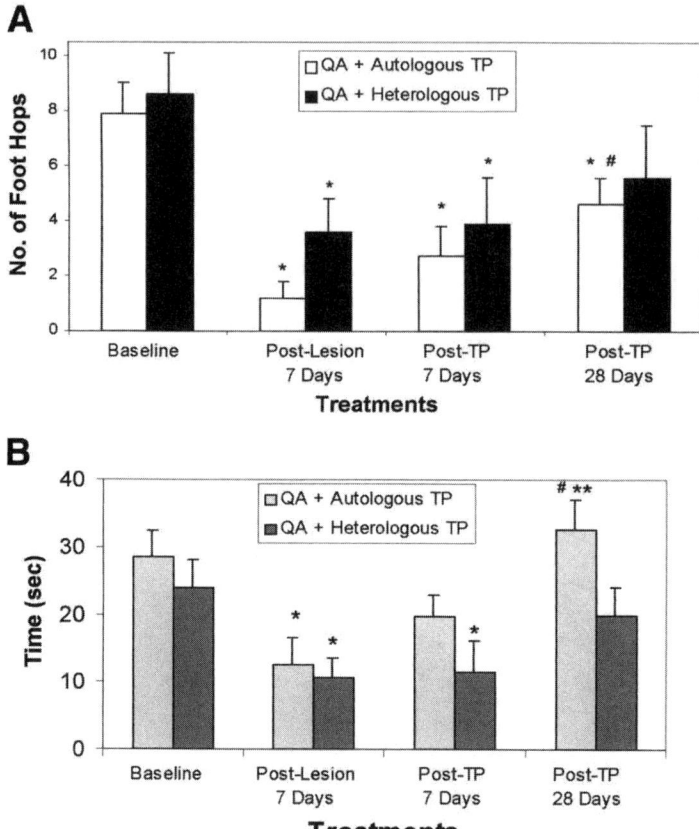

Fig. 4. Ameliorative effects of striatal WBM transplantation on the QA-induced drag test and rotorod performance impairment in rats. **(A)** Effect of autologous (open bar) or heterologous (filled bar) transplantation of WBM cells on the total number of contralateral foot hops is compared at baseline, QA postsurgery day 7, WBM posttransplant day 7, and day 28. Autologous WBM transplantation into the striatum, 7 d following the QA injection, attenuated lesion-induced reduction in contralateral forelimb foot hops 28-d posttransplantation. Heterologous WBM transplantation, in contrast, failed to attenuate the reduced forelimb foot hops on posttransplantation day 28 ($*p < 0.05$ and $^{\#}p < 0.05$, compared to baseline and QA postsurgery day 7, respectively). **(B)** Effect of autologous (light gray bar) or heterologous (dark gray bar) transplantation of WBM cells on latency to fall off the rotorod is compared at baseline, QA postsurgery day 7, WBM posttransplant day 7, and day 28. Autologous WBM transplantation into the striatum, 7 d after QA injection, normalized the lesion-induced reduction in the time the animals were able to remain on an accelerating rotorod during test trials at 7- and 28-d posttransplantation. Heterologous WBM transplants, alternatively failed to reverse the QA-induced impairments on rotorod performance at posttransplantation day 28 ($*p < 0.05$, $**p < 0.05$, and $^{\#}p < 0.05$ vs baseline, posttransplant day 28, and post-QA surgery day 7, respectively).

taining heterologous TPs (Fig. 2F), relative to those containing autologous TPs (Fig. 2G), the dampened ameliorative effects of heterologous transplants may be owing to immunological complications. As such, results of the third study suggest that adult autologous bone marrow TPs can survive and integrate into host tissue, they are therapeutically viable, and they may have less immunological complications than heterologous TPs.

CONCLUSIONS

Collectively, these studies support the findings of other investigators, who have also reported that adult bone marrow–derived stem cells can counteract functional deficits caused by CNS damage (see Table 1). Our work has shown that this effect can be extended to cognitive and motor deficits produced by intrastriatal injections of QA—a neurotoxin that mimics the neuropathology and behavioral problems associated with HD. In all three transplant studies, we observed that the bone marrow stem cell TPs survived and integrated into the host tissue, as had been seen in other studies using TP of MSCs (see Table 1). Nestin-positive cells were also found after 14-d post-TP, as well as GAD-positive cells at 21-d post-TP, but the latter result constituted less than 0.3% of the transplanted cells. In addition, a few of these cells migrated beyond their injection site within the striatum. Our findings support recent evidence provided by of Munoz-Elias and colleagues *(56)*, who observed differentiation, migration, and long-term survival of adult bone marrow stem cells transplanted into embryonic ventricles in rats. Thus, our data suggest that adult bone marrow stem cells possess the plasticity to transdifferentiate into neurons and can successfully integrate with the host tissue and remain metabolically active for over 1-mo posttransplantation. However, the comparison of Munoz-Elias et al. data with ours indicates that neuronal transdifferentiation of MSCs into neurons occurs more readily in a developing brain than in an injured adult brain. Also, the comparison implies that MSCs transplanted in a damaged brain may be missing the level of trophic support needed to transdifferentiate MSCs in greater numbers.

Because less than 1% of the cells expressed neuronal phenotypes, and because the functional recovery observed in all three studies occurred within 1 mo of the TPs, the ameliorative effects of the TPs were probably not because of cell replacement and a reintegration of damaged neuronal circuitry. A more plausible explanation is that the MSC TPs may produce neurotrophic factors or perhaps increase the endogenous production of these factors by the surrounding host tissue. Ample evidence indicates that stem cell transplants are capable of releasing neurotrophic factors, and many members of the neurotrophin family, including nerve growth factor and brain-derived neurotrophic factor, can provide neuroprotection in HD models *(57–59)*.

Interestingly, in all three studies, the TPs were placed just dorsal to the site of the QA injections, but no TP-induced reduction in lesion size was observed. This suggests that the TPs did not exert a neuroprotective effect by rescuing slowly dying neurons. It appears that any ameliorative effects seen in our studies were probably owing to the trophic enhancement of intact, but partially compromised, host tissue. Although the neurotrophic support of this tissue may not be critical for its survival, a potential TP-induced elevation in neurotrophic factors may have resuscitated dysfunctional neurons more efficiently than normal compensatory mechanisms. However, more research is needed to measure the levels of neurotrophic factors in these TP models to better assess the plausibility of this hypothesis. It is still possible that the subtle, but significant, amelioration of QA-induced deficits could be a function of early (but yet incomplete) rewiring of the neural circuitry. More studies using longer-term recovery periods and in vivo electrophysiological measures are required to have a better understanding of the underlying mechanisms of TP-induced recovery.

Comparing the results of these three studies also suggests that autologous MSC TPs may be more efficacious than heterologous MSC TPs in lowering QA-induced behavioral deficits. What appeared to be a more robust functional effect was found, albeit with a different model and using different behavioral parameters, when results from the first study, which involved autologous WBM TPs, were compared to the results of the second study, which included heterologous TPs of presorted MSCs. This effect may have been owing to the fact that the cell-sorting process may have contributed to compromising critical hematopoietic stem cells that could produce beneficial effects, either on their own or by interacting with MSCs when transplanted into the QA-damaged brain. However, the size and morphology of the TP cells were very similar to those observed in our first WBM study, and the transplants were viable and metabolically active at 3-wk post-TP, which suggests that our sorting procedure did not result in gross morphological changes in the TPs. An alternative explanation for the more blunted effects observed in the second study may be that the use of heterologous WBM cells could have produced a subtle immunological response in the host tissue that was mitigated by the use of autologous TP.

Results of the third study provide some evidence that the latter hypothesis is a viable explanation, because direct comparisons of heterologous vs autologous WBM TPs using the same behavioral measures revealed a modest, but significantly more efficacious, effect with the autologous TPs.

This finding is particularly important, considering the use of autologous TPs in HD, despite its potentially greater efficacy, may have limited utility if the mutant gene becomes expressed in the transplanted stem cells. It remains to be determined if and when stem cell transplants containing the mutant HD gene would ever induce cellular dysfunction. However, if heterologous TPs prove to be the only feasible long-term therapy for HD, it may be important to find methods to eliminate or reduce this potential immunological reaction. Clearly, more research is needed to carefully quantify and compare the immunological responses in both heterologous and autologous TPs and to determine if this is indeed the underlying basis for the differential effects in the functional recovery observed in our QA models of HD.

In summary, these data suggest that bone marrow–derived stem cell TPs have significant potential to counteract HD-like cognitive and motor deficits. Our results also indicate that the use of WBM TPs is possibly more effective than presorted MSCs, and that autologous WBM TPs are more efficacious than heterologous WBM TPs. Further study is needed to elucidate the mechanisms underlying the beneficial effects of these TPs and to examine whether heterologous TPs produce an undesirable immunologic response. Research in these areas is critically important to maximize the safety and efficacy of this promising therapeutic approach for HD.

REFERENCES

1. Bjornson, C. R., Rietze, R. L., Reynolds, B. A., et al. (1999) Turning brain into blood: a hematopoietic fate adopted by adult neural stem cells in vivo. *Science* **283,** 534–537.
2. Sanchez-Ramos, J., Song, S., Cardozo-Pelaez, F., et al. (2000) Adult bone marrow stromal cells differentiate into neural cells in vitro. *Exp. Neurol.* **164,** 247–256.
3. Largaespada, D. A. and Verfaillie, C. M. (2000) Pluripotency of mesenchymal stem cells derived from adult marrow. *Nature* **418,** 41–49.
4. Friedenstein, A. J. (1976) Precursor cells of mechanocytes. *Int. Rev. Cytol.* **47,** 327–359.
5. Orlic, D., Kajstura, J., Chimenti, S., et al. (2001) Bone marrow cells regenerate infarcted myocardium. *Nature* **410,** 701–705.
6. Ferrari, G., Cusella-De Angelis, G., Coletta, M., et al. (1998) Muscle regeneration by bone marrow-derived myogenic progenitors. *Science* **279,** 1528–1530.
7. LaBarge, M. A. and Blau, H. M. (2002) Biological progression from adult bone marrow to mononucleate muscle stem cell to multinucleate muscle fiber in response to injury. *Cell* **111,** 589–601.
8. Lagasse, E., Connors, H., Al-Dhalimy, M., et al. (2000) Purified hematopoietic stem cells can differentiate into hepatocytes in vivo. *Nat. Med.* **6,** 1229–1234.
9. Hess, D., Li, L., Martin, M., et al. (2003) Bone marrow-derived stem cells initiate pancreatic regeneration. *Nat. Biotechnol.* **21,** 763–770.

10. Poulsom, R., Forbes, S. J., Hodivala-Dilke, K., et al. (2001) Bone marrow contributes to renal parenchymal turnover and regeneration. *J. Pathol.* **195,** 229–235.
11. Kale, S., Karihaloo, A., Clark, P. R., et al. (2003) Bone marrow stem cells contribute to repair of the ischemically injured renal tubule. *J. Clin. Invest.* **112,** 42–49.
12. Okamoto, R., Yajima, T., Yamazaki, M., et al. (2002) Damaged epithelia regenerated by bone marrow-derived cells in the human gastrointestinal tract. *Nat. Med.* **8,** 1011–1017.
13. Krause, D. S., Theise, N. D., Collector, M. I., et al. (2001) Multi-organ, multi-lineage engraftment by a single bone marrow-derived stem cell. *Cell* **105,** 369–377.
14. Herzog, E. L., Chai, L., and Krause, D. S. (2003) Plasticity of marrow-derived stem cells. *Blood* **102,** 3483–3493.
15. Eglitis, M. A. and Mezey, E. (1997) Hematopoietic cells differentiate into both microglia and macroglia in the brains of adult mice. *Proc. Natl. Acad. Sci. USA* **94,** 4080–4085.
16. Akiyama, Y., Radtke, C., and Kocsis, J. D. (2002) Remyelination of the rat spinal cord by transplantation of identified bone marrow stromal cells. *J. Neurosci.* **22,** 6623–6630.
17. Eglitis, M. A., Dawson, D., Park, K. W., and Mouradian, M. M. (1999) Targeting of marrow-derived astrocytes to the ischemic brain. *Neuroreport* **10,** 1289–1292.
18. Kopen, G. C., Prockop, D. J., and Phinney, D. G. (1999) Marrow stromal cells migrate throughout forebrain and cerebellum, and they differentiate into astrocytes after injection into neonatal mouse brains. *Proc. Natl. Acad. Sci. USA* **96,** 10711–10716.
19. Cogle, C. R., Yachnis, A. T., Laywell, E. D., et al. (2004) Bone marrow transdifferentiation in brain after transplantation: a retrospective study. *Lancet* **363,** 1432–1437.
20. Azizi, S. A., Stokes, D., Augelli, B. J., et al. (1998) Engraftment and migration of human bone marrow stromal cells implanted in the brains of albino rats—similarities to astrocyte grafts. *Proc. Natl. Acad. Sci. USA* **95,** 3908–3913.
21. Brazelton, T. R., Rossi, F. M., Keshet, G. I., and Blau, H. M. (2000) From marrow to brain: expression of neuronal phenotypes in adult mice. *Science* **290,** 1775–1779.
22. Mezey, E., Chandross, K. J., Harta, G., et al. (2000) Turning blood into brain: cells bearing neuronal antigens generated in vivo from bone marrow. *Science* **290,** 1779–1782.
23. Jin, H. K., Carter, J. E., Huntley, G. W., and Schuchman, E. H. (2002) Intracerebral transplantation of mesenchymal stem cells into acid sphingo-myelinase-deficient mice delays the onset of neurological abnormalities and extends their life span. *J. Clin. Invest.* **109,** 1183–1191.
24. Zhao, L. R., Duan, W. M., Reyes, M., et al. (2002) Human bone marrow stem cells exhibit neural phenotypes and ameliorate neurological deficits after grafting into the ischemic brain of rats. *Exp. Neurol.* **174,** 11–20.

25. Lescaudron, L., Unni, D., and Dunbar, G. L. (2003) Autologous adult bone marrow stem cell transplantation in an animal model of huntington's disease: behavioral and morphological outcomes. *Int. J. Neurosci.* **113,** 945–956.
26. Woodbury, D., Schwarz, E. J., Prockop, D. J., and Black, I. B. (2000) Adult rat and human bone marrow stromal cells differentiate into neurons. *J. Neurosci. Res.* **61,** 364–370.
27. Deng, W., Obrocka, M., Fischer, I., and Prockop, D. J. (2001) In vitro differentiation of human marrow stromal cells into early progenitors of neural cells by conditions that increase intracellular cyclic AMP. *Biochem. Biophys. Res. Commun.* **282,** 148–152.
28. Carbonetto, S., Gruver, M. M., and Turner, D. C. (1983) Nerve fiber growth in culture on fibronectin, collagen and glycosaminoglycan substrates. *J. Neurosci.* **3,** 2324–2335.
29. Sakai, T., Johnson, K. J., Murozono, M., et al. (2001) Plasma fibronectin supports neuronal survival and reduces brain injury following transient focal cerebral ischemia but is not essential for skin-wound healing and hemostasis. *Nat. Med.* **7,** 324–330.
30. Eaves, C. J., Cashman, J. D., Kay, R. J., et al. (1991) Mechanisms that regulate the cell cycle status of very primitive hematopoietic cells in long-term human marrow cultures. II. Analysis of positive and negative regulators produced by stromal cells within the adherent layer. *Blood* **78,** 110–117.
31. Majumdar, M. K., Thiede, M. A., Mosca, J. D., et al. (1998) Phenotypic and functional comparison of cultures of marrow-derived mesenchymal stem cells (MSCs) and stromal cells. *J. Cell Physiol.* **176,** 57–66.
32. Kawasaki, H., Mizuseki, K., Nishikawa, S., et al. (2000) Induction of midbrain dopaminergic neurons from ES cells by stromal cell-derived inducing activity. *Neuron* **28,** 31–40.
33. Morizane, A., Takahashi, J., Takagi, Y., et al. (2002) Optimal conditions for in vivo induction of dopaminergic neurons from embryonic stem cells through stromal cell-derived inducing activity. *J. Neurosci. Res.* **69,** 934–939.
34. Schwarz, E. J., Alexander, G. M., Prockop, D. J., and Azizi, S. A. (1999) Multipotential marrow stromal cells transduced to produce L-DOPA: engraftment in a rat model of Parkinson disease. *Hum. Gene Ther.* **10,** 2539–2549.
35. Park, K. W., Eglitis, M. A., and Mouradian, M. M. (2001) Protection of nigral neurons by GDNF-engineered marrow cell transplantation. *Neurosci. Res.* **40,** 315–323.
36. Bartholomew, A., Sturgeon, C., Siatskas, M., et al. (2002) Mesenchymal stem cells suppress lymphocyte proliferation in vitro and prolong skin graft survival in vivo. *Exp. Hematol.* **30,** 42–48.
37. Le Blanc, K., Rasmusson, I., Sundberg, B., et al. (2004) Treatment of severe acute graft-versus-host disease with third party haploidentical mesenchymal stem cells. *Lancet* **363,** 1439–1441.
38. Krampera, M., Glennie, S., Dyson, J., et al. (2003) Bone marrow mesenchymal stem cells inhibit the response of naive and memory antigen-specific T cells to their cognate peptide. *Blood* **101,** 3722–3729.

39. Djouad, F., Plence, P., Bony, C., et al. (2003) Immunosuppressive effect of mesenchymal stem cells favors tumor growth in allogeneic animals. *Blood* **102,** 3837–3844.
40. Hori, S., Takahashi, T., and Sakaguchi, S. (2003) Control of autoimmunity by naturally arising regulatory CD4+ T cells. *Adv. Immunol.* **81,** 331–371.
41. von Herrath, M. G. and Harrison, L. C. (2003) Antigen-induced regulatory T cells in autoimmunity. *Nat. Rev. Immunol.* **3,** 223–232.
42. Bach, J. F. (2003) Autoimmune diseases as the loss of active "self-control". *Ann. NY Acad. Sci.* **998,** 161–177.
43. Babbe, H., Roers, A., Waisman, A., et al. (2000) Clonal expansions of CD8(+) T cells dominate the T cell infiltrate in active multiple sclerosis lesions as shown by micromanipulation and single cell polymerase chain reaction. *J. Exp. Med.* **192,** 393–404.
44. Griffin, D. E., Levine, B., Tyor, W. R., and Irani, D. N. (1992) The immune response in viral encephalitis. *Semin. Immunol.* **4,** 111–119.
45. Cross, A. H., Trotter, J. L., and Lyons, J. (2001) B cells and antibodies in CNS demyelinating disease. *J. Neuroimmunol.* **112,** 1–14.
46. Bruyn, G. W. and von Wolferen, W. J. (1973) Pathogenesis of Huntington's chorea. *Lancet* **1,** 1382.
47. MacMillan, J. C., Morrison, P. J., Nevin, N. C., et al. (1993) Identification of an expanded CAG repeat in the Huntington's disease gene (IT15) in a family reported to have benign hereditary chorea. *J. Med. Genet.* **30,** 1012–1013.
48. Lindvall, O. (1999) Cerebral implantation in movement disorders: state of the art. *Mov. Disord.* **14,** 201–205.
49. Peschanski, M., Defer, G. L., Dethy, S., et al. (1999) The need for phase III studies in experimental surgical treatments of Parkinson's disease. *Adv. Neurol.* **80,** 651–653.
50. Bachoud-Levi, A. C., Remy, P., Nguyen, J. P., et al. (2000) Motor and cognitive improvements in patients with Huntington's disease after neural transplantation. *Lancet* **356,** 1975–1979.
51. Freeman, T. B., Cicchetti, F., Hauser, R. A., et al. (2000) Transplanted fetal striatum in Huntington's disease: phenotypic development and lack of pathology. *Proc. Natl. Acad. Sci. USA* **97,** 13877–13882.
52. Dunbar, G. L., Shear, D. A., Dong, J., and Haik-Cregeur, K. L. (1999) Cognitive and motor deficits produced by acute and chronic administration of 3-nitropropionic acid in rats. In *Mitochondrial Inhibitors and Neurodegenerative Disorders* (Sanberg, P. R., Nishinino, H., and Borlongan, C. V., eds.), Humana Press, Totowa, NJ, pp. 73–92.
53. Shear, D., Dong, J., Haik-Cregier, K. L., et al. (1998) Chronic administration of quinolinic acid in the rat striatum causes spatial learning deficits in a radial arm water maze task. *Exp. Neurol.* **150,** 305–311.
54. Schallert, T., De Ryck, M., Whishaw, I. Q., and Teitelbaum, P. (1979) Excessive bracing reactions and their control by atropine and L-dopa in an animal analog of Parkinsonism. *Exp. Neurol.* **64,** 33–43.
55. Lensch, M. W. and Daley, G. Q. (2004) Origins of mammalian hematopoiesis: in vivo paradigms and in vitro models. *Curr. Top. Dev. Biol.* **60,** 127–196.

56. Munoz-Elias, G., Marcus, A. J., Coyne, T. M., et al. (2004) Adult bone marrow stromal cells in the embryonic brain: engraftment, migration, differentiation, and long-term survival. *J. Neurosci.* **24**, 4585–4595.
57. Alberch, J., Perez-Navarro, E., and Canals, J. M. (2004) Neurotrophic factors in Huntington's disease. *Prog. Brain Res.* **146**, 195–229.
58. Kordower, J. H., Isacson, O., and Emerich, D. F. (1999) Cellular delivery of trophic factors for the treatment of Huntington's disease: is neuroprotection possible? *Exp. Neurol.* **159**, 4–20.
59. Canals, J. M., Pineda, J. R., Torres-Peraza, J. F., et al. (2004) Brain-derived neurotrophic factor regulates the onset and severity of motor dysfunction associated with enkephalinergic neuronal degeneration in Huntington's disease. *J. Neurosci.* **24**, 7727–7739.
60. Ohta, M., Suzuki, Y., Noda, T., et al. (2004) Bone marrow stromal cells infused into the cerebrospinal fluid promote functional recovery of the injured rat spinal cord with reduced cavity formation. *Exp. Neurol.* **187**, 266–278.
61. Jiang, Y., Jahagirdar, B. N., Reinhardt, R. L., et al. (2002) Pluripotency of mesenchymal stem cells derived from adult marrow. *Nature* **418**, 41–49.
62. Wehner, T., Bontert, M., Eyupoglu, I., et al. (2003) Bone marrow-derived cells expressing green fluorescent protein under the control of the glial fibrillary acidic protein promoter do not differentiate into astrocytes in vitro and in vivo. *J. Neurosci.* **23**, 5004–5011.
63. Weimann, J. M., Charlton, C. A., Brazelton, T. R., et al. (2003) Contribution of transplanted bone marrow cells to Purkinje neurons in human adult brains. *Proc. Natl. Acad. Sci. USA* **100**, 2088–2093.
64. Hofstetter, C. P., Schwarz, E. J., Hess, D., et al. (2002) Marrow stromal cells form guiding strands in the injured spinal cord and promote recovery. *Proc. Natl. Acad. Sci. USA* **99**, 2199–2204.
65. Mahmood, A., Lu, D., Yi, L., et al. (2001) Intracranial bone marrow transplantation after traumatic brain injury improving functional outcome in adult rats. *J. Neurosurg.* **94**, 589–595.
66. Cuevas, P., Carceller, F., Dujovny, M., et al. (2002) Peripheral nerve regeneration by bone marrow stromal cells. *Neurol. Res.* **24**, 634–638.
67. Li, Y., Chen, J., Chen, X. G., et al. (2002) Human marrow stromal cell therapy for stroke in rat: neurotrophins and functional recovery. Human marrow stromal cell therapy for stroke in rat: neurotrophins and functional recovery. *Neurology* **59**, 514–523.
68. Li, Y., Chen, J., and Chopp, M. (2001a) Adult bone marrow transplantation after stroke in adult rats. *Cell Transplant.* **10**, 31–40.
69. Li, Y., Chen, J., Wang, L., et al. (2001b) Intracerebral transplantation of bone marrow stromal cells in a 1-methyl-4-phenyl-1,2,3,6-tetrahydropyridine mouse model of Parkinson's disease. *Neurosci. Lett.* **316**, 67–70.
70. Lu, D., Li, Y., Wang, L., et al. (2001) Intraarterial administration of marrow stromal cells in a rat model of traumatic brain injury. *J. Neurotrauma* **18**, 813–819.
71. Mahmood, A., Lu, D., Wang, L., and Chopp, M. (2002) Intracerebral transplantation of marrow stromal cells cultured with neurotrophic factors promotes

functional recovery in adult rats subjected to traumatic brain injury. *J. Neurotrauma* **19**, 1609–1617.

72. Nakano, K., Migita, M., Mochizuki, H., and Shimada, T. (2001) Differentiation of transplanted bone marrow cells in the adult mouse brain. *Transplantation* **71**, 1735–1740.

73. Chen, J., Li, Y., Wang, L., et al. (2001) Therapeutic benefit of intracerebral transplantation of bone marrow stromal cells after cerebral ischemia in rats. *J. Neurol. Sci.* **189**, 49–57.

74. Chen, J., Li, Y., Wang, L., et al. (2001) Therapeutic benefit of intravenous administration of bone marrow stromal cells after cerebral ischemia in rats. *Stroke* **32**, 1005–1011.

75. Chopp, M., Zhang, X. H., Li, Y., et al. (2000) Spinal cord injury in rat: treatment with bone marrow stromal cell transplantation. *Neuroreport* **11**, 3001–3005.

76. Li, Y., Chopp, M., Chen, J., et al. (2000) Intrastriatal transplantation of bone marrow nonhematopoietic cells improves functional recovery after stroke in adult mice. *J. Cereb. Blood Flow Metab.* **20**, 1311–1319.

6

Human Neuroteratocarcinoma Cells as a Neural Progenitor Graft Source for Cell Transplantation in Stroke

Cesario V. Borlongan, Christina Fournier, David C. Hess, and Paul R. Sanberg

ABSTRACT

The use of neuroteratocarcinoma cells for transplantation in stroke has emerged as a strategy for cell replacement therapy and has begun its transition from laboratories to clinical settings. Procurement logistics and novel neuroprotective functions associated with these cells allow their use to serve as an efficacious alternative to the use of fetal cells as donor cell grafts for stroke therapy. However, the optimal transplantation regimen must still be determined. Specifically, the limitations of current stroke management reveal an urgent need to examine the efficacy of experimental treatment options, such as neural transplantation, to develop better therapies. This chapter discusses the characteristics of NT2N cells, the role of the host brain microenvironment and NT2N cell grafts, laboratory research and clinical trials for the intracerebral transplantation of NT2N cells in stroke, the mechanisms underlying the grafts' effects, and NT2N cell grafts and the need for immunosuppression. This chapter also highlights some of the most recent findings regarding NT2N cells.

Key Words: Neuroteratocarcinoma cells; NT2N; intracerebral transplantation; neuroprotection; stroke.

INTRODUCTION

In recent years, basic scientific research efforts have established the efficacy of cell replacement therapy in animal models of neurological disorders; many of these findings have begun their transition to a clinical setting. Limited trials of neural transplantation have already been initiated in chronically ill patients. The use of neuroteratocarcinoma cells for transplantation

From: *Contemporary Neuroscience: Cell Therapy, Stem Cells, and Brain Repair*
Edited by: C. D. Sanberg and P. R. Sanberg © Humana Press Inc., Totowa, NJ

in stroke shows great promise for this transition. Procurement logistics and novel neuroprotective functions relating to this cell line allow neuro-teratocarcinoma cells to be used as donor cell grafts for stroke therapy, but the most suitable transplantation regimen must still be established. The *nonregenerative central nervous system* dogma is arguably a historical part of the past; modern research demonstrates that diseased or aging brain cells can be rescued and their functions can be restored. Cell replacement therapy has emerged as the current translational research trend and may become a promising treatment for various neurological disorders.

To date, stroke is the third leading cause of death, affecting over 500,000 people each year in the United States. In addition, there are currently about 3 million stroke survivors, many of which suffer from significant cognitive and functional disabilities. Rehabilitation therapy has helped some survivors recover, but many patients still experience permanent loss of independent function. The cost for rehabilitation and lost wages is estimated to be $30 billion each year and thus represents a major financial impact on society. Current stroke treatments are typically limited to supportive care and secondary prevention, resulting in only limited improvement in cognitive and motor function. Intravenous tissue plasminogen activator administration has been effective in ameliorating the neurological deficits arising from acute stroke. Yet, this treatment strategy is problematic because of its extremely limited window of efficacy, remaining within 3 h of stroke onset. Current treatments, directed toward acute stroke, are restricted in their application and effectiveness. Unfortunately, no therapy has been proven useful for treating chronic stroke, which is associated with significant morbidity and mortality. Thus, experimental treatments, such as neural transplantation, should be examined to develop better treatment therapies, particularly for chronic stroke.

FETAL NEURAL TRANSPLANTATION

In recent years, delivery of exogenous proteins into the central nervous system (CNS) has utilized a strategy of cellular and gene therapy. Researchers in this field have focused on finding a transplantable and transfectable cellular platform to serve as a local delivery system for gene products of therapeutic value *(1–3)*. A continuous secretion of the gene product may be necessary to achieve a sustained impact from the application of a gene product to affected regions of the CNS. This sustained delivery could be accomplished by transplantation of cells genetically engineered to express the therapeutic protein of interest *(4)*.

Promising laboratory findings in animal models of Parkinson's disease (PD) and Huntington's disease (HD), which were treated with neural transplantation strategies, have formed the scientific basis for clinical trials *(2,5)*. More than 350 PD, HD, and stroke patients have already received intracerebral neural transplantation. However, these patients have demonstrated variable degrees of clinical improvement, owing partly to the low viability of the grafts *(2,6–11)*. Because graft survival is greatly altered by the host immune response, cells that can avoid immunosurveillance, particularly autologous cells (e.g., the transplant recipient's adrenal or stem cells), may limit graft rejection *(12–18)*. Fetal cells persist as the most widely studied graft source for transplantation. Unfortunately, many logistical and ethical issues hinder the use of primary fetal cells in the clinic. Thus, a primary research endeavor in cell transplantation has concentrated on searching for a nonprimary fetal graft source. Various cells have been used as alternatives to primary fetal cells. These include cultured neuronal stem and progenitor cells, cells engineered to secrete neurotransmitters or neurotrophic factors (e.g., immortalized cell lines, fibroblasts, and astrocytes), paraneuronal cells that naturally synthesize neuronal substances and/or have neuron-like properties, and bridge-inducing cell grafts that assist in the physical reconstruction of lost axonal pathways. Typically, when intracerebrally transplanted, these cells have been shown to partially reconstruct the neuronal circuitry and form functional synapses *(19–23)*.

Although transplantation of primary fetal cells may promote the prolonged release of neuronal survival–promoting proteins, the use of cell lines or neural stem cells would be associated with less controversy. These cells are readily available, can be maintained in culture indefinitely, and can be sorted in a homogenous population with their phenotypic features fully characterized. Therefore, establishing human neuronal cell lines or neural stem cells as vehicles for cellular and gene therapy in CNS disorders could have great utility.

GENERATING NEURONS FROM CANCER CELLS

An embryonal carcinoma cell line (NT2 cells), derived from a human teratocarcinoma, can be induced to differentiate into postmitotic neuron-like cells, referred to as *NT2N neurons (1,24–26)*. During a 6-wk retinoic acid (RA) treatment period, NT2 cells, which share many characteristics of neuroepithelial precursor cells, cease expressing neuroepithelial markers and instead develop neuronal markers *(27,28)*. Subsequent exposure to mitotic inhibitors produces more than 99% pure populations of terminally differentiated NT2N neurons *(29)*. Furthermore, NT2N neurons exhibit outgrowth processes and establish functional synapses. Mature NT2N neurons are virtually indistinguishable from terminally differentiated postmitotic embryonic neurons. Notably, these neurons do not divide, and they maintain a neuronal phenotype over a long-term period *(28)*. NT2 cells, unlike other germ cell tumor lines, do not give rise to progeny committed to other well-defined neural or nonneural lineages in response to RA or any other differentiating agent *(27,28)*. Based on this unique property of NT2 cells, they are considered in vitro equivalents to CNS neuronal progenitor cells *(1,30)*. Interestingly, both NT2 cells and NT2N neurons can be genetically engineered, allowing the expression of a gene product of interest in vitro and possibly in vivo, and these applications can be exploited to reveal the cellular and molecular biology of neurons *(1,30)*. It is equally important that both NT2 and NT2N cells can be used as alternative graft sources for transplantation therapy in CNS disorders.

NORMAL HOST BRAIN MICROENVIRONMENT
AND CELL GRAFTS

Pioneering studies using NT2N neurons revealed that purified NT2N neurons survive, mature, and integrate well with the host nervous system following transplantation into the CNS of rodents *(24–26)*. From a developmental cell biology perspective, such a transplantation setup enables the direct examination of growth and maturation of human neuronal cells in an in vivo CNS environment that otherwise could not be fully investigated in an in vitro setting. Compared to human fetal neurons, the use of NT2N neurons offers many advantages. NT2N neurons appear to have a 15% better graft survival, excellent in vitro and in vivo grafted cell homogeneity, and a high degree of host reinnervation *(1,25,26)*. In support of the postmitotic feature of NT2N cells, histological examinations have revealed no observable tumorgenecity, as well as neoplasticity in NT2N cell grafts over prolonged transplantation periods of more than 1 yr *(24–26)*. Because the aforementioned studies used rodents as transplant recipients of NT2N human-derived cells, the observed cross-species graft tolerance required suppressing the host immune response. Surprisingly, some nonimmunosuppressed transplant recipients tolerated the grafts, depending on the transplant target brain area. This observation led to the belief that certain brain sites may be more conducive than others for NT2N cell transplantation.

The role of the host microenvironment in the proliferation and survival of grafted cells is a factor that is recognized but has not yet been fully elucidated. Although it has been established that NT2N cells attained features of fully differentiated neurons following treatment with RA and mitotic inhibitors, and these cells do not revert to a neoplastic state after transplantation, concerns abound regarding the possibility that quiescent mitotic capacity remains in grafted NT2N neurons that may be stimulated by the host microenvironment. Studies examining transplantation of NT2N neurons do not indicate that these grafted cells reacquire mitotic features, at least when they are transplanted into the striatum. In contrast, evidence exists that proliferation and survival of parent NT2 cells are affected by the host microenvironment *(24–26)*. The neoplastic potential of grafted NT2 cells has been explored by grafting these cells into different regions of the brains of subacute combined immunodeficient (SCID) mice and nude mice *(24–26)*. The anatomical site into which the NT2 cells were implanted significantly influenced the survival, proliferation, and differentiation of NT2 cells.

Histological results revealed that the NT2 cells continued to proliferate and undergo an apoptotic-like cell death, with minimal capacity to differentiate into neurons following implantation into the subarachnoid space and superficial neocortex. However, when NT2 cells were transplanted in the lateral ventricles, liver, and muscle, the grafted cells rapidly progressed into bulky and lethal tumors within 10 wk after transplantation. The observed tumorgeneic and neoplastic state of grafted NT2 cells was in sharp contrast to the phenotypic features displayed by NT2 cells transplanted into the caudoputamen of SCID mice. Caudoputaminal grafted NT2 cells stopped proliferating, showed no evidence of necrosis or apoptosis, and did not form tumors after more than 20-wk posttransplantation. Furthermore, neuronal phenotypic markers demonstrate that the majority of NT2 cell grafts in the caudoputamen differentiate into postmitotic immature, neuron-like cells. The marked differential histopathological effects produced by the caudoputamen and other brain transplant target sites suggest that the choice of host microenvironment for transplantation of NT2 cells critically influences the eventual survival, proliferation, and differentiation fate of grafted cells. These observations support the notion that the host microenvironment (in this case, the caudoputamen) may promote signaling molecules or other cues, such as cell–cell contacts, which are capable of regulating the fate of grafted NT2 cells. Importantly, RA has been present in both developing and adult rodent striatum tissues *(31)*, and RA is believed to potentially influence cell fate *(32,33)*. As mentioned above, RA and other mitotic inhibitors are primarily used as factors in the differentiation process of NT2 cells into NT2N in vitro. Accordingly, the presence of RA in the striatum likely influences the nontumorgenic fate of NT2 cells following transplantation into this brain area. Although the effects of the host microenvironment appear limited to NT2 cells, more in-depth examinations are warranted to investigate the influence of specific brain transplant target sites on survival, proliferation, and differentiation of NT2N neurons.

INJURED HOST BRAIN MICROENVIRONMENT AND CELL GRAFTS

Because progressive neurodegeneration ensues after the onset of many neurological disorders, as exemplified in stroke, critically identifying a conducive host microenvironment seems to be a prerequisite for successful cell transplantation therapy. In stroke animal models, the reported NT2N neuronal graft survival rate of 15% *(34)* is a bit higher, compared to fetal cell grafts, but this rate is still low, considering that ischemic stroke is not limited to a specific cell population. Moreover, stratified ischemic zones exist, i.e., the predominantly necrotic core and apoptotic penumbra.

To produce therapeutic effects following an ischemic stroke, either via pharmacologic treatment or cell replacement therapy, the consensus is to target the ischemic penumbra, rather than the core. Such preferential penumbral rescue is logical, because apoptotic cell death that accompanies the penumbra may be potentially reversed, as opposed to the necrosis inherent in the core. Targeting secondary cell death mechanisms, as in the case of ischemia-induced apoptosis, suggests that the ischemic penumbra seems to be a more conducive host microenvironment than the ischemic core.

Nonetheless, the brain damage that accompanies stroke, regardless of the location in the penumbra or core, is characterized by the degeneration of many cell populations and brain structures. Accordingly, a greater number of cells with high viability and increased survival ability must be transplanted into the ischemic regions. Despite the moderately conducive nature of the ischemic penumbra, the extensive brain area encompassing this region may require multiple brain targets to repair the damaged neuroanatomical circuitry. Multiple intracerebral transplantations may not be feasible, however, because of the trauma associated with such an invasive surgical procedure. In addition, different types of donor cells may have to be transplanted, considering that many cell populations are destroyed because of stroke. NT2N neurons have been shown to differentiate into dopaminergic and GABAergic neuron-like cells *(35,36)*. Alternatively, RA-naive NT2 cells may possess some multipotent properties, such as those attributed to neural stem cells, and these features may be potentially advantageous for generating different cell populations. Indeed, transplanted human neural stem cells can mature into the phenotype of cells that are undergoing cell death in the brains of animals that were introduced to neuronal injury *(37)*. Interestingly, NT2 cells can differentiate into neurons when transplanted into the caudoputamen. Thus, at least for focal caudoputaminal stroke, both NT2N and NT2 cell grafts may be beneficial. The pluripotent features of NT2 cells, and the highly differentiated neuronal-like characteristics of NT2N cells, may be directed toward specific stroke types to enhance their therapeutic effects. For example, a large striatal stroke may benefit from the multipotent NT2 cell grafts, whereas a highly localized striatal stroke may appeal to the differentiated NT2N cell grafts. These hypotheses have not yet been tested in the laboratory.

PRECLINICAL INTRACEREBRAL TRANSPLANTATION OF NT2N CELLS IN STROKE

The rodent model of middle cerebral artery occlusion replicates many pathophysiological changes seen in clinical cerebral ischemia, leading to studies of treatment strategies for stroke. The potential efficacy of neural transplantation of NT2N neurons to correct the neurobehavioral deficits associated with cerebral ischemia have been elucidated by the use of this model. Data have shown that ischemia-induced behavioral dysfunctions are ameliorated by NT2N neuronal grafts as early as 1-mo posttransplantation *(38–40).* Compared to the transplantation of fetal striatal cells, which were previously shown to reverse motor abnormalities in stroke rats, NT2N neuronal grafts induced a significantly greater robust recovery. These pioneering data provides justification for the use of NT2N neurons for transplantation therapy to circumvent the logistical and ethical concerns inherent with the use of fetal striatal cells.

Subsequent studies using this human neuronal cell line have revealed that pretransplantation viability and posttransplantation survival of NT2N neurons are highly correlated with the functional recovery of transplanted stroke animals *(39).* These observations suggest that the positive behavioral effects seen in transplanted stroke animals can be attributed to viable and functional NT2N neuronal grafts. During the pretransplantation period, NT2N neuron viability counts revealed a variable range of 52–95%. Within-subject comparisons of pretransplantation cell viability and subsequent behavioral changes in transplanted stroke animals revealed that a high cell viability just prior to transplantation surgery correlated highly with a robust and sustained functional improvement in transplant recipients. In addition, the histological analysis of grafted brains revealed a positive correlation between the number of surviving NT2N neurons and the degree of functional recovery. Similar correlational data regarding fetal tissue transplantation has been reported between pretransplantation viability, or posttransplantation survival, of grafted cells and behavioral outcome.

In support of the aforementioned positive correlations between motor recovery and neuronal regeneration in stroke animals, dose-dependent functional effects of NT2N neuronal grafts have also been indicated *(34)*. Stroke animals that received 40, 80, or 160×10^3 NT2N neurons dose-dependently exhibited performance improvements in both the passive avoidance and elevated body swing tests. Moreover, dose-dependent survival of NT2N neuronal grafts was observed; grafts of 80 or 160×10^3 NT2N neurons demonstrated a 12–15% survival of NT2N neurons in the graft, whereas grafts of 40×10^3 NT2N neurons demonstrated only a 5% survival. NT2N neuronal grafts may have been affected by progressive stroke, suggesting again the influence of the host microenvironment. Correlational analyses revealed that ischemic animals that received 80 or 160×10^3 NT2N neurons produced a significantly better amelioration of behavioral dysfunctions, compared to those that received lower dosages of NT2N neurons. In concert with the earlier speculation that varying extent of stroke brain damage would require manipulation of the number of donor cells, this study demonstrated that transplantation of more viable NT2N neurons is required to rescue larger stroke-induced brain damage *(34)*. Determining the optimal dosage of NT2N neurons for a given stroke case should be viewed in terms of NT2N cell viability at pre- and posttransplantation, as well as the extent of stroke brain areas.

The issue of NT2N neuron procurement feasibility arises when considering such devastating diseases as stroke. Because a very narrow therapeutic window exists following a stroke episode, the immediate availability of NT2N cells must be considered. Laboratory studies have shown that NT2N neuronal grafts still produce robust functional recovery at 1 mo poststroke *(38–40)*. Clinical trials also revealed that transplanted patients who had a stroke at least 6 mo prior to NT2N neuronal transplantation displayed some improvement. Although transplantation therapy could potentially reverse chronic stroke, better functional outcomes may be achieved if treatment is initiated acutely (<3 h) poststroke, when brain damage would presumably still be limited. Thus, a strategy incorporating immediate availability and transplantation of NT2N neurons is deemed more effective for stroke therapy.

Before advancing NT2N cell transplantation in stroke, another important factor that should be closely examined in the laboratory is the severe inflammatory glial response that accompanies the early stages of the disease. Controversy surrounds whether such a response works for or against graft survival *(41,42)*. For example, during the early poststroke period, cytokines and inflammatory signals are highly elevated *(43,44)*, which could be harmful to grafted cells. Therefore, transplantation in acute stroke could be detrimental to grafted cells. However, evidence also indicates that chemo-attractants may be produced by glial cells or macrophages following stroke, and these cues may guide grafted cells to the site of injury *(45–47)*. These data suggest that early transplantation at a stage of high glial response may aid in cell graft trafficking toward appropriate stroke target sites. Accordingly, the glial response/inflammation may produce both inhibitory and facilitatory effects during the early period after stroke, particularly impacting the migration of grafted cells, and these effects warrant further investigation. Yet, in studies that used the chronic stroke model, NT2N cell grafts have been shown to migrate away from the original transplant site *(38–40)*. Stimulating NT2N cells to migrate in acute and chronic stroke may be a challenge, considering the large extent of brain damage inherent in the disease. Modulating the glial response/inflammation may be beneficial to NT2N cell graft migration.

Because handling freshly cultured NT2N neurons would not be altogether feasible in the clinic, cryopreserved NT2N neurons are recommended. In the laboratory setting, cryopreservation of NT2N neurons did not produce any significant deleterious effects on cell viability prior to, or after, transplantation in animals that suffered from stroke *(39)*. The sustained viability of NT2N neurons following cryopreservation fares much better than fetal cells, which display a significant decline in viability after cryopreservation, rendering them nontransplantable. Cryopreservation of NT2N neurons therefore allows immediate cell transplantation in acute stroke. Whether enhanced functional outcomes can be achieved with an acute stroke transplantation regimen needs to be examined in the laboratory.

NT2N CELL GRAFTS AND IMMUNOSUPPRESSION

Although the brain is perceived as an immunoprivileged site, neuronal graft rejection still persists. Indeed, the sustained motor and cognitive improvements noted in NT2N neuronal transplanted animals were only observed with systemic administration of cyclosporin A (CsA) immunosuppression *(38–40)*. In contrast, behavioral recovery in the nonimmunosuppressed animals transplanted with NT2N neurons began to diminish by about 2-mo posttransplantation. Moreover, histological analysis revealed surviving NT2N cells in the brains of immunosuppressed transplanted animals but not in nonimmunosuppressed transplanted animals. The near absence of visible grafts in nonimmunosuppressed animals transplanted with NT2N neurons suggests host immunological rejection of the grafts, as observed previously *(48)*. Nonetheless, the magnitude of the behavioral recovery produced by NT2N cell grafts in rats that did not receive CsA was greater than that seen in animals with ischemia-induced brain injury, followed by injections of rat fetal cerebellar cells or medium alone. This suggests that NT2N cell grafts promote behavioral effects at early time periods posttransplantation, even in the absence of immunosuppression. However, these nonimmunosuppressed animals, despite displaying significant improvements when compared to control animals at 6-mo posttransplantation, still remained impaired versus immunosuppressed animals that received NT2N cells. Accordingly, immunosuppression with CsA enhanced the survival of grafted NT2N cells; this finding is consistent with a previous study *(48)*. Furthermore, histological examinations revealed many surviving NT2N cells in immunosuppressed transplanted animals that exhibited a robust functional recovery for more than 6-mo posttransplantation. Based on these results, the need for chronic immunosuppressive therapy as an adjunct to the transplantation of human NT2N cells in rats appears necessary to obtain optimal and sustained functional improvement, as well as prolonged graft survival. Interestingly, no evidence from these studies, or any other previous reports, has been found to demonstrate that transplanted NT2N cells, with or without immunosuppression, have any deleterious effects on the host brain *(24–26)*.

However, in future clinical trials involving neural transplantation of NT2N cells, chronic immunosuppression, which is normally required for successful xenografting, may not be necessary, because NT2N neurons are human-derived cells. Some studies of human fetal cell transplantation for PD have, in fact, found that the absence of immunosuppression does not deleteriously affect the survival of fetal cell grafts and their ability to produce clinical improvement *(49,50)*. In addition, preliminary data suggest that NT2N neurons may have immunosuppressive properties *(51)*. Recent studies have implied that stem cells only minimally elicit an immune response and may even secrete their own immunosuppressant factors following intracerebral transplantation *(37)*. Similar immunosuppressant factors may also be secreted by NT2N cells, which may exert localized immunosuppression within the transplant site, allowing them to circumvent host immunosurveillance. Thus, long-term systemic immunosuppression may not be necessary in humans. Nevertheless, because recent studies have indicated that immunosuppressants and their analogs exert neuroprotective effects *(2,41)*, adjunct immunosuppression with NT2N cell transplantation could be considered for enhanced graft survival and functional effects.

SPECULATIVE MECHANISMS UNDERLYING FUNCTIONAL EFFECTS OF NT2N CELL GRAFTS

The observation of acute behavioral effects in nonimmunosuppressed NT2N transplanted stroke animals suggests that the neurotrophic effects of transplanted NT2N cells may have mediated functional recovery, at least during the early posttransplantation period. This finding supports neuronal rescue via neurotrophic factor therapy. In many preclinical and clinical studies of neural transplantation, the use of neurotrophic factors has been shown to significantly enhance the survival rate of grafted cells *(5,52,53)*. Direct infusion of neurotrophic factors alone, or their use as a transplant facilitator by pretreating donor cells or coadministration during and after neural transplantation therapy, has been efficacious in CNS animal models. Hence, administration of neurotrophic factors may serve as another adjunct to neural transplantation. One of the most potent neurotrophic factors is the glial-cell line–derived neurotrophic factor (GDNF). Encouraging laboratory results have been reported in neural transplantation of GDNF-secreting fetal kidney cells for stroke *(54,55)*. Interestingly, an in vitro study has demonstrated that NT2N cells respond positively to putative neurotrophic factors secreted by an immortalized human fetal astrocyte cell line *(56)*.

The functional improvement seen in stroke animals was initially ascribed to neurotrophic effects of NT2N cells to the injured area *(38)*. However, this study did not demonstrate any direct evidence that neuroprotection was indeed a function of NT2N neuronal grafts. The first suggestion that neuroprotection by NT2N neurons could be mediated by a neurotrophic factor mechanism was reported recently in a study showing that NT2N neurons are positive for GDNF mRNA *(57)*. Because GDNF has been found to be neuroprotective for stroke animals, the indication that NT2N neurons can exert GDNF expression offers a mechanistic explanation for the observed neuroprotection by NT2N neuronal grafts in stroke. In fact, although a similar pattern of behavioral recovery was seen in animals that received NT2N neurons, as well as in those that received fetal striatal transplants, the NT2N-transplanted animals showed a more robust recovery at 1-mo posttransplantation. This effect of NT2N neurons was also evident in transplanted nonimmunosuppressed animals. Considering no evidence has been reported of neural transplants replacing lost host brain tissue at this early posttransplantation stage, the observed functional effects may be owing to the release of trophic factors from the grafted NT2N neurons. Another indication of trophic factors mediating NT2N's action is that the effective dose of transplanted NT2N neurons required to produce functional recovery was 10 times less than that of transplanted striatal cells. Therefore, at the early posttransplantation period, connectivity with, or repair of, the stroke brain by NT2N neurons does not account for the functional recovery; instead, it is possible that trophic factors secreted by NT2N neurons enable stroke animals to display functional improvement.

At a later posttransplantation period, an alternative mechanism may underlie the behavioral recovery produced by NT2N neuronal grafts; NT2N cells might have replaced the degenerated host brain cells. After transplantation into nude mice, NT2N cells can integrate and change the phenotype into neurons similar to the target neurons, such as striatal neurons (24–26,28). NT2N neurons can become striatal-like neurons and may also be capable of secreting neurochemicals or even performing functions of lost striatal cells of the host brain. Indeed, NT2N cells can be stimulated through the application of neurotrophic factors, such as acidic fibroblast growth factor and activating factors (e.g., catecholamines or forskolin) to express the rate-limiting enzyme in catecholamine biosynthesis: tyrosine hydroxylase (TH; 58,59). Notably, in vitro studies have shown that the percentage of TH-positive neuron-like cells in the NT2N cells treated with RA, cocultured with striatal extracts, exceeded the percentage of TH-positive cells induced in sister cultures exposed to RA alone by more than 10-fold (24–26). Interestingly, many behavioral dysfunctions seen in a stroke model are dopamine-mediated behaviors (38–40). Thus, the possibility that NT2N cell grafts can be induced by the host microenvironment, particularly the remaining host striatal neurons or the whole host striatum itself, to secrete dopamine would contribute greatly to the amelioration of ischemia-induced behavioral deficits.

The host microenvironment of the adult mouse striatum appears to have the potential ability to induce grafted NT2 cells to differentiate progressively into fully mature, adult CNS neurons (24–26). The striatum may exert similar neuronal differentiation effects on grafted NT2N neurons. This differentiation of NT2N following transplantation is important, especially if it is necessary to target specific disease types that entail degeneration of different cell populations. For example, subsets of NT2N cells have been shown to express neuronal markers for dopaminergic and GABAergic neurons (60,61), which would be an appropriate cell graft source for PD and HD, respectively. Because multiple cell populations are affected by stroke, the ability of NT2N cells to differentiate into many cell types will be advantageous. Furthermore, if it is possible to recreate the microenvironment characteristic of the striatum in other brain areas (e.g., the cortex or hippocampus, which are also damaged in stroke), then this strategy could extend the efficacy of NT2N cell transplantation to a variety of stroke types and other neurological disorders. Multiple mechanisms may mediate the neuro-behavioral benefits produced by NT2N neuronal grafts, and these mechanisms warrant further research.

TRANSPLANTATION OF NT2N CELLS IN STROKE PATIENTS

The preclinical studies (see above) demonstrating successful implantation of human-derived NT2N neurons into rat brains paved the way for limited clinical trials. The target stroke patients chosen were at a chronic stage because laboratory data indicated the possibility of reversing motor symptoms associated with a stable stroke. The Food and Drug Administration approved phase I clinical trials of transplantation of NT2N neurons to evaluate this therapy in patients with stable stroke. NT2N cells were transplanted into patients with basal ganglia stroke and fixed motor deficits, including 12 patients ages 44–75 yr, with an infarct of 6 mo to 6 yr who were stable for at least 2 mo *(62)*. Serial evaluations at 12–18 mo showed no adverse cell-related serologic or imaging-defined effects. These results suggest that transplantation of NT2N cells is feasible in patients with motor infarction.

The intracranial transplantation of certain stem cell lines has been shown to induce tumor formation when transplants were targeted to the cortex *(63)*. However, no evidence was found of neoplastic formation from NT2N transplants into the striatum *(24)*. The presence of RA in the striatum may have aided in the suppression of tumor formation following NT2 cell grafts and may also have facilitated further differentiation of NT2N cell grafts into neuronal lineages. Consistent with this evidence, the above report demonstrated no serious adverse events in transplanted stroke patients at 1 yr posttransplantation. Thus, it appears that grafted NT2N cells do not exhibit neoplasticity, thereby preventing any tumor formation. However, considering the intrastraital transplantation of NT2N cells has been the preclinical and clinical method that has demonstrated a consistent nontumorgenic outcome following NT2N cell grafts, extending the transplant target sites to other brain areas outside the striatum must be done with caution. At this time, future transplantation trials should be limited to targeting the striatum, focusing only on striatal stroke patients. A strategy needs to be developed that makes the graft material nonresponsive to tumor formation cues from the host microenvironment or that suppresses the host microenvironment from releasing these signals before proceeding with extrastriatal NT2N cell transplantation.

Additional challenges in the field of neural transplantation include demonstration of graft viability and functional effects. A subsequent clinical report evaluated the function of NT2N-transplanted cells using positron emission tomography (PET) *(64)*. Uptake of fluorodeoxyglucose (FDG) was measured at baseline and at 6 and 12 months after transplantation of NT2N neurons. At 6-mo posttransplantation, 7 of 11 patients showed more than 10% increase in FDG uptake in the area of cell implantation; this increase correlated with clinical improvement, as measured by stroke scale values. In a recent study that reported the first postmortem brain in an NT2N-transplanted patient at 2 yr posttransplantation *(65)*, histological examination revealed neurofilament immunoreactive neurons resembling those seen in NT2N neurons in vitro. The observed NT2N cell graft survival in this patient suggests that these transplanted cells mediated functional outcome. The PET and histological data from transplanted stroke patients allow some comparisons with long-term graft survival of fetal ventral mesencephalic neurons in PD patients who also died from causes unrelated to the transplants. In both stroke and PD patients, robust graft survival was seen using PET scans and was accompanied by the expression of neuronal phenotypes in grafted cells postmortem. No overt side effects from the transplants were observed, indicating that the grafts did not exacerbate disease progression. These parallel clinical outcomes seen in stroke and PD patients support the use of NT2N cells as an efficacious alternative to fetal cells.

The optimal number of transplanted cells necessary to achieve improvements in functional outcome remains to be determined. In the previous experiment, there was no observed difference in outcome between transplantation of 2 million or 6 million cells. Interestingly, increased uptake of FDG persisted in only three patients at the 12-mo posttransplant evaluation *(64)*. However, clinical improvement was maintained in six of the seven patients who showed an initial increase in FDG uptake. This finding suggests that improvement in the clinical examination may be mediated by a factor that requires only functional effects of transplanted cells during a critical time period; even a limited number of NT2N cells may promote some degree of functional recovery. This notion of minimally required viable cell grafts for functional effects is also true for fetal cell transplantation, at least in PD, where it has been suggested that as few as 300 dopaminergic neurons could exert behavioral recovery *(66)*. However, because recurrent stroke episodes may likely ensue following the initial stroke, transplanting more viable NT2N cells may be required for long-term improvement. Transplantation of a higher number of viable cells may also be needed for other neurological disorders, including PD, if grafted cells are affected by ongoing neurodegeneration.

One advantage of transplanting NT2N neurons into humans may be the circumvention of host immunosuppression. Although transplantation of human NT2N neurons appeared to be well tolerated by immunosuppressed rats, these human-derived cells may not require immunosuppression in stroke patients. Furthermore, because these are cultured cells, examination for possible infectious diseases can be performed ahead of the scheduled transplant surgery. Therefore, a more efficient transplantation protocol can be achieved with the use of these cells, compared to using fetal cells.

In addition to avoiding the ethical concerns surrounding the use of fetal cells, transplantation of such clonal cells as NT2N neurons allows a logistical advantage of conducting neural transplantation in a wider therapeutic window after stroke. As discussed above, the success of treating cerebral ischemia depends highly on the timing of intervention; thus, the ready availability of clone cells as a graft source would significantly reduce the time between the ischemic event and the therapeutic intervention. Nonetheless, the robust recovery of animals with a stable stroke following transplantation of NT2N neurons suggests the possibility of treating stroke patients, even with a long delay after a stroke episode.

Many reservations must still be acknowledged regarding this pioneering clinical trial of NT2N neuron transplantation *(67)*. The preceding clinical trial is an open-label study and was therefore not designed to prove efficacy. At best, the results from this phase I study revealed that NT2N neuronal transplantation is feasible and safe. The follow-up studies suggest no malignant tumor formation over moderate posttransplantation periods of 2 yr. Continuous monitoring of the transplanted patients at longer time periods is in progress and should reveal further safety issues associated with the therapy. To optimize the protocol to achieve effective and consistent improved clinical outcomes, carefully designed laboratory studies and limited clinical trials should be considered. Subsequent research may determine the optimal number of cells needed to reach significant improvement and should help the development of imaging techniques that would allow characterization of grafted cells to assess viability, migration, differentiation, and graft–host integration over the NT2N graft maturation period.

Table 1
Summary of N2T Cell/NT2N Neuron Transplantation Studies

Study	References
(1) Purified NT2N neurons survive, mature, and integrate well with the host nervous system following transplantation into the CNS of rodents	(24–26)
(2) NT2N neurons have better graft survival than human fetal neurons (15%), excellent in vitro and in vivo grafted cell homogeneity, and a high degree of host reinnervation; prolonged NT2N cell grafts have no observable tumorgenecity or neoplasticity	(24–26,63)
(3) Proliferation and survival of parent NT2 cells affected by host microenvironment; anatomical site into which NT2 cells are implanted significantly influences survival, proliferation, and differentiation of NT2 cells; NT2 cells survive and differentiate into neurons when transplanted into the caudoputamen	(24–26)
(4) NT2N neurons differentiate into dopaminergic and GABAergic neuron-like cells	(35) Kondo et al., 1997
(5) Ischemia-induced behavioral dysfunctions ameliorated by NT2N neuronal grafts as early as 1-mo posttransplantation, pretransplantation viability and posttransplantation survival of NT2N neurons correlates highly with functional recovery of transplanted stroke animals	(38–40)
(6) Functional effects and survival of NT2N neuronal grafts are dose-dependent; transplantation of more viable NT2N neurons is required to rescue larger stroke-induced brain damage	(34)
(7) NT2N neuronal grafts still produce robust functional recovery at 1-mo poststroke	(38–40)
(8) Cytokines and inflammatory signals are highly elevated at early poststroke period, which could be harmful to grafted cells	(43–44)
(9) Chemoattractants produced by glial cells/macrophages following stroke may guide grafted cells to site of injury	Chopp 1999, 2000, 2001
(10) NT2N cell grafts migrate away from original transplant site in chronic stroke model	(38–40)
(11) Cryopreservation of NT2N neurons has no deleterious effects on cell viability prior to and after transplantation in stroke animals	(39)

(12) Sustained motor and cognitive improvements and cell survival in NT2N neuronal transplanted animals only observed with systemic CsA immunosuppression (38–40)

(13) Host rejection of NT2N grafts in nonimmunosuppressed animals. These animals significantly improve, compared to control animals at 6-mo posttransplantation but remain impaired vs immunosuppressed animals that receive NT2N cells; immunosuppression with CsA enhances graft survival (48)

(14) No evidence that transplanted NT2N cells with or without immunosuppression have deleterious effects on host brain; NT2N neurons may have immunosuppressive properties (24–26,51)

(15) Functional improvement in stroke animals ascribed to neurotrophic effects of NT2N cells. Neuroprotection by NT2N neurons could be mediated by a neurotrophic factor mechanism; NT2N neurons have GDNF mRNA (38,57)

(16) After transplantation into nude mice, NT2N cells can integrate and change phenotype into neurons similar to target neurons. Percentages of TH-positive neuron-like cells in NT2N cells treated with RA, cocultured with striatal extracts, exceeds percentage of TH-positive cells induced in sister cultures exposed to RA alone. Host microenvironment of adult mouse striatum has potential ability to induce grafted NT2 cells to differentiate progressively into fully mature adult CNS neurons (24–26,28)

(17) NT2N cells can be stimulated through application of neurotrophic factors and activating factors to express TH. Subsets of NT2N cells express neuronal markers for dopaminergic and GABAergic neurons (58–61)

(18) NT2N cells transplanted in patients with basal ganglia stroke and fixed motor deficits; serial evaluations show no adverse cell-related serologic or imaging-defined effects (62)

(19) No evidence of neoplastic formation from NT2N transplants into striatum. No serious adverse events in transplanted stroke patients at 1-yr posttransplantation (24)

(20) PET and histological data from NT2N-transplanted stroke patients shows uptake increase in FDG in area of cell implantation and neurofilament immunoreactive neurons that resembled those seen in NT2N neurons in vitro. Clinical improvement was maintained in six of seven patients who showed initial increase in uptake of FDG (64,65)

CONCLUSION

Although the transplantation of NT2N cells has great potential for therapeutic efficacy in CNS disorders, concerns still remain regarding NT2N transplantation in a clinical setting. One recent study that raised such concerns found that embryonic cortical neurons and NT2N have different network properties. Neurons derived from the human NT2 cell line were found to form networks with a clustered neuritic architecture in vitro, whereas primary dissociated embryonic rat cortical neurons displayed a more homogenous cell assembly. Also, NT2N neurons showed a mostly uncorrelated firing pattern, in contrast to the primary dissociated embryonic rat cortical neurons that displayed highly synchronized bursting. These findings bring forth additional issues that need to be considered before NT2N neurons are used clinically in CNS grafting *(68)*.

Although some valid concerns remain regarding NT2N cell transplantation, many recent studies have made further progress in demonstrating the efficacy of NT2N cell transplants and elucidating the mechanisms that underlie the effects of these grafts (Table 1). One recent study has shown that defined populations of genetically modified human NT2N neurons are practical and effective for stable ex vivo gene delivery into the CNS. This study successfully displayed stable, efficient, and nontoxic gene transfer into undifferentiated NT2 cells using a pseudotyped lentiviral vector. NT2 cells were differentiated into NT2N neurons via treatment with RA, then transplanted into the striatum of adult nude mice. Transduced NT2N neurons survived and continued to express the reporter gene for long-term time points in vivo. Transplantation of NT2N neurons that were genetically modified to express nerve growth factor also significantly attenuated cognitive dysfunction following traumatic brain injury in mice *(69)*.

Another recent study examined whether lithium treatment of NT2N neurons increases TH expression when cells are transplanted into the striatum of hemiparkinsonian rats. The histological analysis indicated a much better cell survival in the group treated briefly with lithium, thus providing an option for enhancing NT2N graft survival in future transplantation studies *(70)*.

In addition, NT2N neurons induced by RA were found to express the Nurr1 receptor, which has been shown to be essential for the development, differentiation, and survival of midbrain dopamine neurons. This study also confirmed the coexpression of Nurr1 and TH in NT2N neurons. These findings suggest that Nurr1 may be important during the development of NT2N neurons and could also be involved in their differentiation into the dopaminergic phenotype *(71)*.

The use of NT2N cells for transplantation has also been evaluated recently in the laboratory to ameliorate spinal cord injury. The histological data from one such study revealed that graft survival in rats receiving transplants was displayed in 66.7% of the surviving grafted animals. Fiber outgrowth was observed in both rostral and caudal directions bridging the lesion. These results suggest that NT2N grafts may be able to structurally reconnect the proximal and distal spinal cord across the region of injury, thus presenting the future possibility of extending the clinical use of these transplants to spinal cord injury *(72)*.

A laboratory study found that transplantation of NT2N neurons could be an effective means of reestablishing electrical connectivity in the injured spinal cord. Rats were given a complete spinal cord contusion injury, producing a complete loss of motor-evoked potentials, then selected rats underwent transplantation with NT2N cells within the contusion site either immediately after injury or at a delayed point 2 wk following injury. Rats receiving delayed transplants exhibited a significant functional recovery, seen by the return of motor-evoked potentials, as well as a modest improvement of motor function, again implying the capability of NT2N cell transplantation to restore function in spinal cord injury *(73)*.

REFERENCES

1. Trojanowski, J. Q., Kleppner, S. R., Hartley, R. S., et al. (1997) Transfectable and transplantable postmitotic human neurons: a potential "platform" for gene therapy of nervous system diseases. *Exp. Neurol.* **144,** 92–97.
2. Borlongan, C. V., Sanberg, P. R., and Freeman, T. B. (1999) Neural transplantation for neurodegenerative disorders. *Lancet* **353,** SI29–SI30.
3. Nishino, H. and Borlongan, C. V. (2000) Restoration of function by neural transplantation in the ischemic brain. *Prog. Brain Res.* **127,** 461–476.
4. Johnston, R. E., Dillon-Carter, O., Freed, W. J., and Borlongan, C. V. (2001) Trophic factor secreting kidney cell lines: in vitro characterization and functional effects following transplantation in ischemic rats. *Brain Res.* **900,** 268–276.
5. Alexi, T., Borlongan, C. V., Faull, R. L., et al. (2000) Neuroprotective strategies for basal ganglia degeneration: Parkinson's and Huntington's diseases. *Prog. Neurobiol.* **60,** 409–470.
6. Freeman, T. B., Cicchetti, F., Hauser, R. A., et al. (2000) Transplanted fetal striatum in Huntington's disease: phenotypic development and lack of pathology. *Proc. Natl. Acad. Sci.* **97,** 13877–13882.
7. Kordower, J. H. and Sortwell, C. E. (2000) Neuropathology of fetal nigra transplants for Parkinson's disease. *Prog. Brain Res.* **127,** 333–344.
8. Bjorklund, A. (1992) Dopaminergic transplants in experimental parkinsonism: cellular mechanisms of graft-induced functional recovery. *Curr. Opin. Neurobiol.* **2,** 683–689.

9. Mahalik, T. J., Hahn, W. E., Clayton, G. H., and Owens, G. P. (1994) Programmed cell death in developing grafts of fetal substantia nigra. *Exp. Neurol.* **129,** 27–36.
10. Nikkhah, G., Bentlage, C., Cunningham, M. G., and Bjorklund, A. (1994) Intranigral fetal dopamine grafts induce behavioral compensation in the rat Parkinson model. *J. Neurosci.* **14,** 3449–3461.
11. Nikkhah, G., Cunningham, M. G., Jodicke, A., et al. (1994) Improved graft survival and striatal reinnervation by microtransplantation of fetal nigral cell suspensions in the rat Parkinson model. *Brain Res.* **633,** 133–143.
12. Freed, W. J., Adinol, A. M., Laskin, J. D., and Geller, H. M. (1989) Transplantation of B16/C3 melanoma cells into the brains of rats and mice. *Brain Res.* **485,** 349–362.
13. Kawaja, M. D., Fagan, A. M., Firestein, B. L., and Gage, F. H. (1991) Intracerebral grafting of cultured autologous skin fibroblasts into the rat striatum: an assessment of graft size and ultrastructure. *J. Comp. Neurol.* **307,** 695–706.
14. Nishino, H., Hashitani, T., and Kumazaki, M. (1990) Phenotypic plasticity of grafted catecholaminergic cells in the dopamine-depleted caudate nucleus in the rat. *Neurosci. Res. Suppl.* **13,** S54–S60.
15. Schueler, S. B., Ortega, J. D., Sagen, J., and Kordower, J. H. (1993) Robust survival of isolated bovine adrenal chromaffin cells following intrastriatal transplantation: a novel hypothesis of adrenal graft viability. *J. Neurosci.* **13,** 4496–4510.
16. Emerich, D. F., Winn, S. R., Christenson, L., et al. (1992) A novel approach to neural transplantation in Parkinson's disease: use of polymer-encapsulated cell therapy. *Neurosci. Biobehav. Rev.* **16,** 437–447.
17. Studer, L., Tabar, V., and McKay, R. D. (1998) Transplantation of expanded mesencephalic precursors leads to recovery in parkinsonian rats. *Nat. Neurosci.* **1,** 290–295.
18. Svendsen, C. N., Caldwell, M. A., Shen, J., et al. (1997) Long-term survival of human central nervous system progenitor cells transplanted into a rat model of Parkinson's disease. *Exp. Neurol.* **148,** 135–146.
19. Freund, T. F., Bolam, J. P., Bjorklund, A., et al. (1985) Efferent synaptic connections of grafted dopaminergic neurons reinnervating the host neostriatum: a tyrosine hydroxylase immunocytochemical study. *J. Neurosci.* **5,** 603–616.
20. Mahalik, T. J., Finger, T. E., StroÈmberg, I., and Olson, L. (1985) Substantia nigra transplants into denervated striatum of the rat: ultrastructure of graft and host interconnections. *J. Comp. Neurol.* **240,** 60–70.
21. Faull, R. L. M., Waldvogel, H. J., Nicholson, L. F. B., et al. (1995) Huntington's disease and neural transplantation: GABAA receptor changes in the basal ganglia in Huntington's disease, in the human brain and in the quinolinic acid lesioned rat model of the disease following fetal neuron transplants. In *Neurotransmitters in the Human Brain* (Tracey, D. J., ed.), Plenum Press, New York, pp. 173–197.
22. Hoffer, B. and Olson, L. (1997) Treatment strategies for neurodegenerative diseases based on trophic factors and cell transplantation techniques. *J. Neural. Transm. Suppl.* **49,** 1–10.

23. Isacson, O., Costantini, L., Schumacher, J. M., et al. (2001) Cell implantation therapies for Parkinson's disease using neural stem, transgenic or xenogeneic donor cells. *Parkinsonism Relat. Disord.* **7**, 205–212.
24. Kleppner, S. R., Robinson, K. A., Trojanowski, J. Q., and Lee, V. M. (1995) Transplanted human neurons derived from a teratocarcinoma cell line (NTera-2) mature, integrate, and survive for over 1 year in the nude mouse brain. *J. Comp. Neurol.* **357**, 618–632.
25. Miyazono, M., Lee, V. M., and Trojanowski, J. Q. (1995) Proliferation, cell death, and neuronal differentiation in transplanted human embryonal carcinoma (NTera2) cells depend on the graft site in nude and severe combined immuno-deficient mice. *Lab. Invest.* **73**, 273–283.
26. Miyazono, M., Nowell, P. C., Finan, J. L., et al. (1996) Long-term integration and neuronal differentiation of human embryonal carcinoma cells (NTera-2) transplanted into the caudoputamen of nude mice. *J. Comp. Neurol.* **376**, 603–613.
27. Pleasure, S. J., Page, C., and Lee, V. M. (1992) Pure, postmitotic, polarized human neurons derived from NTera 2 cells provide a system for expressing exogenous proteins in terminally differentiated neurons. *J. Neurosci.* **12**, 1802–1815.
28. Pleasure, S. J. and Lee, V. M. (1993) NTera 2 cells: a human cell line which displays characteristics expected of a human committed neuronal progenitor cell. *J. Neurosci. Res.* **35**, 585–602.
29. Lee, V. M. and Andrews, P. W. (1986) Differentiation of NTERA-2 clonal human embryonal carcinoma cells into neurons involves the induction of all three neurofilament proteins. *J. Neurosci.* **6**, 514–521.
30. Lee, V. M., Hartley, R. S., and Trojanowski, J. Q. (2000) Neurobiology of human neurons (NT2N) grafted into mouse spinal cord: implications for improving therapy of spinal cord injury. *Prog. Brain Res.* **128**, 299–307.
31. Zetterstrom, R. H., Simon, A., Giacobini, M. M., et al. (1994) Localization of cellular retinoid-binding proteins suggests specific roles for retinoids in the adult central nervous system. *Neuroscience* **62**, 899–918.
32. Faiella, A., Zappavigna, V., Mavilio, F., and Boncinelli, E. (1994) Inhibition of retinoic acid-induced activation of 3' human HOXB genes by antisense oligonucleotides affects sequential activation of genes located upstream in the four HOX clusters. *Proc. Natl. Acad. Sci. USA* **91**, 5335–5339.
33. Faiella, A., Wernig, M., Consalez, G. G., et al. (2000) A mouse model for valproate teratogenicity: parental effects, homeotic transformations, and altered HOX expression. *Hum. Mol. Genet.* **9**, 227–236.
34. Saporta, S., Borlongan, C. V., and Sanberg, P. R. (1999) Neural transplantation of human neuroteratocarcinoma (hNT) neurons into ischemic rats. A quantitative dose-response analysis of cell survival and behavioral recovery. *Neuroscience* **91**, 519–525.
35. Zigova, T., Barroso, L. F., Willing, A. E., et al. (2000) Dopaminergic phenotype of hNT cells in vitro. *Brain Res. Dev. Brain Res.* **122**, 87–90.
36. Matsuoka, T., Kondoh, T., Tamaki, N., and Nishizaki, T. (1997) The GABAA receptor is expressed in human neurons derived from a teratocarcinoma cell line. *Biochem. Biophys. Res. Commun.* **237**, 719–723.

37. Flax, J. D., Aurora, S., Yang, C., et al. (1998) Engraftable human neural stem cells respond to developmental cues, replace neurons, and express foreign genes. *Nat. Biotechnol.* **16,** 1033–1039.
38. Borlongan, C. V., Tajima, Y., Trojanowski, J. Q., et al. (1998) Transplantation of cryopreserved human embryonal carcinoma-derived neurons (NT2N cells) promotes functional recovery in ischemic rats. *Exp. Neurol.* **149,** 310–321.
39. Borlongan, C. V., Tajima, Y., Trojanowski, J. Q., et al. (1998) Cerebral ischemia and CNS transplantation: differential effects of grafted fetal rat striatal cells and human neurons derived from a clonal cell line. *Neuroreport* **9,** 3703–3709.
40. Borlongan, C. V., Saporta, S., Poulos, S. G., et al. (1998) Viability and survival of hNT neurons determine degree of functional recovery in grafted ischemic rats. *Neuroreport* **9,** 2837–2842.
41. Borlongan, C. V. (2000) Transplantation therapy for Parkinson's disease. *Expert Opin. Investig. Drugs* **9,** 2319–2330.
42. Chang, C. F., Morales, M., Chou, J., et al. (2002) Bone morphogenetic proteins are involved in fetal kidney tissue transplantation-induced neuroprotection in stroke rats. *Neuropharmacology* **43,** 418–426.
43. Carroll, J. E., Hess, D. C., Howard, E. F., and Hill, W. D. (2000) Is nuclear factor-kappaB a good treatment target in brain ischemia/reperfusion injury? *Neuroreport* **11,** 1–4.
44. Hill, W. D., Hess, D. C., Carroll, J. E., et al. (2001) The NF-kappaB inhibitor diethyldithiocarbamate (DDTC) increases brain cell death in a transient middle cerebral artery occlusion model of ischemia. *Brain Res. Bull.* **55,** 375–386.
45. Mahmood, A., Lu, D., Wang, L., et al. (2001) Treatment of traumatic brain injury in female rats with intravenous administration of bone marrow stromal cells. *Neurosurgery* **49,** 1196–1203.
46. Li, Y., Chen, J., and Chopp, M. (2002) Cell proliferation and differentiation from ependymal, subependymal and choroid plexus cells in response to stroke in rats. *J. Neurol. Sci.* **193,** 137–146.
47. Chen, J., Li, Y., Wang, L., et al. (2001) Therapeutic benefit of intravenous administration of bone marrow stromal cells after cerebral ischemia in rats. *Stroke* **32,** 1005–1011.
48. Trojanowski, J. Q., Mantione, J. R., Lee, J. H., et al. (1993) Neurons derived from a teratocarcinoma cell line establish molecular and structural polarity following transplantation into the rodent brain. *Exp. Neurol.* **122,** 283–294.
49. Henderson, B. T. H., Clough, C. G., Hughes, R. C., et al. (1991) Implantation of human ventral mesencephalon to the right caudate nucleus in advanced Parkinson's disease. *Arch. Neurol.* **48,** 822–827.
50. Freed, C. R., Breeze, R. E., Rosenberg, N. L., et al. (1992) Survival of implanted fetal dopamine cells and neurologic improvement 12 to 46 months after transplantation for Parkinson's disease. *N. Engl. J. Med.* **327,** 1549–1555.
51. Sarnacki, P. G., Engelman, R. W., Chang, Y., et al. (1998) Immunosuppression by hNT neurons and supernatant. *Exp. Neurol.* **153,** 386.
52. Connor, B. and Dragunow, M. (1998) The role of neuronal growth factors in neurodegenerative disorders of the human brain. *Brain Res. Rev.* **27,** 1–39.

53. Hughes, P. E., Alexi, T., Walton, M., et al. (1999) Activity and injury-dependent expression of inducible transcription factors, growth factors, and apoptosis-related genes within the central nervous system. *Prog. Neurobiol.* **57,** 421–450.

54. Tomac, A. C., Grinberg, A., Huang, S. P., et al. (2000) Glial cell line-derived neurotrophic factor receptor alpha1 availability regulates glial cell line-derived neurotrophic factor signaling: evidence from mice carrying one or two mutated alleles. *Neuroscience* **95,** 1011–1023.

55. Chiang, Y. H., Lin, S. Z., Borlongan, C. V., et al. (1999) Transplantation of fetal kidney tissue reduces cerebral infarction induced by middle cerebral artery ligation. *J. Cereb. Blood Flow Metab.* **19,** 1329–1335.

56. Tornatore, C., Baker-Cairns, B., Yadid, G., et al. (1996) Expression of tyrosine hydroxylase in an immortalized human fetal astrocyte cell line; in vitro characterization and engraftment into the rodent striatum. *Cell Transplant.* **5,** 145–163.

57. Lin, S. Z., Wang, Y., Hoffer, B. J., et al. (1999) Transplantation of hNT cells in adult rats with cortical stroke. *Soc. Neurosci. Abstr.* **25,** 1344.

58. Schinstine, M., Stull, N. D., and Iacovitti, L. (1996) Induction of tyrosine hydroxylase in hNT neurons. *Soc. Neurosci. Abstr.* **22,** 1959.

59. Iacovitti, L. and Stull, N. D. (1997) Expression of tyrosine hydroxylase in newly differentiated neurons from a human cell line (hNT). *Neuroreport* **8,** 1471–1474.

60. Zigova, T., Willing, A. E., Tedesco, E. M., et al. (1999) Lithium chloride induces the expression of tyrosine hydroxylase in hNT neurons. *Exp. Neurol.* **157,** 251–258.

61. Hurlbert, M. S., Gianani, R. I., Hutt, C., et al. (1999) Neural transplantation of hNT neurons for Huntington's disease. *Cell Transplant.* **8,** 143–151.

62. Kondziolka, D., Wechsler, L., Goldstein, S., et al. (2000) Transplantation of cultured human neuronal cells for patients with stroke. *Neurology* **55,** 565–569.

63. Trojanowski, J. Q., Kleppner, S. R., Hartley, R. S., et al. (1997) Transfectable and transplantable postmitotic human neurons: a potential "platform" for gene therapy of nervous system diseases. *Exp. Neurol.* **144,** 92–97.

64. Meltzer, C. C., Kondziolka, D., Villemagne, V. L., et al. (2001) Serial {^{18}F} fluorodeoxyglucose positron emission tomography after human neuronal implantation for stroke. *Neurosurgery* **49,** 586–592.

65. Nelson, P. T., Kondziolka, D., Wechsler, L., et al. (2002) Clonal human (hNT) neuron grafts for stroke therapy: neuropathology in a patient 27 months after implantation. *Am. J. Path.* **160,** 1201–1206.

66. Brundin, P., Dunnett, S., Bjorklund, A., and Nikkhah, G. (2001) Transplanted dopaminergic neurons: more or less? *Nat. Med.* **7,** 512–513.

67. Zivin, J. A. (2000) Cell transplant therapy for stroke: Hope or hype. *Neurology* **55,** 467.

68. Gortz, P., Fleischer, W., Rosenbaum, C., et al. (2004) Neuronal network properties of human teratocarcinoma cell line-derived neurons. *Brain Res.* **1018,** 18–25.

69. Watson, D. J., Longhi, L., Lee, E. B., et al. (2003) Genetically modified NT2N human neuronal cells mediate long-term gene expression as CNS grafts in vivo and improve functional cognitive outcome following experimental traumatic brain injury. *J. Neuropathol. Exp. Neurol.* **62,** 368–380.

70. Willing, A. E., Zigova, T., Milliken, M., et al. (2002) Lithium exposure enhances survival of NT2N cells (hNT neurons) in the hemiparkinsonian rat. *Eur. J. Neurosci.* **16,** 2271–2278.

71. Misiuta, I. E., Anderson, L., McGrogan, M. P., et al. (2003) The transcription factor Nurr1 in human NT2 cells and hNT neurons. *Brain Res. Dev. Brain Res.* **145,** 107–115.

72. Christie, S. D., Sadi, D., and Mendez, I. (2004) Intraspinal transplantation of hNT neurons in the lesioned adult rat spinal cord. *Can. J. Neurol. Sci.* **31,** 87–96.

73. Saporta, S., Makoui, A. S., Willing, A. E., et al. (2002) Functional recovery after complete contusion injury to the spinal cord and transplantation of human neuroteratocarcinoma neurons in rats. *J. Neurosurg.* **97(1 Suppl),** 63–68.

7

Therapeutic Applications of Bone Marrow–Derived Stem Cells in Neurologic Injury and Disease

C. Dirk Keene, Xilma R. Ortiz-Gonzalez, Yuehua Jiang, Catherine M. Verfaillie, and Walter C. Low

ABSTRACT

Stem cells exhibit the unique properties of continual self-renewal and the ability to differentiate into a variety of cell types with the appropriate inductive cues. In the past hematopoietic stem cells were identified in bone marrow that have the capability to differentiate into cells that are part of the blood system. The cells therefore are restricted to the types of cells they can become. Recently, other types of stem cells such as mesenchymal stem cells and multipotent adult progenitor cells have been isolated from bone marrow that have the ability to differentiate into a broader range of cell types. The potential of these bone marrow-derived stem cells to be used in treating neurological disorders is the focus of this chapter.

Key Words: Stem cells; hematopoietic stem cells; marrow stromal cells; mesenchymal stem cells; multipotent adult progenitor cells.

INTRODUCTION

The recent discovery of the neural potential of bone marrow–derived stem cells (BMSCs) has enabled scientists and clinicians to envision novel transplantation therapies for central nervous system (CNS) injury and disease. Although such treatments are, at present, mostly theoretical or experimental, recent advances portend possible future applications of BMSCs in neurological medicine. This chapter discusses the clinical applications of BMSCs in relation to neurologic injury and disease. Then, the different types and classifications of BMSCs are reviewed. Because it is unclear how many different BMSC populations naturally exist, such distinctions may involve

From: *Contemporary Neuroscience: Cell Therapy, Stem Cells, and Brain Repair*
Edited by: C. D. Sanberg and P. R. Sanberg © Humana Press Inc., Totowa, NJ

the same basal stem cells that are independently isolated and characterized by variable means. Alternatively, the disparate BMSC types may comprise distinctly different stem cell populations, each with unique properties and developmental potencies. Next, the chapter describes proposed modalities for the use of BMSCs in the treatment of neurologic disorders and reviews experimental progress toward therapeutic implementation for each type of BMSC. Finally, this chapter outlines the concepts and studies necessary to translate BMSC research into clinical applications for the treatment of CNS maladies. Although the chapter does not reference or discuss the clinical potential of stem cells derived from embryonic, neural, or umbilical cord sources, the applications discussed are applicable to these populations as well. The goal of this chapter is to provide a reference for researchers to understand the current status of BMSCs and to provide novel ideas regarding the future potential of BMSCs in the treatment of patients with neurologic injury and disease.

POTENTIAL CLINICAL APPLICATIONS OF BMSCs IN NEUROLOGIC INJURY AND DISEASE

Since the discovery that a subpopulation of cells within the adult bone marrow may be able to differentiate into neural lineages, BMSCs have become an appealing alternative to traditional donor populations, including fetal neural tissue, neural stem cells, and embryonic stem cells, which are encumbered by significant limitations. First and foremost, utilization of fetal brain and embryonic tissues is fraught with ethical concerns. Procurement of embryos for embryonic stem cells, fetuses for fetal brain, and neural tissue for neural stem cells is limited by supply and tissue accessibility. Embryonic stem cell–derived transplants are prone to the formation of tumors, primarily teratomas. Finally, by definition, each of these tissues beget allogeneic grafts, which are susceptible to immune rejection in the host CNS. BMSCs are much more ethically acceptable than fetal or embryonic tissues and, to date, no donor-derived neoplasms have been reported in experimental models or in clinical trials. In addition, adult BMSCs are readily available and can be obtained from the patient's own marrow in most cases, thus minimizing the risk of immune rejection. Thus, BMSCs provide a tantalizing alternative to current cell- and tissue-based treatment modalities for neurologic injury and disease.

Most pathophysiologic processes involving the CNS are amenable to treatment with BMSCs in some manner. These include (but are not limited to) neurodegenerative conditions, vascular or traumatic insult, inherited protein and enzyme deficiencies, autoimmune disorders, and neoplastic processes. Indeed, each entity results in the death or dysfunction of neurons and/or glia within the CNS. BMSC transplant therapy can be envisioned in one of two principle ways: cell replacement and gene delivery.

Cell replacement mechanisms are self-evident. Although the prevention of pathologic processes is the ultimate goal, replacing lost CNS tissues with autologous grafts is an ideal alternative for conditions in which these measures are not available. Parkinson's disease (PD) is the classic example of a disease in which a specific cell population becomes dysfunctional and dies by a process that is poorly understood and not yet preventable. Moreover, there is longstanding experimental and clinical evidence that supports the idea that replacement of dopaminergic neurons can ameliorate functional disabilities. Thus, BMSC-derived dopaminergic neurons could have significant therapeutic potential. Whether the BMSCs are induced to undergo neural differentiation prior to transplantation, rather than naturally *in situ*, will need to be carefully evaluated, but cell replacement strategies using BMSCs hold major promise. Indeed, Huntington's disease, spinocerebellar ataxia, stroke, traumatic brain injury, and many other conditions, in addition to PD, may also be amenable to cell replacement strategies.

The use of BMSCs as vectors for gene delivery encompasses a wide array of applications. However, two general systems are envisioned that involve unmodified cells and genetically engineered BMSCs. Stem cells produce various neurotrophins, cytokines, and cell adhesion molecules that could provide neuroprotective or neurotrophic support for CNS injury and disease. A principle use is associated with genetic disease, in which a protein or enzyme deficiency, such as lysosomal storage diseases, is present. Hurler's syndrome, for example, results from the deletion or mutation of the α-L-iduronidase gene that renders the individual incapable of processing and metabolizing glycosaminoglycans, which are eventually toxic to the CNS. Transplantation and engraftment of allogeneic bone marrow in the CNS should reconstitute the deficient enzyme, resulting in decreased neurotoxicity and improved clinical outcomes. In acute neurologic insult, e.g., stroke or trauma, implantation of stem cells that normally produce various, and as yet unidentified, neurotrophins and cytokines could mitigate the degenerative processes that occur in these disorders. As a direct extension of this, autologous BMSCs that can be isolated, purified, and cultured for extended periods can be modified to express certain genes. These include genes that are deficient in the host, genes to modify or counteract the expression of toxic proteins, and suicide genes for the treatment of neoplastic processes. For instance, BMSCs can be engineered to express glial-derived neurotrophic factor (GDNF), which is beneficial in animal models of PD. Thus, BMSCs represent potentially diverse utility as gene delivery vectors, regardless of their capability for neural differentiation.

CLASSIFICATION AND NEURAL POTENTIAL OF BMSCs

Distinct populations of stem and progenitor cells have long been recognized within the bone marrow. The discovery of colony-forming units (CFU) by Friedenstein et al. *(1)*, in addition to the isolation and characterization of hematopoietic stem cells (HSCs), demonstrated the mesodermal and hematopoietic potential of the bone marrow. Studies within the last 5–7 yr have demonstrated the existence of cells within the bone marrow that can differentiate into neural lineages. Although the specific identity of this precursor cell remains unclear, numerous isolation and characterization attempts have been conducted. Technological innovations have enabled the characterization of multiple subpopulations of stem cells within bone marrow, and heterogeneous culture techniques have resulted in the isolation of various potentially disparate stem cell populations. The existence of a single or multiple distinct BMSC populations continues to be debated, but BMSC diversity can be classified into four main populations with variable characteristics and neural potential, including whole bone marrow, HSCs, mesenchymal stem cells/marrow stromal cells (MSCs), and multipotent adult progenitor cells (MAPCs). Each population is presented according to isolation parameters, phenotypes, demonstrated neural differentiation, and preclinical studies.

WHOLE BONE MARROW

Bone marrow transplantation (BMT) has been utilized for decades to repopulate cells in patients who received whole-body irradiation/chemotherapy in the treatment of hematopoietic malignancies. However, only after the discovery that bone marrow cells transplanted into the brains of adult mice expressed neural antigens, such as glial fibrillary acidic protein (GFAP) *(2)*, did investigators begin to systematically examine the neural differentiation of whole bone marrow in the CNS. After transplanting whole bone marrow systemically, Brazelton et al. *(3)* and Mezey et al. *(4)* independently discovered donor-derived cells that expressed neural antigens within the brains of lethally irradiated or genetically myelodeficient mice. Corti et al. *(5)* found β-III-tubulin (TuJ1)–, neurofilament–, and neuronal nuclei-specific antigen (NeuN)–positive donor cells in spinal cord and sensory ganglia of mice 3 mo after BMT, as well as GFAP-positive cells in the spinal cord. Moreover, analysis of donor cell ploidy by quantitative, computer-assisted analysis of 4',6-diamidino-2-phenylindole fluorescence intensity revealed no evidence of cell fusion. After direct CNS injection of bone marrow, Nakano et al. *(6)* discovered cells that expressed markers for oligodendrocytes, astrocytes, and microglia (but not neurons). In addition, Priller et al. *(7)* and Weimann et al. *(8)* reported donor-derived cerebellar Purkinje neurons after systemic injection of whole bone marrow in rodents and humans, respectively. Hudson et al. *(9)* indicated that most bone marrow cells retained their hematopoietic identity in vitro; however, there was expression of nestin, TuJ1, and GFAP, as well as migration along established parenchymal pathways, by bone marrow cells transplanted directly into the murine subventricular zone. Mezey et al. *(10)* found donor-derived neurons in the hippocampus and cortex in a postmortem analysis of BMT patients. Finally, Cogle et al. *(11)* found Y chromosome–positive microglia in the brains of three female BMT recipients and detected no fusion sex chromosome phenotype in donor cells (XXY, XXXY). Although no male neurons were detected in BMT recipients who died within weeks of transplantation, donor-derived neurons, astrocytes, and microglia were detected in the hippocampus of a patient who died 6 yr after the first of two BMT procedures.

The mechanism of these findings continues to be debated, and only Corti et al. *(5)* and Cogle et al. *(11)* address host–donor cell fusion. Indeed, Castro et al. *(12)* did not observe neural engraftment after whole bone marrow or HSC transplantation, whereas differentiation in other tissues was present. Wehner et al. *(13)* observed no evidence of astrocytic differentiation after BMT with GFAP promoter–controlled enhanced green fluorescent protein–expressing cells. Thus, despite that support for neural migration and differentiation of systemically administered whole bone marrow exists, it seems to be a rare occurrence.

Multiple studies have examined the efficacy of whole BMT in experimental models of neurologic injury and disease. Mahmood et al. *(14)* demonstrated improved functional outcomes, as well as NeuN, microtubule-associated protein (MAP), and GFAP labeling of donor marrow cells transplanted into the contusion periphery in a rat model of traumatic brain injury. Whole marrow has been transplanted into the spinal cord of rats with radiation-induced demyelinating injury and was associated with increased remyelination in a Schwann cell pattern that was not mediated by HSCs *(15)*. In the *twitcher* mutant mouse model of genetic demyelinating disease, such as globoid cell leukodystrophy, Yagi et al. *(16)* demonstrated improved function with intraperitoneal BMT. Yet, although numerous donor cells were found in the brains of these mice, no evidence of neural differentiation was noted. In both of these models, it is unclear whether BMT contributes directly to cell replacement, but there is significant anatomic and functional improvement, and the underlying mechanism is currently unknown.

Despite that neural differentiation of whole bone marrow seems to be infrequent, especially with systemic transplantation, BMT may have a role in gene delivery and therapy. Park et al. *(17)* found improved motor function and increased nigral dopaminergic neurons in mice that received intravenous GDNF-engineered BMT, followed by methylphenyl-tetrahydropyridine (MPTP) lesions, compared to untreated controls, suggesting neuroprotection by BMT-mediated growth factor delivery. In a classic gene therapy paradigm using retroviral gene transfer, whole bone marrow was engineered to overexpress arylsulfatase A (ASA)—the enzyme deficient in metachromatic leukodystrophy—and then administered intravenously to ASA knockout mice. These mice exhibited partial correction of altered lipid metabolism and demonstrated mild improvement in behavioral and neuropathological measures *(18)*. However, the authors conclude that ASA expression or delivery was insufficient to significantly reverse or delay the course of the experimental disease. Using whole bone marrow grown in conditioned neural stem cell media, Lee et al. *(19)* found that bone marrow–derived neural stem-like cells migrated toward sites of CNS freeze injury and tumor, could be effectively transduced to express tumor necrosis factor-related apoptosis-inducing ligand and, when transplanted near the glioblastoma, significantly increased tumor cell apoptosis and host survival.

Whole BMT for CNS injury has been most extensively studied in ischemic stroke. Hess et al. *(20)* and Zhang et al. *(21)* have examined the role of bone marrow–derived endothelial precursor cells in revascularization after cerebral infarction. They found donor-derived endothelial cells during vasculogenesis in mice that received middle cerebral artery occlusion (MCAO) following systemic administration of whole bone marrow. Although Hess et al. *(20)* did find rare NeuN-expressing donor cells in the striatum of lesioned mice, neither group studied whether this mitigated functional deficits.

However, other groups have studied behavioral recovery. Initially, Li et al. *(22)* found NeuN- and GFAP-expressing cells in the ischemic penumbra after intracerebral transplantation of whole bone marrow. In a follow-up study, Chen et al. *(23)* found NeuN-, MAP-2-, and GFAP-positive donor cells, decreased donor cell apoptosis, and improved adhesive removal ability in mice that received intracerebral transplantation of whole bone marrow coinjected with the antiapoptotic caspase inhibitor Z-Val-Ala-DL-Asp-fluoromethylketone (Z-VAD). Z-VAD alone did not improve functional recovery, which suggests that improved survival of bone marrow cells mediated this effect. Iihoshi et al. *(24)* found improved functional outcomes and reduced lesion volumes in rats treated with intravenous BMT *after*

MCAO, and demonstrated maximal effects if treatment was initiated within 3 h of the infarction. These authors also report CNS migration and neuronal and astrocytic differentiation of donor cells in lesioned mice. Finally, Chen et al. *(25)* demonstrated that whole bone marrow supplemented with brain-derived neurotrophic factor (BDNF), and transplanted into the periphery of MCAO lesions, resulted in increased neural differentiation of transplanted cells and improved sensorimotor function. This suggests that enhanced neural differentiation of donor cells leads to improved neuropathological and functional outcomes.

Phenotypic neural differentiation of donor cells was found in numerous studies; however, transplantation of whole bone marrow for treatment of CNS disease may be suboptimal for several reasons. First and foremost, access to the CNS is limited with systemic transplantation, but it is likely greater in CNS injury secondary to breakdown of the blood-brain barrier. Also, the stem cell component in these populations is a small fraction of the total number of cells transplanted, and cells capable of crossing lineage boundaries traditionally established for bone marrow are probably very rare in whole bone marrow. Although direct CNS injection and experimental administration in animal models of CNS disease improves delivery, the dilutional effect of transplanting whole bone marrow, rather than purified BMSC components, remains problematic. To this end, multiple research teams, using various methodologies, have isolated and expanded stem cell populations that exist within the bone marrow. These BMSCs may be more advantageous, as larger numbers of multipotent, or even pluripotent, stem cells can be delivered to the CNS.

HSCs

HSCs are a vital component of BMT, because this population is responsible for reconstitution of the host hematopoietic system. HSCs can be isolated by depleting bone marrow of lineage-committed cells, e.g., lymphocytes and granulocytes, and subsequently deriving colonies that grow from such "lin" cells by specific CFU assays. Other protocols positively select for specific markers known to be highly expressed by HSCs, such as CD45 and CD34, using immunomagnetic or fluorescent-activated cell sorting (FACS) methods. Another FACS-based purification strategy relies on the ability of HSCs to exclude Hoechst dye (owing to expression of the ABCG2 transporter) and generate a "side population" that is highly enriched for HSCs *(26,27)*. In experiments described previously, it is unclear whether HSCs contribute to the neural differentiation of donor cells after BMT. Indeed, there exists only one report of neural differentiation of HSCs in vitro, and these cells expressed nestin and GFAP but did not become neurons or oligodendrocytes *(28)*. In vivo, Eglitis and Mezey *(29)* found that donor-derived cells in the CNS expressed neural markers for microglia (F4/80) and astrocytes (GFAP) following intravenous HSC transplantation. However, Wagers et al. *(30)* found that clonal HSC transplantation did not lead to donor-derived CNS engraftment, and Ono et al. *(31,32)* report significant donor-derived CNS microglia, but they did not find any neurons in mice that received intravenous HSC transplants weeks earlier. Hess et al. *(33)* also found donor-derived CNS microglia, in addition to perivascular cells associated with small and large blood vessels, in mice transplanted with rigorously selected, clonal populations of HSCs. However, after MCAO, infrequent NeuN-positive donor cells were detected, as well as increased frequency of donor-derived microglia and perivascular cells. Considering multiple methods exist for the isolation of HSCs, contamination of HSC grafts by non-HSC entities may contribute to neural engraftment, and HSCs may be incapable of neural differentiation. In other words, it is unclear whether there exists within the HSC population a pluripotent precursor contaminant capable of neural differentiation or if HSCs can cross lineage boundaries to become cells of the CNS. Nevertheless, several groups have begun to examine whether HSCs can be useful in the treatment of neurologic disease.

In multiple sclerosis, HSC transplantation might have therapeutic value by suppressing the inflammatory component of this progressively debilitating disease (for review, see ref. *34*). The potential use of HSCs as a source for cell replacement/repair in CNS disorders is also being examined. Bonilla et al. *(35)* indicated that c-kit$^+$ (CD117$^+$)–enriched HSC transplanted into neonatal mice brains expressed mostly oligodendrocytic but also neuronal and astrocytic markers in vivo. Vitry et al. *(36)* found that primordial HSCs expressed nestin, polysialic acid-neural cell adhesion molecule, TuJ1, and GFAP in hematopoietic conditions. However, the cells did not adopt a neural fate when cocultured with neural precursors or in neural stem cell conditions. These cells derived microglia in vivo but failed to form myelinating cells in the *Shiverer* mouse model of hypomyelination. Sasaki et al. *(15)* found that a CD34$^+$ HSC population transplanted in a demyelinated rat spinal cord survived but failed to remyelinate fibers within the lesion. However, remyelination did occur after transplantation of acutely isolated whole bone marrow cells (containing mixed MSCs, HSCs, and so on), suggesting that cells other than HSCs in the bone marrow are able to mediate remyelination. One potential explanation for the discrepancy between these studies is that HSCs used by Sasaki et al. *(15)* were positively selected by immunomagnetic means, but the cells used by Bonilla et al. *(35)* were selected by FACS. Only 20% of the population transplanted was actually c-kit+. Thus, there is limited evidence that naive HSCs can be effectively utilized to treat primary disorders of the CNS.

However, genetically modified HSCs have been successfully utilized to treat genetic enzyme deficiency. Leimig et al. *(37)* engineered hematopoietic progenitor cells to express protective protein/cathepsin A (PPCA), which is deficient in galactosialidosis. When administered to PPCA-deficient mice via tail vein injection, engineered cells expressed PPCA. PPCA was internalized into host cells, and this mediated significant reduction in vacuolizations and pathologic storage material and preserved cerebellar Purkinje cell architecture and population. The identity of hematopoietic progenitor cells, compared to classic HSCs, is unclear. Yet, this study demonstrates the potential efficacy of enzyme delivery with bone marrow–derived hematopoietic cells. Extensive research is required to determine the potential use of HSCs versus alternative BMSC populations for the treatment of neurologic injury and disease. Several of these alternative populations within the bone marrow have also been studied, with more encouraging results.

MESENCHYMAL STEM CELLS/MARROW STROMAL CELLS

The bone marrow stroma contains a heterogeneous cell population, consisting of HSCs and non-HSCs that can be isolated and expanded by multiple variable means *(38)*. Typically, MSCs are isolated from bone marrow aspirates via adherence to plastic culture vessels in high-serum media (10–20%). MSC, originally described by Friedenstein in 1974 *(1)* as CFU with fibroblastic morphology, were shown to differentiate into cartilage, bone, and adipose tissues in vitro. Since this time, various isolation and culture techniques have arisen, which purport to purify select stem cell populations collectively referred to as *mesenchymal stem cells*. However, because distinctly different methods are used, it is uncertain whether various reports of MSC plasticity and transdifferentiation can be viewed as utilizing the same population of stem cells. Additionally, this heterogeneity is likely responsible, at least in part, for disparate and often inconsistent reports of the neural plasticity of MSC cultures. Until a characteristic expression pattern or antigenic fingerprint is found that positively identifies the various MSC components, it cannot be assumed that MSCs represent or comprise a uniform population of stem cells. Indeed, most MSC culture methods yield morphologically heterogeneous cell populations, which likely exhibit variable plasticity in vitro and in vivo. Furthermore, MSC isolation and culture techniques likely exclude other subpopulations of BMSCs, including HSCs and MAPCs (see below). Nevertheless, interesting results specifically investigating the neural differentiation of MSCs have been reported.

In 1998, Azizi et al. *(2)* reported astrocytic differentiation in vivo after intrastriatal transplantation of MSCs in mice. However, they did not find evidence of neuronal or oligodendrocytic differentiation. Soon after, Kopen et al. *(39)* reported MSC expression of GFAP and neurofilament after intraventricular transplantation into neonatal mice. These initial reports stimulated investigations into the neural potential of MSCs. In 2000, Sanchez-Ramos et al. *(40)* and Woodbury et al. *(41)* independently demonstrated MSC expression in vitro of glial and neuronal antigens using significantly different induction methods. By supplying basic fibroblast growth factor (bFGF) to cultured MSCs, Muñoz-Elias et al. *(42)* found expression of tau, NSE, TUC-4 (TOAD/ Ulip/CRMP-4), and NeuN concomitantly with decreased fibronectin production and demonstrated this differentiation was not owing to cell selection via apoptosis. Later, this group showed donor cell engraftment, migration to neocortex, hippocampus, rostral migratory stream and olfactory bulbs, and neuronal differentiation of MSCs transplanted intraventricularly into E15.5 fetal rat ventricles *(43)*. Electrically active neural cells were generated from plastic adherent "size-sieved" human bone marrow by Hung et al. *(44)*. Dezawa et al. *(45)* showed that MSCs acquired neural progenitor cell antigen expression, including glutamate transporter, phosphoglycerate dehydrogenase, and nestin after endosome-mediated *Notch* intracellular domain gene transfer.

In addition, subsequent administration of forskolin, bFGF, and ciliary neurotrophic factor induced 95% of the cells to express MAP-2. These cells exhibited delayed rectifier potassium currents and maintained lower resting membrane potentials. After further addition of BDNF and nerve growth factor (NGF), these cells also exhibited tetrodotoxin-sensitive inward currents, likely mediated by voltage-gated sodium channels. Finally, treatment with glial-derived neurotrophic factor (GDNF) resulted in tyrosine hydroxylase (TH) expression in 41% of cells. In addition to primary neural differentiation of MSCs, Lou et al. *(46)* reported increased neuronal differentiation of mesencephalic neural stem cells cocultured with MSCs, and this effect was reproduced with MSC-conditioned media. Although there is strong evidence that MSCs can fuse with embryonic stem cells *(47)* in vitro, along with hepatocytes, cardiac muscle, and Purkinje cells in vivo *(48)*, most reports of neuronal differentiation in vitro occurred without the presence of coculture, and no reports of cell fusion with non-Purkinje neurons or astrocytes exist. Further investigations into the potential contribution of fusion in MSC neural differentiation are required, and rigorous preclinical studies are necessary to determine the potential for, and efficacy of, MSC transplantation for the treatment of neurologic disease. To date, studies have yielded encouraging results.

MSCs have been transplanted into animal models of stroke, traumatic brain and spinal cord injury, PD, and genetic disorders. CNS injury promotes many molecular signaling events that could affect stem cell survival and/or differentiation, which might explain why most reports of neural differentiation of MSCs in the CNS derive from studies in injury models, rather than the intact brain. Indeed, an injured and diseased CNS may be the most appropriate milieu in which to study neural differentiation of MSCs, but it is not clear whether neural differentiation of MSCs will be required in all cases to mediate clinical improvement.

For instance, intracerebral transplantation of MSC transduced to express acid sphingomyelinase (ASM) into an ASM knockout mouse model of Niemann-Pick disease resulted in MSC survival and migration away from graft site, decreased cerebral sphingomyelin, delayed Purkinje cell loss that was inversely proportional to distance from MSC graft, and increased survival *(49)*. In follow-up experiments, Jin et al. *(50)* demonstrated near-normal ASM levels, significantly improved neuropathological findings, major reductions in cerebral sphingomyelin, normal cerebellar function, and near-normal Purkinje cell populations after intravenous plus intracerebral transplantation of human MSC-expressing ASM. In addition, many Purkinje cells also expressed human ASM, likely because of donor-derived Purkinje cells or host-donor cell fusion. Thus, MSC grafts could mediate a therapeutic benefit in metabolic diseases via enzyme replacement ability, neural differentiation and replacement, or both.

Several studies have examined the use of MSCs in PD. Schwarz et al. *(51)* transduced MSCs with TH and GTP cyclohydrolase I, resulting in levodopa (L-DOPA)–secreting MSCs. After intrastriatal transplantation into 6-hydroxydopamine (6-OHDA)–lesioned rats, a temporary reduction in apomorphine-induced rotations was observed despite that neural differentiation of grafted MSCs was not reported. Although the cells survived and engrafted for at least 3 mo, transient transgene expression was likely responsible for the temporary behavioral improvement. In the MPTP mouse model of PD, intrastriatally transplanted MSCs survived and engrafted, expressed TH, and mitigated sensorimotor impairment, as measured by Rota-Rod *(22)*. Finally, transplantation in rats of TH-expressing MSCs transduced with *Notch* intracellular domain and treated with multiple growth factors, including GDNF, resulted in improved apomorphine-induced rotational asymmetry and step- and paw-reaching tests in the 6-OHDA lesion model of PD *(45)*.

Acute CNS injury may be highly amenable to treatment with MSC transplantation; extensive studies have examined MSC treatment potential in spinal cord injury, CNS trauma, and stroke. Using magnetic resonance tracking, Jendelova et al. *(52)* found that intravenously injected iron oxide nanoparticle–labeled MSCs preferentially migrated to the site of balloon-induced spinal cord compression lesion. Chopp et al. *(53)* reported donor cell expression of NeuN, increased oligodendrocytic expression of Rip (a marker for functional myelinating cells), and improved functional outcomes after MSC transplantation into rat spinal cord 1 wk after traumatic cord injury. Wu et al. *(54)* expanded on these findings by demonstrating reduced lesion cavities and improved functional recovery in rats with spinal cord contusion following MSC transplantation. In vitro, they also showed that MSC cocultured with spinal cord–derived neural stem cells resulted in increased neural stem cell process formation and chemotaxis toward MSCs. Similar anatomic and functional improvement was observed in vivo after transplantation of MSC into the cerebrospinal fluid of spinal cord–contused rats *(55)*. Green fluorescent protein (GFP)–labeled MSCs grafted directly into a X-irradiation–induced demyelinated spinal cord lesion were found to express the oligodendrocyte marker myelin basic protein, mediate remyelination, and improve action potential conduction velocity *(56)*. Finally, undifferentiated and culture-manipulated MSCs improved transected sciatic nerve regeneration after transplantation into the distal stump *(57,58)*.

MSCs have also been evaluated in the treatment of traumatic brain injury. Several studies indicate improved functional outcomes and donor cell expression of neuronal and astrocytic markers after intracerebral *(59)*, intraarterial *(60)*, and intravenous *(61–63)* administration of rodent MSCs in adult rats with traumatic brain injury. Treatment efficacy was improved with pretreatment of rodent MSCs with BDNF and NGF *(59)*. Furthermore, similar efficacy has been reported with intravenous administration of human MSCs after traumatic brain injury in rats. Donor cells expressed neuronal (TuJ1) and astrocytic (GFAP) markers *(64)*. MSC transplantation led to behavioral recovery following traumatic brain injury, but it is unlikely mediated solely by replacement of injured and dead neurons.

Chen et al. *(65)* demonstrated time-dependent release of several growth factors, including BDNF, NGF, and vascular endothelial growth factor, after exposure of MSCs to cerebral tissue extracted from traumatic brain–injured animals. If similar growth factor release occurs in vivo, this would likely improve neuronal survival and recovery. Mahmood et al. *(66)* found increased intracerebral concentrations of NGF and BDNF in injured brain tissue after intravenous transplantation of MSCs, compared to noninjected controls. The authors reported MSC migration into the injured CNS tissue, but it is unclear whether the growth factors are donor- or host cell-derived. In any case, MSC transplantation mediates increased growth factor expression, which likely contributes to functional recovery in traumatic brain injury.

MSC treatment of ischemic stroke has also been extensively evaluated. MSCs transplanted intracerebrally were shown to migrate preferentially toward cortical photochemical lesions in a model of thrombotic stroke using in vivo magnetic resonance tracking of MSC labeled with iron oxide nanoparticles *(52)*. Several studies have shown functional improvement in rodents with transient MCAO after intrastriatal *(67)*, intracarotid *(22)*, or intracerebral *(68)* MSC transplantation, and this effect was improved with intracerebral transplantation of MSCs pretreated with NGF *(69)*. Moreover, in interface culture with ischemic cerebral tissue, MSC migration was enhanced, and likely mediated, by inflammatory chemotactic agents, e.g., monocyte chemotactic protein-1, interleukin-8, and macrophage inflammatory protein-1 *(70,71)*.

As with traumatic brain injury, the mechanism of recovery mediated by MSC in stroke is probably multifactorial. Although neural differentiation of grafted cells is identified, the time course and degree of differentiation cannot explain the recovery based on cell replacement alone. In vitro studies have shown increased human MSC growth factor production in the presence of ischemic rat brain extracts *(65)*. In the presence of rodent MSCs, ischemic hippocampal slice cultures suffered less degeneration and exhibited more neurite extension than control cultures *(72)*. Chen et al. *(73)* found MSCs mediated reduction in penumbral apoptosis and increased 5-bromo-2-deoxyuridine (BrdU)-positive host cells in the subventricular zone, implying increased endogenous cell proliferation with intravenous MSC transplantation for ischemic stroke. Finally, intravenous administration of MSCs, with or without the nitric oxide donor diethyltetraamine-NONOate (DETA/NONOate), improved functional recovery with evidence of induction and, in the presence of DETA/NONOate, enhanced angiogenesis *(74,75)*. Moreover, doublecortin-positive donor cells were identified, implicating enhanced neurogenesis as well *(75)*. Thus, MSC-mediated recovery from stroke is likely owing to MSC secretion of neurotrophic factors, MSC induced host neurotrophic factor expression, increased host neurogenesis, and cell replacement.

Overall, growing evidence indicates that MSC transplantation in lesion models of human disease can ameliorate functional deficits. However, the contribution of neural differentiation of grafted cells, and subsequent neuronal and glial cell replacement in functional recovery, is unclear. Future MSC studies will be important to delineate the mechanisms by which transplants mediate recovery. In addition, considering the highly variable isolation and culture methods by which MSCs are obtained and expanded, and in view of the heterogeneous nature of MSC cultures and transplants, it will be important to determine the identity of the cell subpopulation(s) that contribute to multipotency in MSC grafts. By identifying and characterizing specific bone marrow stem cell types, scientists and clinicians can ensure reproducible graft tissues and can best predict clinical outcomes in various conditions. HSCs are perhaps the best characterized population of BMSCs, but it is questionable whether they retain properties that will be clinically useful for the treatment of neurologic disease. However, a population of BMSCs, conservatively referred to as MAPCs, has been identified that may be an ideal tissue source for BMSC neural transplantation.

MAPCs

Reyes et al. *(76,77)* originally identified human MAPCs as a population of stem cells that initially copurified with MSCs and have subsequently been isolated and expanded from mouse, rat, dog, pig, and rhesus monkey bone marrow, as well as murine brain and muscle *(78)*. However, unlike MSCs which rely on adherence to plastic, MAPCs are selected using ficoll separation, followed by depletion of $CD45^+$ and glycophorin A^+ cells and subsequent very low-density culture in media supplemented with specific growth factors (epidermal growth factor, platelet-derived growth factor, and, in rodents, leukemia inhibitory factor) and that contain minimal (2%) or no serum. Indeed, because MAPCs differentiate and do not retain stem cell properties in high-serum conditions, the high-serum (10–20%) conditions required by MSC cultures likely exclude MAPCs shortly after plating. Human MAPCs are well characterized, consistently expressing CD10, CD13, CD49b, CD49d, CDw90, and Flk1, with no expression of CD34, CD36, CD38, CD44, CD45, HLA-DR, and HLA-type I, even after up to 80 (human MAPCs) or 150 (murine MAPCs) population doublings (for review of MAPC culture conditions and phenotype, see refs. *76,79–81)*. Undifferentiated MAPCs exhibit extensive self-renewal and (with very few exceptions) retain cytogenetic stability in culture. These cells consistently express active telomerase and exhibit telomere stability *(82)*. This abundantly reproducible, highly stable, and well-characterized phenotype allows researchers to provide reliable MAPC preparations for experiments in vitro and in vivo, which is currently not possible with MSCs, making them an ideal tissue source for neural transplantation.

MAPCs exhibit dramatic potential for differentiation. Indeed, clonal populations of MAPCs, confirmed by sequencing DNA flanking retroviral insertion sites, differentiate into representative tissues from each embryonic germ layer. MAPCs differentiate into mesodermal tissues that are traditionally associated with MSCs, including osteoblasts, chondroblasts, myoblasts, and adipocytes. However, unlike MSCs, MAPCs can also become endothelial cells *(76,77,83)*. Endodermal differentiation of MAPCs has been shown with the production of functional hepatocytes *(81)*, and ectodermal differentiation is identified by the production of neurons, astrocytes, and oligodendrocytes (see Fig. 1). Indeed, MAPCs injected intravenously engrafted in multiple organ systems, e.g., hematopoietic, pulmonary, and gastrointestinal tract (intestines and liver) *(79)*. After MAPC transplantation into blastocyst-stage embryos, further studies revealed that chimerism was identified in up to 80% of mice and most (if not all) somatic tissues were derived partly from β-galactosidase–expressing donor cells (see Fig. 2). Therefore, thus far, MAPCs are the only adult bone marrow–derived cells that have been shown to be pluripotent both in vivo and vitro.

Human MAPCs were initially found to generate neural cells consistent with all three neural lineages, including astrocytes, oligodendrocytes, and neurons in vitro upon exposure to bFGF *(77)*. It also appeared that MAPCs responded to cytokines, e.g., BDNF and GDNF, to generate predominantly dopaminergic and GABAergic neuronal phenotypes. Murine MAPCs were subsequently found to exhibit similar neuroectodermal potential in vitro in the absence of coculture *(79)*. Using an induction protocol that draws from embryonic midbrain development (including FGF-8 and *sonic hedgehog*), murine MAPCs generate a phenotype reminiscent of dopaminergic neurons, such as the appropriate sequential expression of nestin, nurr-1, neurofilament, tubulin, TH, and DAT. Furthermore, when these predominantly dopaminergic-like cells were cocultured with fetal astrocytes, the genetically labeled MAPC-derived cells acquired anatomical and electrophysiological properties of midbrain dopaminergic neurons. Some of these properties are functional voltage-gated sodium channels, spontaneous spiking behavior, and inducible action potential generation (see Fig. 3), as well as compartmentalization of dendritic and axonal compartments *(80)*. Thus, under physiologically appropriate conditions, MAPCs differentiate into functional dopaminergic neuron-like cells that could potentially be a substitute for fetal or embryonic stem cell–derived neural grafts for PD. However, neural differentiation in vitro must be confirmed in vivo prior to any conclusions concerning clinical applicability.

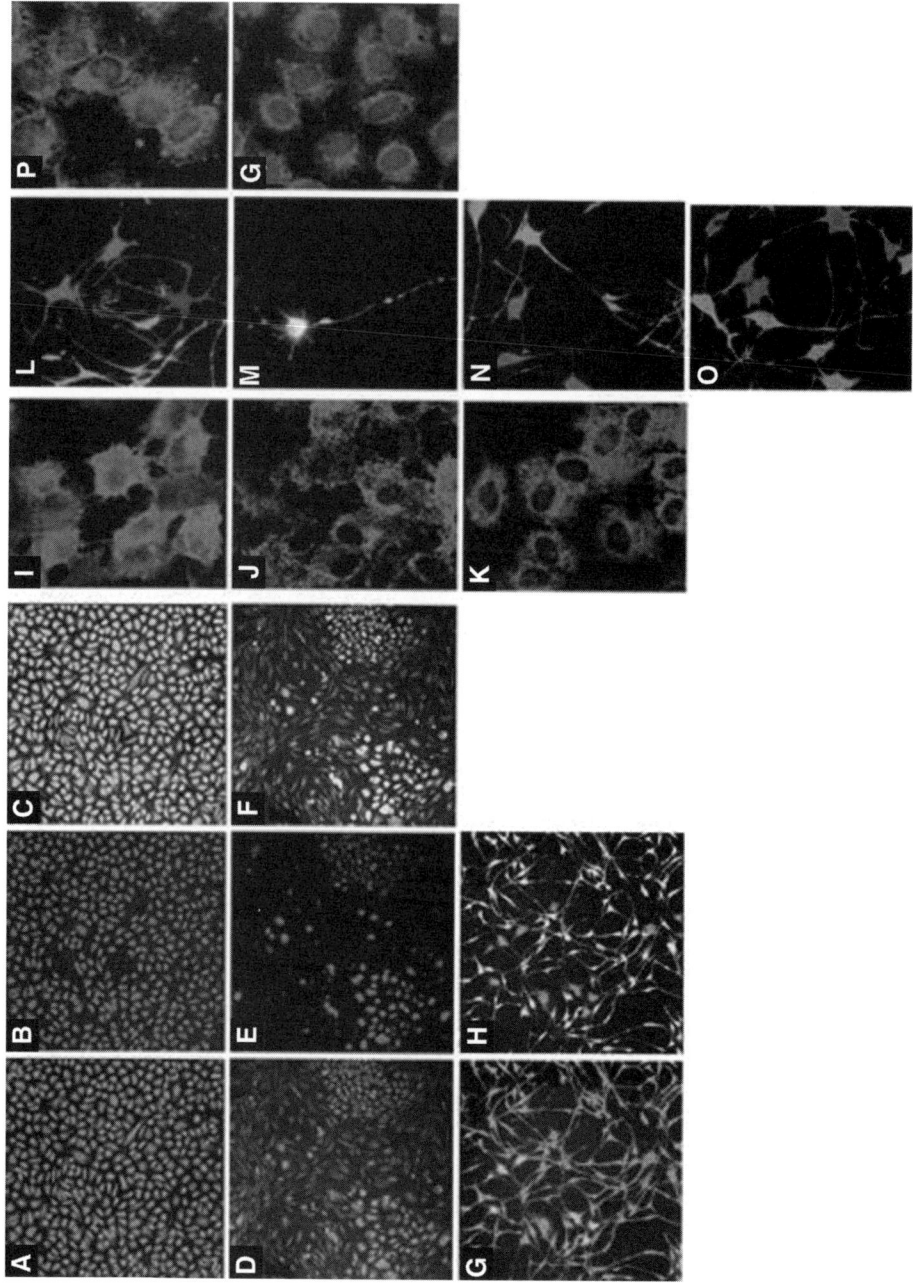

Fig. 1. In vitro differentiation of mMAPC to endothelium, neuroectoderm, and endoderm. (**A–H**) Clonal eGFP+ mMAPC treated with vascular endothelial growth factor (VEGF) (**A–C**), FGF-4 + HGF (**D–F**), or bFGF (**G,H**) for 14 d. Cultures were stained with anti-vWF labeled with Cy3 (**B**), antialbumin-Cy3 (**E**), anti-GFAP-Cy5 (**G**), or anti-NF200-Cy3 (**H**). **C**, **F**, **G**, and **H** show an overlay of Cy3 or Cy5 staining with eGFP. **A**, **C**, **D**, **F**, **G**, and **H** show that 100% of cells were eGFP-positive (×10 magnification). (**I–K**) eGFP+ mMAPC treated with VEGF for 14 d were stained with antibodies against CD31-Cy3 (**I**), Flk-1-Cy3 (**J**), or vWF-Cy3 (**K**) (×60 magnification). (**L–O**) ROSA26 mMAPC treated with bFGF for 14 d were stained with antibodies against GFAP-Cy3, FITC-labeled NF200, and anti-GalC-Cy5 (**L**) (×20 magnification). Alternatively, ROSA26 mMAPC treated sequentially with bFGF, FGF-8, and BDNF were stained with antibodies against MAP2-Cy3 and FITC-labeled anti-Tau (**M**), GABA-Cy3 and FITC-labeled serotonin (**O**) Cy3 and FITC-labeled DDC (**N**), or TH-Cy3 and FITC-labeled serotonin (**O**) (×40 magnification). (**P,Q**) eGFP+ mMAPC treated with FGF-4 and HGF for 14 d stained with antibodies against CK18-Cy3 (**P**) or albumin-Cy3 and HNF1-Cy5 (**Q**) (×60 magnification). Color staining: Cy3, red; FITC, green; Cy5, blue. Double positive for FITC and Cy3, yellow. In **Q**, the nucleus area is positive for HNP1-Cy5 and weakly positive for Cy3 (purple). Magnification: **A–H**, ×10; **I–K**, ×60; **L**, ×20; **M–O**, ×40; **P**, **Q**, ×60. Reprinted with permission from ref. 79.

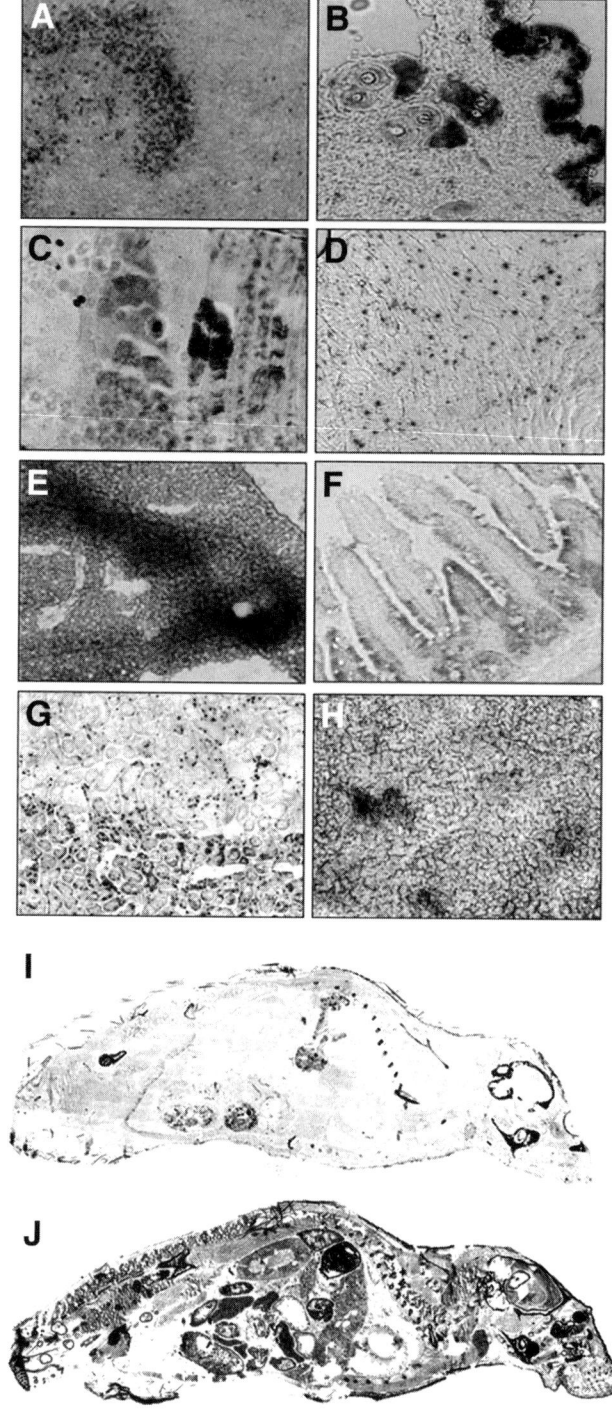

Fig. 2.

Analysis of the brains of adult chimeric mice that were transplanted with a single (lac-Z labeled) MAPC at the blastocyst stage *(79)* revealed evidence of extensive neural differentiation of MAPCs in vivo *(84)*. Histochemical analysis for β-galactosidase in the chimeric brains revealed extensive contribution of MAPC-derived progeny throughout the entire brain, including the cortex, striatum, hippocampus, substantia nigra, and cerebellum (see Fig. 4). Double and triple immunofluorescent-labeling studies revealed that MAPC-derived neurons (NeuN-bGal) and astrocytes (GFAP-bGal) contributed to each cellular layer of the cortex, the pyramidal and granule cell layers, stratum oriens, stratum moleculare, in the hilus in the hippocampus, Purkinje and granule cell layers in the cerebellum, and matrix and stroma within the caudate and putamen (see Fig. 5). Chimeric brains were anatomically indistinct from controls, exhibited appropriate architectural morphology, and contained normal neurotransmitter localization and associations, such as the presence of TH[+] processes and MAPC-derived GABAergic neurons in the neostriatum (see Fig. 6). In addition, the chimeric mice exhibited no gross behavioral or neuroanatomical abnormalities, further supporting the functionality of the MAPC-derived neural cells. Thus, MAPCs can become functional neurons and glia in vitro and in vivo under developmentally appropriate environmental conditions and may therefore provide an ideal alternative source of cells for treatment of neurologic disease. However, investigation into the clinical potential of MAPCs is in its earliest stages.

The first study to evaluate the therapeutic potential of MAPCs was conducted in rodents that had undergone ischemic stroke. Zhao et al. *(85)* transplanted human MAPC into the ischemic penumbra of rats that had undergone MCAO 1 wk earlier. There was significant amelioration of sensorimotor deficits with the limb placement test at 2- and 6-wk posttransplantation (see Fig. 7). Although cortical and subcortical MAPC-derived cells expressed various neural markers, including neurofilament, β-III tubulin, NSE, GFAP, and galactocerebroside, they were infrequent and exhibited little evidence of dendritic arborization and connectivity *(85)*. Thus, MAPC-

---

Fig. 2. *(previous page)* Chimerism detection by X-gal staining and anti-β-gal staining in animals generated from blastocysts microinjected with a single ROSA26 MAPC. **(A–H)** Images from X-gal-stained individual organs from a 45% chimeric mouse, determined by Q-polymerase chain reaction for Neo on tail clip. Tissue sections were from the brain **(A)**, skin **(B)**, skeletal muscle **(C)**, myocardium **(D)**, liver **(E)**, small intestine **(F)**, kidney **(G)**, and spleen **(H,I,J)**. Images from an X-gal-stained section through a mouse that was not chimeric **(I)** or was 45% chimeric **(J)**. Magnification: ×20. Reprinted with permission from ref. *79*.

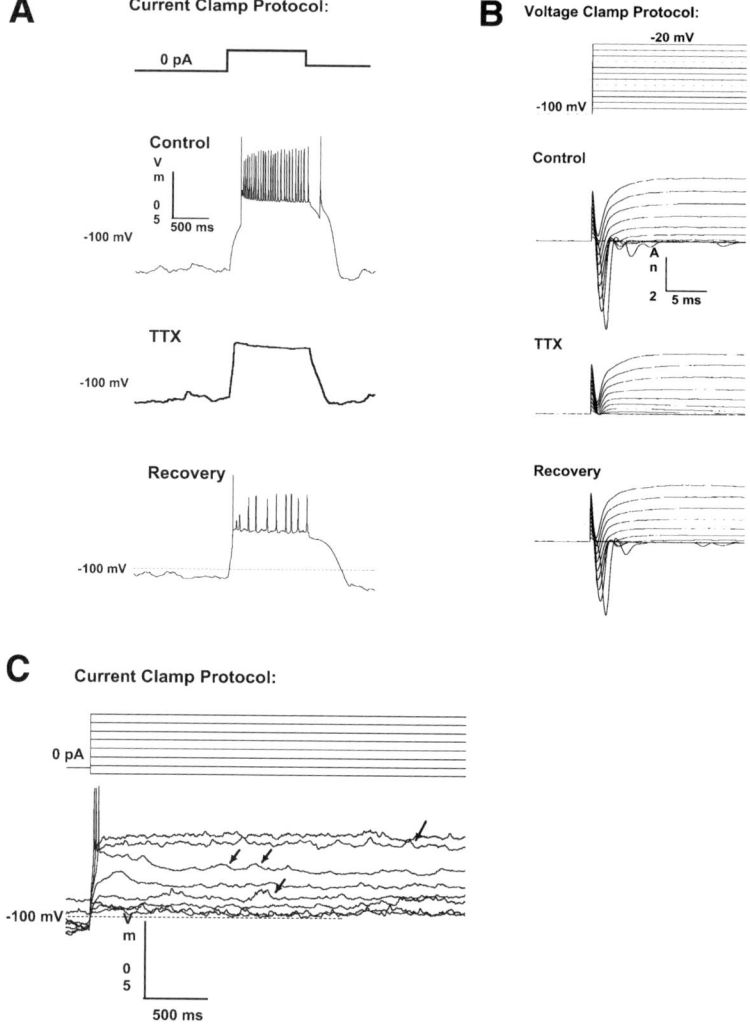

Fig. 3. Spiking behavior and voltage-gated currents from MAPCs in coculture with fetal mouse brain astrocytes. (**A**) Current-clamp recordings from a MAPC that had been cocultured with astrocytes for 8 d. Illustrated in the bottom three panels are the voltage responses elicited by the current-injection protocol shown (top, a 17-pA current-injection step). The repetitive spiking recorded in this cell was blocked reversibly by tetrodotoxin (TTX). The current injection protocol reports the current injected relative to a negative DC current that was injected into the cell to "hold" it near −100 to −130 mV. (**B**) Voltage-clamp recordings of leak-subtracted currents from the same cell shown in A. (Top) The voltage-clamp protocol used to elicit the families of currents shown in the bottom three panels. A large transient inward current was evident that could be blocked reversibly by TTX. (**C**) Current clamp records obtained from a MAPC that had been in culture with astrocytes for 8 d. In this example, the cell produced only one spike in response to depolarizing current injections (Δ pA = 7). The arrows point to synaptic potentials. Reprinted with permission from ref. *80*.

mediated functional improvement was likely not from the replacement of dead and dying neurons but could be a function of MAPC-mediated neurotrophic support or transplant induction of endogenous neuroreparative mechanisms (e.g., mobilization of host neural stem cells). Indeed, neurotrophic effects of stem cell transplants are likely to mediate such behavioral recovery, which is consistent with reports in other CNS lesion models using neural stem cell grafts *(84,85)*. Optimal transplantation therapies would recapitulate damaged or dysfunctional neuroanatomic pathways via reparative and replacement mechanisms, but several factors may have impacted neural differentiation in this study. For instance, it is not known whether appropriate cues are present in adult rodent parenchyma to induce neural differentiation of human MAPCs. No neural engraftment was identified in adult mice that received intravenous murine MAPCs *(79)*. Also, although rats in this study received cyclosporin A throughout the study, more rigorous immunosuppression may be required to ensure adequate xenograft survival. Finally, neural differentiation was examined 2- and 6-wk posttransplantation, which enabled the expression of neuronal and glial antigens to be identified but may be insufficient for neuritic extension and maturation of human MAPC-derived cells in the rodent CNS. Long-term studies are in progress that address MAPC differentiation and the functional effects of intracerebral transplantation of xenogeneic, allogeneic, and syngeneic MAPCs in neonatal and adult immunosuppressed rodents.

MAPCs represent a population of pluripotent BMSCs that are well-characterized and can be routinely isolated, expanded, and differentiated into mesodermal, endodermal, and neuroectodermal lineages. MAPCs are readily transducible and exhibit stable haplotypes over extensive passages. MAPCs differentiate into various neural lineages using developmentally appropriate induction protocols and have full capacity to integrate into most, if not all, components of the rodent CNS under the appropriate conditions. It is unclear whether MAPCs normally exist in the adult organism or if their pluripotency is derived from molecular induction in culture; however, in either case, their potential clinical impact is significant. If stimulated and mobilized, endogenous MAPCs could be manipulated in injury and disease to mitigate and/or repair diseased or damaged tissues. Understanding pathways that artificially create pluripotency in vitro would greatly enhance our knowledge and ability to manipulate cells and tissues for future therapeutic interventions. MAPCs represent a homogeneous cell population that does not require fetal bovine serum for maintenance and expansion, and multiple antigenic and genetic markers can be utilized to assess the "stemness" of the MAPCs. Thus, these cells exhibit significant advantages over traditional BMSCs that

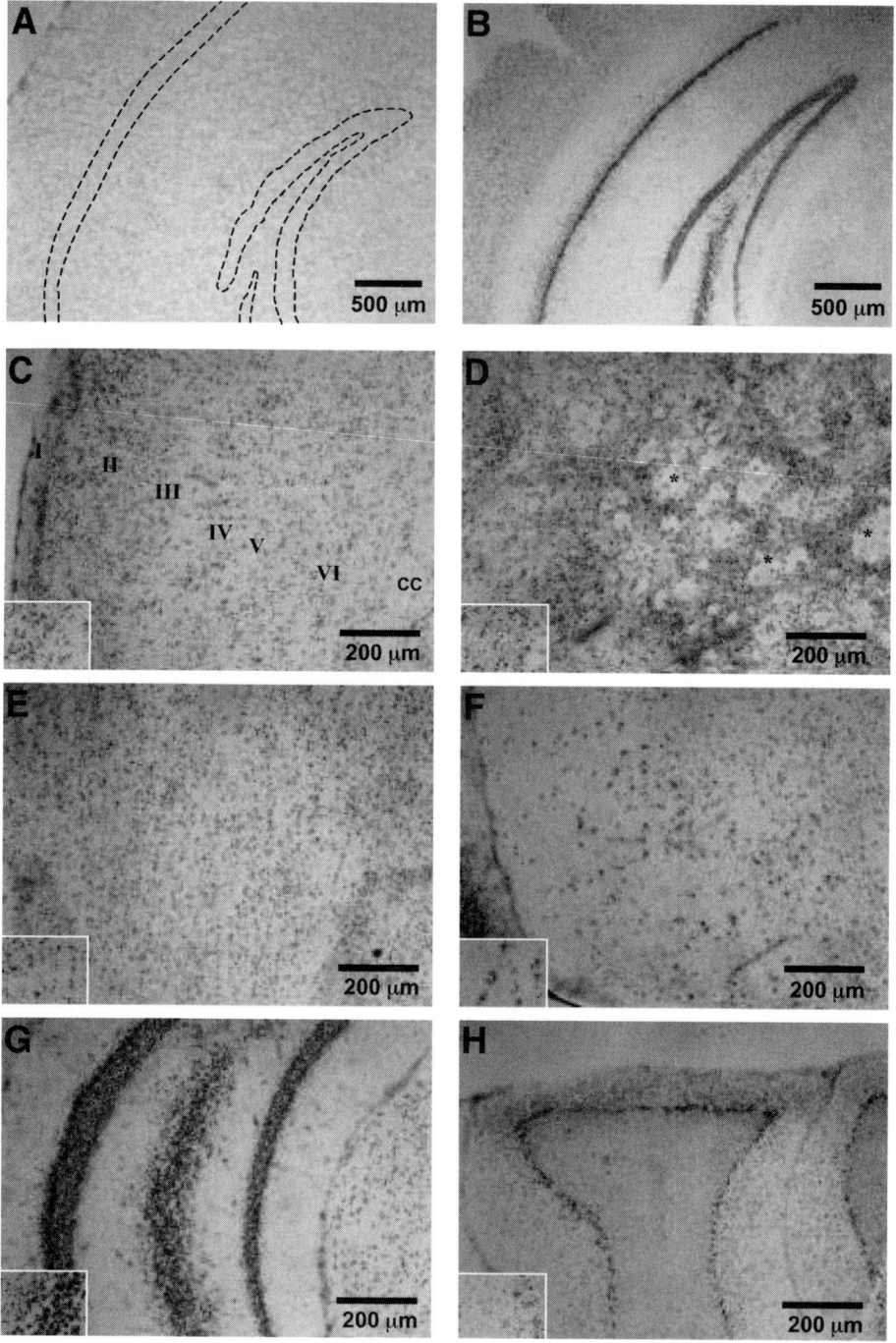

Fig. 4.

are heterogeneous, variably characterized, require animal serum for growth, and may not possess the neural potency of MAPCs. Studies investigating the efficacy of MAPCs to treat a variety of neurological disorders are underway, and MAPCs may represent an ideal alternative to traditional sources of tissue for neural transplantation.

DISCUSSION

The clinical applicability of BMSCs in neurologic injury and disease remains to be established. Multiple avenues of research will be necessary prior to its integration into patient care. These remaining issues include, but are not limited to, discovery of the mechanism and control of neural differentiation, evaluation of neural function in vivo, rigorous characterization of BMSC phenotype, and standardization of BMSC isolation, purification, and expansion. Additional concerns are optimal BMSC phenotype in various potential applications, optimal site and timing of BMSC transplantation, safety, evaluation of immune reaction to allogeneic grafts, feasibility of autologous BMSC transplantation, and efficacy of BMSC transplantation in each individual condition.

The identity of BMSCs in vivo remains unclear. Do pluripotent stem cells continue to populate the human marrow throughout life? If so, what is their role in normal human physiology? This continues to be debated, and it has yet to be conclusively demonstrated that MSC and MAPC multipotency and pluripotency do not result from manipulation in culture, resulting in artifactual plasticity. Although it is tempting to believe biological organisms retain a primitive population of stem cells with remarkable regenerative capacity, would it matter to the patient with debilitating stroke whether their treatment

---

Fig. 4. *(previous page)* β-galactosidase histochemistry reveals significant engraftment throughout the brains of single-MAPC–injected chimeric mice. β-Gal activity is undetectable in MAPC-injected, nonchimeric mouse hippocampus (**A**). (Granule and pyramidal cell layers, and hilus are delineated with superimposed dashed lines for comparison with chimeric hippocampus (**B**), in which X-gal labeling is strikingly dense.) MAPC-derived cells were found in every area of the brain examined. Their widespread distribution is shown in each of the (numbered) layers of the cortex (**C**), in the neuronal and fiber tracts (*) of the striatum (**D**), in the cells of the medial septal nucleus (**E**), and in the substantia nigra (**F**). Higher power examination of the hippocampus reveals increased MAPC-derived cells in neuronal layers, such as the stratum granulosum and hilus (**G**). MAPC contribution to the cerebellum was also very extensive, with increased histochemical density in the granule and Purkinje cell layers (**H**). Inset photomicrographs (**C–H**) show higher power views of each associated tissue. Reprinted with permission from ref. *84.*

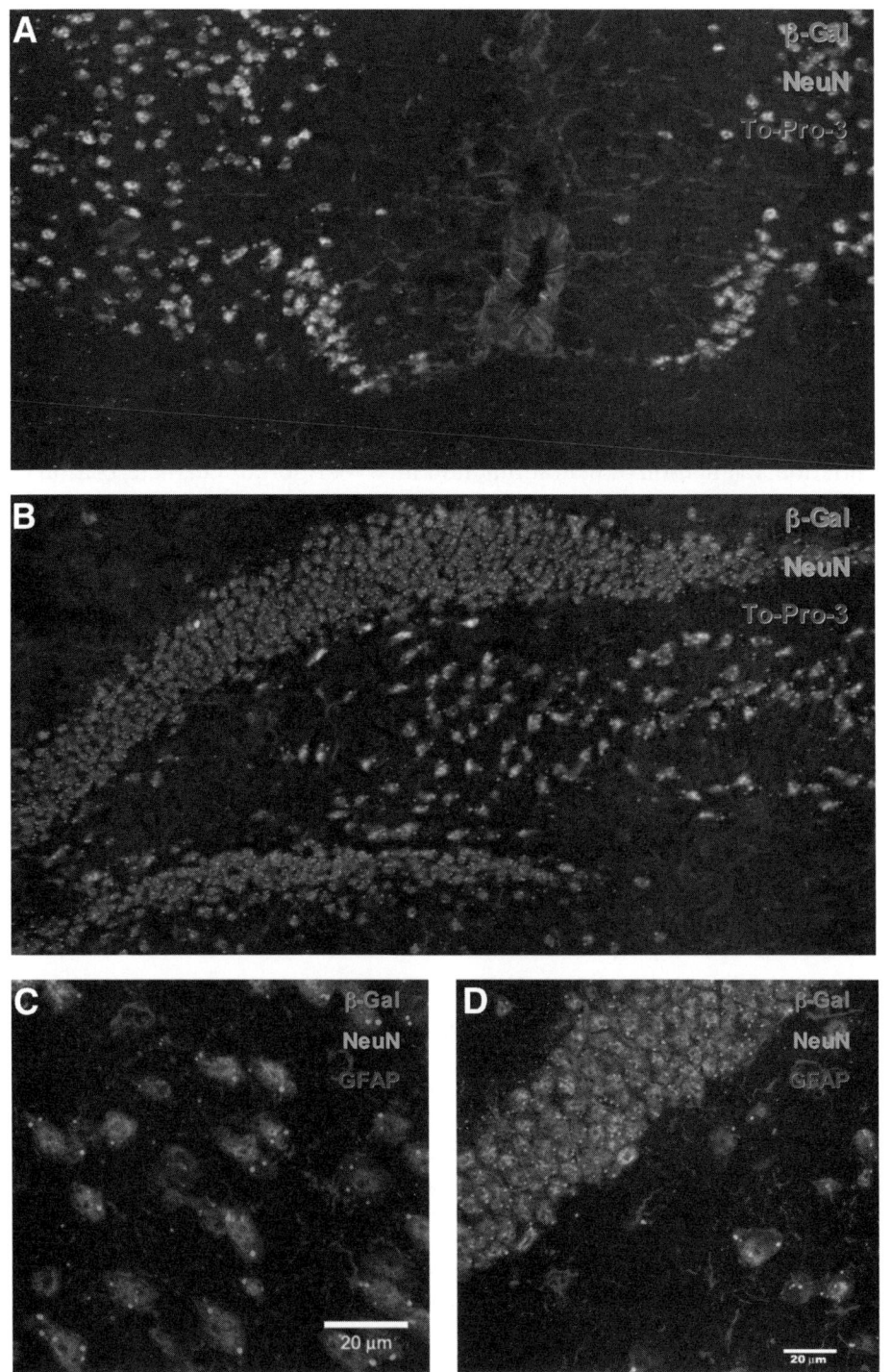

Fig. 5.

resulted from true biological plasticity or culture artifact? Yet, it is important to determine the molecular mechanisms responsible for maintaining plasticity and for neural differentiation to better manipulate the system and to ensure the safety of patients receiving BMSC treatments. Continued evaluation of the molecular, antigenic, and morphological phenotype of various stem cell components within whole bone marrow is vital. To eventually ensure consistent graft properties and a reliable degree of clinical efficacy, specific signature characteristics must be discovered. To establish product consistency, standardized isolation, selection, and expansion criteria need to be implemented. Significant advances in this regard have been accomplished with HSCs and MAPCs, but continued characterization and standardization will be critical to accurately determining clinical potential and efficacy.

Perhaps the most debated issue surrounding the neural potential of BMSCs is the mechanism by which these cells acquire neuronal and glial characteristics. There is continued concern surrounding the contribution of fusion of BMSCs with other cells in coculture or in vivo. Terada et al. *(47)* demonstrated that MSCs could undergo fusion with embryonic stem cells, albeit at a very low frequency. This is unlikely to be solely responsible for neural differentiation of BMSCs, considering neural differentiation occurs without the presence of coculture in multiple studies *(40,51,77–81)*. However, until it is convincingly examined through intense chromosomal analysis of terminally differentiated BMSCs following intracerebral transplantation, the role of fusion in BMSC neural differentiation cannot be negated.

Another pressing issue is whether the observed plasticity results in functional neurons or oligodendrocytes capable of adequately replacing lost or dysfunctional cells in such conditions as PD. For instance, it is not sufficient to demonstrate that BMSCs differentiate to express NeuN and TH. At a minimum, these cells must secrete dopamine in the caudate nucleus and

---

Fig. 5. *(previous page)* MAPC-derived neurons and astrocytes in cortex and hippocampus. Confocal microscopy of triple-label immunofluorescence staining in brain from Rosa-26 MAPC and WT chimera reveals β-galactosidase–positive inclusions throughout the cortex **(A)** and dentate gyrus and hilus of hippocampus **(B)**. Higher magnification reveals extensive colabeling NeuN and β-galactosidase and less colabeling of GFAP and β-galactosidase in midline cortex **(C)** and in the granule cell layer of the dentate gyrus **(D)**. No cells were observed that colabeled with GFAP and NeuN. Thus, extensive MAPC-derived neuronal (NeuN+) engraftment is observed, with limited MAPC contribution to glial (GFAP+) cell types. Secondary fluorescent labels are β-gal/Cy3, NeuN/Alexa488, and GFAP/Cy5. Modified and reprinted from refs. *79* and *84*.

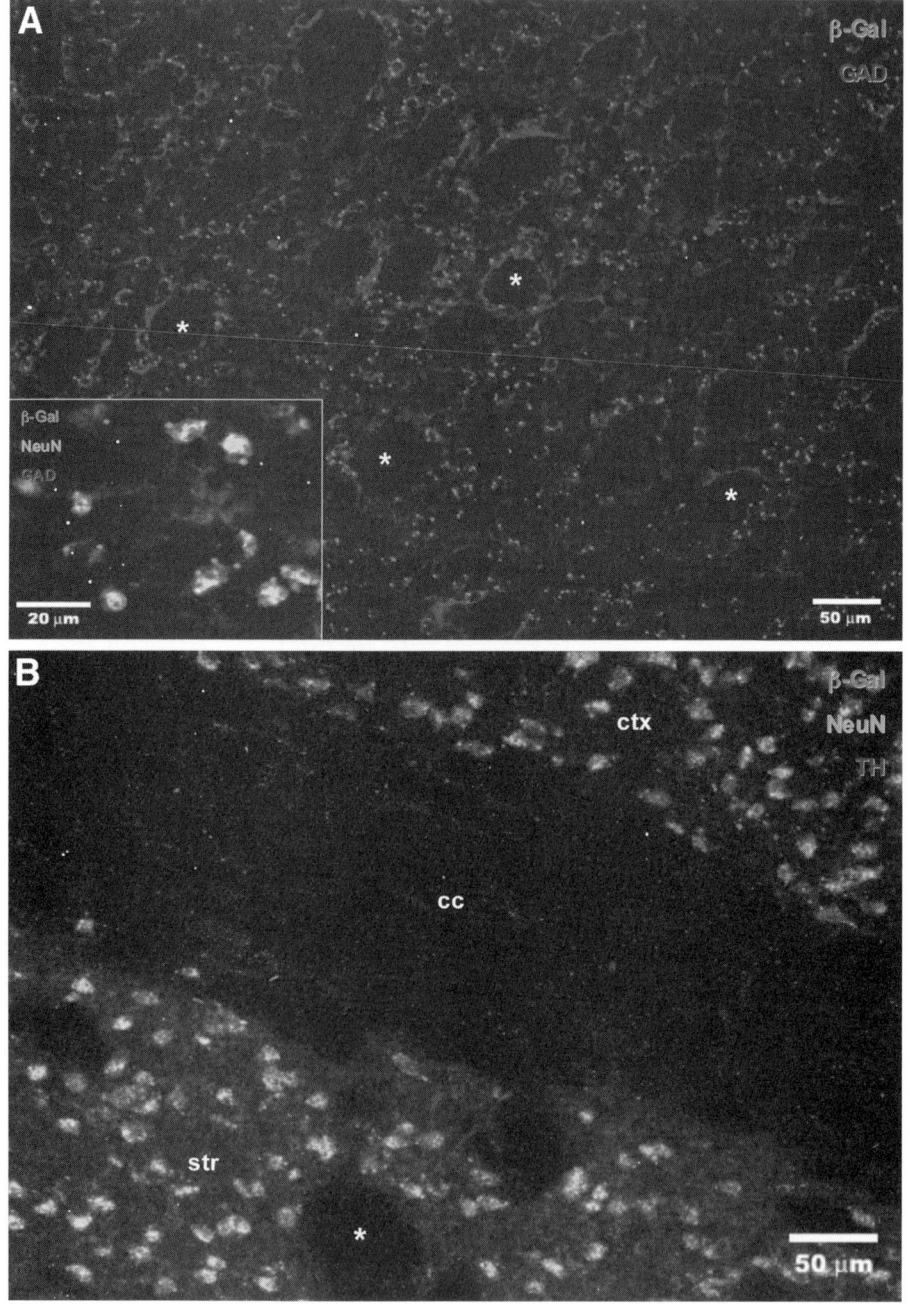

Fig. 6.

putamen; optimally, BMSCs would be transplanted into the nigral compartments and recapitulate the nigrostriatal circuitry, establish appropriate pre- and postsynaptic connectivity, and regulate striatothalamic output in a manner similar to an intact neuroanatomical system. This has yet to be conclusively demonstrated. As found in fetal neural graft studies, the nigrostriatal dopaminergic system is forgiving, but most other cell replacement strategies will require functionally integrated donor cells to provide effective therapy. Experiments that indicate this are difficult and labor-intensive but must be performed to help understand the mechanism by which BMSCs provide, or fail to provide, clinical efficacy in various conditions.

Although encouraging reports of the benefit of BMSCs in experimental neurologic disease exist, the safety of each BMSC entity must be established. A principle concern with BMSC transplantation involves the potential implantation of neoplastic or precancerous cells. To date, no reports of tumors arising from transplanted BMSCs exist, but it will be critical to understand the biology of BMSC plasticity to ensure patient safety. For instance, certain populations of BMSCs can be cultured and expanded extensively, and whether this increases the chance of inducing genetic mutations that can lead to malignant transformation must be determined. Furthermore, as new, more complex manipulations are developed and implemented, the safety of each variant must be ascertained. In addition to malignancy, other risks must be evaluated. These include graft vs. host disease (unlikely except with whole bone marrow or HSCs) and transmission of such pathogens as HIV or HCV. (Diligent screening should alleviate this concern.) Other risks to be examined are inappropriate engraftment or differentiation, such as the overexpression of dopamine in PD, resulting in tardive dyskinesias, and (although unlikely) stimulation of host autoimmunity or graft rejection. Some concerns surrounding BMSC transplantation relate to standard surgical risk, e.g., hemorrhage and infection. The cost–benefit

---

Fig. 6. *(previous page)* Site-specific neurotransmitter expression in the striatum is anatomically normal and demonstrates specific GABAergic phenotype of some MAPC-derived striatal neurons. Low-power striatal image shows normal GAD immunoreactivity among widespread β-Gal labeling (**A**), including MAPC-derived GABAergic neurons (**A**, inset), which are appropriately positioned within striosomal/matrix areas. TH was also appropriately expressed in chimeric striatum (str) that exhibited extensive β-gal immunoreactivity and was not expressed in corpus callosum (cc) or cortical (ctx) regions (**B**). Striatal fiber tracts (*) were normally distributed and proportioned in chimeric brains (**A,B**). β-gal /Cy3 (**A,B**), NeuN/ Alexa 488 (**A** inset, **B**), GAD/Cy5 (**A**), and TH/Cy5 (**B**). Reprinted with permission from ref. *84.*

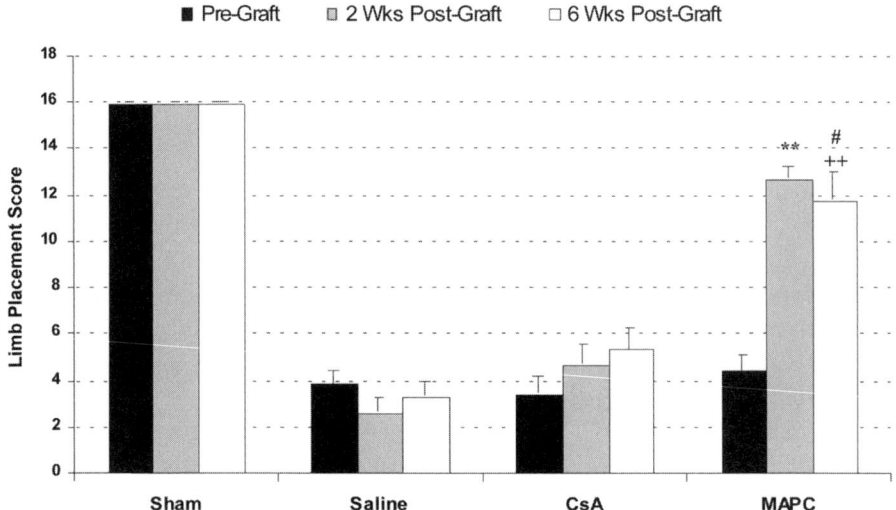

Fig. 7. MAPC-mediated mitigation of ischemia-induced neurologic deficits evaluated with limb placement test (scale 0, severe sensorimotor dysfunction, to 16, normal). At 2-wk posttransplantation, neurologic deficits were significantly ameliorated in rats with MAPC transplants, compared to controls receiving saline (**$p < 0.01$) or cyclosporin A (CsA) (**$p < 0.01$). At 6-wk posttransplantation, MAPC-treated animals continued to perform significantly better than saline (++$p < 0.01$) and CsA controls (++$p < 0.01$). Before transplantation, rats that would receive MAPC grafts were not statistically different from controls (mean = 4.5; $p > 0.05$), whereas grafted animals showed a major functional improvement, compared to pregraft performance 2- and 6-wk posttransplantation. The improved function by hMAPC transplantation was seen at 6 wk postgrafting (**$p < 0.01$; #$p = 0.03$, respectively). Groups: Sham ($n = 4$), sham ischemic rats with sham transplants; saline ($n = 14$), ischemic rats with intracerebral phosphate-buffered saline (PBS) injections and intraperitoneal (ip) saline injections; CsA ($n = 8$), ischemic rats with intracerebral PBS injections and ip CsA injections; MAPC ($n = 15$), ischemic rats with intracerebral hMAPC grafts and ip CsA injections. Nonparametric technique was used for statistical analysis. Data were presented as mean ± SEM. Adapted and reprinted with permission from ref. *85*.

ratio of these risks can only be determined once treatment efficacy is established. Finally, culture methods that require the coincubation of nonhuman cells and tissues for support or differentiation, nonhuman serum, or other animal-derived media reagents should be closely evaluated because of the risk of transmission of xenopathogens. Every attempt should be made to limit the amount of the use of these reagents in culture.

Ultimately, clinical efficacy must be rigorously compared to current standards of medical care for each condition. Initial preclinical studies involving transplantation in rodent and, in some cases, nonhuman primate models of human neurologic disease, are in progress and cite encouraging results thus far. However, to proceed to clinical trials, standardized isolation and culture methods must be optimized for preclinical benefit. For instance, it is unclear whether significant BMSC expansion affects the neural potential of the cells. Optimal cell preparations, including cell density, concentration protocols, and volumes, must be determined regarding efficacy of treatment. It must be determined whether naïve or differentiated cells are more effective. Protective factors, such as growth factors and antiapoptotic reagents, should be evaluated for inclusion with grafted cells to improve graft survival. Direct CNS injection should be compared with systemic administration to determine the optimal route of administration. Surgical techniques, such as approach, number and location of transplant sites, and avoidance of complications (e.g., intracranial hemorrhage), need to be optimized. Patient selection and exclusionary criteria must be developed and modified over time. Such factors as age, comorbidities, severity of illness, latency from injury or disease onset, and immunological phenotype must be established. Careful adherence to accepted and relevant outcome measures must be rigorously followed, and appropriate placebo or sham controls must be incorporated into definitive clinical trials in a prospective, double-blind manner to ensure valid conclusions of efficacy.

CONCLUSION

Many steps need to be taken prior to routine therapeutic transplantation of BMSCs for neurologic disease; yet, substantial progress toward this goal has been made. The utilization of bone marrow stem cells in the treatment of neurologic injury and disease will soon be tested clinically. Although substantial basic and preclinical research should be conducted to identify optimal cell types and administration strategies, BMSCs present enticing promise for use in the treatment of traumatic, vascular, neurodegenerative, and metabolic CNS processes. Future innovative strategies will likely uncover new BMSC populations and more efficacious means of treating insults to the human CNS.

REFERENCES

1. Friedenstein, A. J., Deriglasova, U. F., Kulagina, N. N., et al. (1974) Precursors for fibroblasts in different populations of hematopoietic cells as detected by the in vitro colony assay method *Exp. Hematol.* **2,** 83–92.
2. Azizi, S. A., Stokes, D., Augelli, B. J., et al. (1998) Engraftment and migration of human bone marrow stromal cells implanted in the brains of albino rats-similarities to astrocyte grafts. *Proc. Natl. Acad. Sci. USA* **95,** 3908–3913.

3. Brazelton, T. R., Rossi, F. M. V., Keshet, G. I., and Blau, H. M. (2000) From marrow to brain: expression of neuronal phenotypes in adult mice. *Science* **290,** 1775–2122.

4. Mezey, E., Chandross, K. J., Harta, G., et al. Turning blood into brain; cells bearing neuronal antigens generated in vivo from bone marrow. *Science* **290,** 1779–1781. 2000.

5. Corti, S., Locatelli, F., Donadoni, C., et al. (2002) Neuroectodermal and microglial differentiation of bone marrow cells in the mouse spinal cord and sensory ganglia. *J. Neurosci. Res.* **70,** 721–733.

6. Nakano, K., Migita, M., Mochizuki, H., and Shimada, T. (2001) Differentiation of transplanted bone marrow cells in the adult mouse brain. *Transplantation* **71,** 1735–1740.

7. Priller, J., Persons, D. A., Klett, F. F., et al. (2001) Neogenesis of cerebellar Purkinje neurons from gene-marked bone marrow cells in vivo. *J. Cell. Biol.* **155,** 733–738.

8. Weimann, J. M., Charlton, C. A., Brazelton, T. R., et al. (2003) Contribution of transplanted bone marrow cells to Purkinje neurons in human adult brains. *Proc. Natl. Acad. Sci. USA* **100,** 2088–2093.

9. Hudson, J. E., Chen, N., Song, S., et al. (2004) Green fluorescent protein bone marrow cells express hematopoietic and neural antigens in culture and migrate within the neonatal rat brain. *J. Neurosci. Res.* **76,** 255–264.

10. Mezey, E., Key, S., Vogelsang, G., et al. (2003) *Proc. Natl. Acad. Sci. USA* **100,** 1634–1369.

11. Cogle, C. R., Yachnis, A. T., Laywell, E. D., et al. (2004) Bone marrow transdifferentiation in brain after transplantation: a retrospective study. *Lancet* **363,** 1432–1437.

12. Castro, R. F., Jackson, K. A., Goodell, M. A., et al. (2002) Failure of bone marrow cells to transdifferentiate into neural cells in vivo. *Science* **297,** 1299.

13. Wehner, T., Bontert, M., Eyupoglu, I., et al. (2003) Bone marrow-derived cells expressing green fluorescent protein under the control of the glial fibrillary acidic protein promoter do not differentiate into astrocytes in vitro and in vivo. *J. Neurosci.* **23,** 5004–5011.

14. Mahmood, A., Lu, D., Wang, L., et al. (2001) Treatment of traumatic brain injury in female rats with intravenous administration of bone marrow stromal cells. *Neurosurgery* **49,** 1196–1203.

15. Sasaki, M., Honmou, O., Akiyama, Y., et al. (2001) Transplantation of an acutely isolated bone marrow fraction repairs demyelinated adult rat spinal cord axons. *Glia* **35,** 26–34.

16. Yagi, T., McMahon, E. J., Takikita, S., et al. (2004) Fate of donor hematopoietic cells in demyelinating mutant mouse, twitcher, following transplantation of GFP+ bone marrow cells. *Neurobiol. Dis.* **16,** 98–109.

17. Park, K. W., Eglitis, M. A., and Mouradian, M. M. (2001) Protection of nigral neurons by GDNF-engineered marrow cell transplantation. *Neurosci. Res.* **40,** 315–323.

18. Matzner, U., Hartmann, D., Lullmann-Rauch, R., et al. (2002) Bone marrow stem cell-based gene transfer in a mouse model for metachromatic leukodystrophy: effects on visceral and nervous system disease manifestations. *Gene Ther.* **9,** 53–63.
19. Lee, J., Elkahloun, A. G., Messina, S. A., et al. (2003) Cellular and genetic characterization of human adult bone marrow-derived neural stem-like cells: a potential antiglioma cellular vector. *Cancer Res.* **63,** 8877–8889.
20. Hess, D. C., Hill, W. D., Martin-Studdard, A., et al. (2002) Bone marrow as a source of endothelial cells and NeuN-expressing cells After stroke. *Stroke* **33,** 1362–1368.
21. Zhang, Z. G., Zhang, L., Jiang, Q., and Chopp, M. (2002) Bone marrow-derived endothelial progenitor cells participate in cerebral neovascularization after focal cerebral ischemia in the adult mouse. *Circ. Res.* **90,** 284–288.
22. Li, Y., Chen, J., Wang, L., et al. (2001) Intracerebral transplantation of bone marrow stromal cells in a 1-methyl-4-phenyl-1,2,3,6-tetrahydropyridine mouse model of Parkinson's disease. *Neurosci. Lett.* **316,** 67–70.
23. Chen, J., Li, Y., Wang, L., Lu, M., and Chopp, M. (2002) Caspase inhibition by Z-VAD increases the survival of grafted bone marrow cells and improves functional outcome after MCAo in rats. *J. Neurol. Sci.* **199,** 17–24.
24. Iihoshi, S., Honmou, O., Houkin, K., et al. (2004) A therapeutic window for intravenous administration of autologous bone marrow after cerebral ischemia in adult rats. *Brain Res.* **1007,** 1–9.
25. Chen, J., Li, Y., and Chopp, M. (2000) Intracerebral transplantation of bone marrow with BDNF after MCAo in rat. *Neuropharmacology* **39,** 711–716.
26. Pallavicini, M. G., Summers, L. J., Dean, P. N., and Gray, J. W. (1985) Enrichment of murine hemopoietic clonogenic cells by multivariate analyses and sorting. *Exp. Hematol.* **13,** 1173–1181.
27. Scharenberg, C. W., Harkey, M. A., and Torok-Storb, B. (2002) The ABCG2 transporter is an efficient Hoechst 33342 efflux pump and is preferentially expressed by immature human hematopoietic progenitors. *Blood* **99,** 507–512.
28. Hao, H. N., Zhao, J., Thomas, R. L., et al. (2003) Fetal human hematopoietic stem cells can differentiate sequentially into neural stem cells and then astrocytes in vitro. *J. Hematother. Stem Cell Res.* **12,** 23–32.
29. Eglitis, M. A. and Mezeley, E. (1997) Hematopoietic cells differentiate into both microglia and macroglia of adult mice. *Proc. Natl. Acad. Sci. USA* **94,** 4080–4085.
30. Wagers, A. J., Sherwood, R. I., Christensen, J. L., and Weissman, I. L. (2002) Little evidence for developmental plasticity of adult hematopoietic stem cells. *Science* **297,** 2256–2259.
31. Ono, K., Takii, T., Onozaki, K., et al. (1999) Migration of exogenous immature hematopoietic cells into adult mouse brain parenchyma under GFP-expressing bone marrow chimera. *Biochem. Biophys. Res. Commun.* **262,** 610–614.
32. Ono, K., Yoshihara, K., Suzuki, H., et al. (2003) Preservation of hematopoietic properties in transplanted bone marrow cells in the brain. *J. Neurosci. Res.* **72,** 503–507.

33. Hess, D. C., Abe, T., Hill, W. D., et al. (2004) Hematopoietic origin of micro-glial and perivascular cells in brain. *Exp. Neurol.* **186,** 134–144.
34. Muraro, P. A., Cassiani, I. R., and Martin, R. (2003) Hematopoietic stem cell transplantation for multiple sclerosis: current status and future challenges. *Curr. Opin. Neurol.* **16,** 299–305.
35. Bonilla, S., Alarcon, P., Villaverde, R., et al. (2002) Haematopoietic progenitor cells from adult bone marrow differentiate into cells that express oligodendroglial antigens in the neonatal mouse brain. *Eur. J. Neurosci.* **15,** 575–582.
36. Vitry, S., Bertrand, J. Y., Cumano, A., and Dubois-Dalcq, M. (2003) Primordial hematopoietic stem cells generate microglia but not myelin-forming cells in a neural environment. *J. Neurosci.* **23,** 10724–10731.
37. Leimig, T., Mann, L., Martin, M. P., et al. (2002) Functional amelioration of murine galactosialidosis by genetically modified bone marrow hematopoietic progenitor cells. *Blood* **99,** 3169–3178.
38. Fibbe, W. E. and Noort, W. A. (2003) Mesenchymal stem cells and hematopoietic stem cell transplantation. *Ann. NY Acad. Sci.* **996,** 235–244.
39. Kopen, G. C., Prockop, D. J., and Phinney, D. G. (1999) Marrow stromal cells migrate throughout forebrain and cerebellum, and they differentiate into astrocytes after injection into neonatal mouse brains. *Proc. Natl. Acad. Sci. USA* **96,** 10711–10716.
40. Sanchez-Ramos, J., Song, S., Cardozo-Pelaez, F., et al. (2000) Adult bone marrow stromal cells differentiate into neural cells in vitro. *Exp. Neurol.* **164,** 247–256.
41. Woodbury, D., Schwarx, E. J., Prockop, D. J., and Black, I. B. (2000) Adult rat and human bone marrow stromal cells differentiate into neurons. *J. Neurosci. Res.* **61,** 364–370.
42. Munoz-Elias, G., Woodbury, D., and Black, I. B. (2003) Marrow stromal cells, mitosis, and neuronal differentiation: stem cell and precursor functions. *Stem Cells* **21,** 437–448.
43. Munoz-Elias, G., Marcus, A. J., Coyne, T. M., et al. (2004) Adult bone marrow stromal cells in the embryonic brain: engraftment, migration, differentiation, and long-term survival. *J. Neurosci.* **24,** 4585–4595.
44. Hung, S. C., Cheng, H., Pan, C. Y., et al. (2002) In vitro differentiation of size-sieved stem cells into electrically active neural cells. *Stem Cells* **20,** 522–529.
45. Dezawa, M., Kanno, H., Hoshino, M., et al. (2004) Specific induction of neuronal cells from bone marrow stromal cells and application for autologous transplantation. *J. Clin. Invest.* **113,** 1701–1710.
46. Lou, S., Gu, P., Chen, F., et al. (2003) The effect of bone marrow stromal cells on neuronal differentiation of mesencephalic neural stem cells in Sprague-Dawley rats. *Brain Res.* **968,** 114–121.
47. Terada, N., Hamazaki, T., Oka, M., et al. (2002) Bone marrow cells adopt the phenotype of other cells by spontaneous cell fusion. *Nature* **416,** 542–545.
48. Alvarez-Dolado, M., Pardal, R., Garcia-Verdugo, J. M., et al. (2003) Fusion of bone-marrow-derived cells with Purkinje neurons, cardiomyocytes and hepatocytes. *Nature* **425,** 968–973.

49. Jin, H. K., Carter, J. E., Huntley, G. W., and Schuchman, E. H. (2002) Intracerebral transplantation of mesenchymal stem cells into acid sphingomyelinase-deficient mice delays the onset of neurological abnormalities and extends their life span. *J. Clin. Invest.* **109**, 1183–1191.
50. Jin, H. K. and Schuchman, E. H. (2003) Ex vivo gene therapy using bone marrow-derived cells: combined effects of intracerebral and intravenous transplantation in a mouse model of Niemann-Pick disease. *Mol. Ther.* **8**, 876–885.
51. Schwarz, E. J., Alexander, G. M., Prockop, D. J., and Azizi, S. A. (1999) Multipotential marrow stromal cells transduced to produce L-DOPA: engraftment in a rat model of Parkinson disease. *Hum. Gene Ther.* **10**, 2539–2549.
52. Jendelova, P., Herynek, V., Urdzikova, L., et al. (2004) Magnetic resonance tracking of transplanted bone marrow and embryonic stem cells labeled by iron oxide nanoparticles in rat brain and spinal cord. *J. Neurosci. Res.* **76**, 232–243.
53. Chopp, M., Zhang, X. H., Li, Y., et al. (2000) Spinal cord injury in rat: treatment with bone marrow stromal cell transplantation. *NeuroReport* **11**, 3001–3005.
54. Wu, S., Suzuki, Y., Ejiri, Y., et al. (2003) Bone marrow stromal cells enhance differentiation of cocultured neurosphere cells and promote regeneration of injured spinal cord. *J. Neurosci. Res.* **72**, 343–351.
55. Ohta, M., Suzuki, Y., Noda, T., et al. (2004) Bone marrow stromal cells infused into the cerebrospinal fluid promote functional recovery of the injured rat spinal cord with reduced cavity formation. *Exp. Neurol.* **187**, 266–278.
56. Akiyama, Y., Radtke, C., and Kocsis, J. D. (2002) Remyelination of the rat spinal cord by transplantation of identified bone marrow stromal cells. *J. Neurosci.* **22**, 6623–6630.
57. Dezawa, M., Takahashi, I., Esaki, M., et al. (2001) Sciatic nerve regeneration in rats induced by transplantation of in vitro differentiated bone-marrow stromal cells. *Eur. J. Neurosci.* **14**, 1771–1776.
58. Cuevas, P., Carceller, F., Dujovny, M., et al. (2002) Peripheral nerve regeneration by bone marrow stromal cells. *Neurol. Res.* **24**, 634–638.
59. Mahmood, A., Lu, D., Wang, L., and Chopp, M. (2002) Intracerebral transplantation of marrow stromal cells cultured with neurotrophic factors promotes functional recovery in adult rats subjected to traumatic brain injury. *J. Neurotrauma* **19**, 1609–1617.
60. Lu, D., Li, Y., Wang, L., et al. (2001) Intraarterial administration of marrow stromal cells in a rat model of traumatic brain injury. *J. Neurotrauma* **18**, 813–819.
61. Lu, D., Mahmood, A., Wang, L., et al. (2001) Adult bone marrow stromal cells administered intravenously to rats after traumatic brain injury migrate into brain and improve neurological outcome. *NeuroReport* **12**, 559–563.
62. Mahmood, A., Lu, D., Yi, L., et al. (2001) Intracranial bone marrow transplantation after traumatic brain injury improving functional outcome in adult rats. *J. Neurosurg.* **94**, 589–595.
63. Lu, D., Li, Y., Mahmood, A., et al. (2002) Neural and marrow-derived stromal cell sphere transplantation in a rat model of traumatic brain injury. *J. Neurosurg.* **97**, 935–940.

64. Mahmood, A., Lu, D., Lu, M., and Chopp, M. (2003) Treatment of traumatic brain injury in adult rats with intravenous administration of human bone marrow stromal cells. *Neurosurgery* **53**, 697–702.

65. Chen, X., Li, Y., Wang, L., et al. (2002) Ischemic rat brain extracts induce human marrow stromal cell growth factor production. *Neuropathology* **22**, 275–279.

66. Mahmood, A., Lu, D., and Chopp, M. (2004) Intravenous administration of marrow stromal cells (MSCs) increases the expression of growth factors in rat brain after traumatic brain injury. *J. Neurotrauma* **21**, 33–39.

67. Li, Y., Chopp, M., Chen, J., et al. (2000) Intrastriatal transplantation of bone marrow nonhematopoietic cells improves functional recovery after stroke in adult mice. *J. Cereb. Blood Flow Metab.* **20**, 1311–1319.

68. Chen, J., Li, Y., Wang, L., et al. (2001a) Therapeutic benefit of intracerebral transplantation of bone marrow stromal cells after cerebral ischemia in rats. *J. Neurol. Sci.* **189**, 49–57.

69. Chen, J., Li, Y., Wang, L., et al. (2001b) Therapeutic benefit of intravenous administration of bone marrow stromal cells after cerebral ischemia in rats. *Stroke* **32**, 1005–1011.

70. Wang, L., Li, Y., Chen, J., et al. (2002) Ischemic cerebral tissue and MCP-1 enhance rat bone marrow stromal cell migration in interface culture. *Exp. Hematol.* **30**, 831–836.

71. Wang, L., Li, Y., Chen, X., et al. (2002) MCP-1, MIP-1, IL-8 and ischemic cerebral tissue enhance human bone marrow stromal cell migration in interface culture. *Hematology* **7**, 113–117.

72. Zhong, C., Qin, Z., Zhong, C. J., et al. (2003) Neuroprotective effects of bone marrow stromal cells on rat organotypic hippocampal slice culture model of cerebral ischemia. *Neurosci. Lett.* **342**, 93–96.

73. Chen, J., Li, Y., Katakowski, M., et al. (2003) Intravenous bone marrow stromal cell therapy reduces apoptosis and promotes endogenous cell proliferation after stroke in female rat. *J. Neurosci. Res.* **73**, 778–786.

74. Chen, J., Zhang, Z. G., Li, Y., et al. (2003) Intravenous administration of human bone marrow stromal cells induces angiogenesis in the ischemic boundary zone after stroke in rats. *Circ. Res.* **92**, 692–699.

75. Chen, J., Li, Y., Zhang, R., et al. (2004) Combination therapy of stroke in rats with a nitric oxide donor and human bone marrow stromal cells enhances angiogenesis and neurogenesis. *Brain Res.* **1005**, 21–28.

76. Reyes, M., Lund, T., Lenvik, T., et al. (2001) Purification and ex vivo expansion of postnatal human marrow mesodermal progenitor cells. *Blood* **98**, 2615–2625.

77. Reyes, M. and Verfaillie, C. M. (2001) Characterization of multipotent adult progenitor cells, a subpopulation of mesenchymal stem cells. *Ann. NY Acad. Sci.* **938**, 231–233.

78. Jiang, Y., Vaessen, B., Lenvik, T., et al. (2002) Multipotent progenitor cells can be isolated from postnatal murine bone marrow, muscle, and brain. *Exp. Hematol.* **30**, 896–904.

79. Jiang, Y., Jahagirdar, B. N., Reinhardt, R. L., et al. (2002) Pluripotency of mesenchymal stem cells derived from adult marrow. *Nature* **418,** 41–49.
80. Jiang, Y., Henderson, D., Blackstad, M., et al. (2003) Neuroectodermal differentiation from mouse multipotent adult progenitor cells. *Proc. Natl. Acad. Sci. USA* **100(Suppl 1),** 11854–11860.
81. Schwartz, R. E., Reyes, M., Koodie, L., et al. (2002) Multipotent adult progenitor cells from bone marrow differentiate into functional hepatocyte-like cells. *J. Clin. Invest.* **109,** 1291–1302.
82. Verfaillie, C. M., Schwartz, R., Reyes, M., and Jiang, Y. (2003) Unexpected potential of adult stem cells. *Ann. NY Acad. Sci.* **996,** 231–234.
83. Reyes, M., Dudek, A., Jahagirdar, B., et al. (2002) Origin of endothelial progenitors in human postnatal bone marrow. *J. Clin. Invest.* **109,** 337–346.
84. Keene, C. D., Ortiz-Gonzalez, X. R., Jiang, Y., et al. (2003) Neural differentiation and incorporation of bone marrow-derived multipotent adult progenitor cells after single cell transplantation into blastocyst stage mouse embryos. *Cell Transplant.* **12,** 201–213.
83. Zhao, L. R., Duan, W. M., Reyes, M., et al. (2002) Human bone marrow stem cells exhibit neural phenotypes and ameliorate neurological deficits after grafting into the ischemic brain of rats. *Exp. Neurol.* **174,** 11–20.
84. Lu, P., Jones, L. L., Snyder, E. Y., and Tuszynski, M. H. (2003) Neural stem cells constitutively secrete neurotrophic factors and promote extensive host axonal growth after spinal cord injury. *Exp. Neurol.* **181,** 115–129.
85. Hagan, M., Wennersten, A., Meijer, X., et al. (2003) Neuroprotection by human neural progenitor cells after experimental contusion in rats. *Neurosci. Lett.* **351,** 149–152.

8

Cell Therapy for Models of Pain and Traumatic Brain Injury

Mary Eaton and Jacqueline Sagen

ABSTRACT

This chapter reviews the cell transplantation strategies that have been explored as potential options in the treatment of pain and traumatic brain injury (TBI). As the goals of these two therapeutic targets are widely disparate, approaches have evolved along distinctive paths. Thus, although the provision of a local cellular source of pharmacologic analgesic molecules may be most appropriate in the management of chronic pain, central nervous system (CNS) repair after traumatic injury will most likely require replacement of lost neural populations and reestablishment of appropriate neurocircuitry to be maximally effective. The development of cellular strategies that would replace, or be used as, an adjunct to current clinical treatments for neuropathic pain have progressed over the past 30 yr. There are a variety of useful surgical and pharmacologic interventions, including electric stimulation, implantable mechanical pumps, and a myriad of drugs for pain relief, but cell and molecular technologies are a new frontier in pain medicine.

The earliest cell therapy studies for pain relief tested adrenal chromaffin cells from rat or bovine sources that were placed in the subarachnoid space and functioned as cell minipumps, secreting a cocktail of antinociceptive agents around the spinal cord for peripheral nerve injury, inflammatory, or arthritic pain. These animal, and later clinical, studies suggested that the spinal intrathecal space was a safe and accessible location for cell grafts. A major problem remained: a lack of homogeneous expandable cell source to supply the antinociceptive agents. Cell lines that are either naturally immortalized or can be reversibly immortalized are the next phase for a practical homogenous source. These technologies have been modeled with various murine cell lines, where cells are transplanted that downregulate their proliferative or oncogenic phenotype either before or after transplant.

From: *Contemporary Neuroscience: Cell Therapy, Stem Cells, and Brain Repair*
Edited by: C. D. Sanberg and P. R. Sanberg © Humana Press Inc., Totowa, NJ

Most recently, cell lines for pain have used molecular switches to remove the oncogenic sequence before grafting, to take advantage of the proliferative property induced by an oncogene for expansion, followed by its removal before grafting to make it safe, without the danger of tumor formation in the host. An alternate approach for current human cell lines is the use of neural or adrenal precursors, where their antinociceptive properties are induced by in vitro treatment with molecules that drive the cells to an irreversible neural or chromaffin phenotype. Although such human cell lines are at an early stage of investigation, their potential for clinical antinociception are enormous against the daunting problem of neuropathic spinal cord injury (SCI) pain.

For TBI, early transplants using fetal neural tissue has more recently led to stem cell transplants and various manipulations, including the generation of immortalized neural stem cell lines, use of neurally differentiated embryonic stem cells and neurons derived from a human teratocarcinoma cell line, and nonneural sources, e.g., bone marrow and umbilical cord. Although the ideal goal of these grafts would be to replace neural cells lost to injury and to reintegrate within the host CNS circuitry, it is more likely that their beneficial effects (when observed) are attributable to the provision of trophic support. Thus, a promising approach in transplantation strategies for improved therapeutic outcomes following TBI may be to utilize grafted cells engineered to produce appropriate neuroprotective and trophic agents as vehicles for delivery to damaged CNS sites, similar to the approach taken in the cell-based delivery for pain management.

Key Words: Immortalized; adrenal; transplant; subarachnoid; cell minipumps; NT2; stem cells.

INTRODUCTION

The transplantation of cells into the CNS for therapeutic purposes can be envisioned with several increasingly demanding goals in mind: (1) provision of structural support or permissive conduits for regenerating axons; (2) local and sustained provision of therapeutic molecules, such as pharmacologic agents and neurotrophic factors; and (3) replacement of lost cellular populations and reconstruction of local neuronal circuitry. To a large extent, the demands of the particular therapeutic application will be the key factor in determining the goals of the transplant paradigm, and this guides both the selection of optimal cell type(s) and ideal parameters (e.g., graft site, timing) for transplantation. The current chapter explores the wide variety of approaches in neural transplantation for the therapeutic management of pain and traumatic brain injury (TBI). Overlap in the goals of these two indications can be envisioned, e.g., the provision of neurotrophic or neuroprotective molecules for TBI or replacement of lost inhibitory neurocircuitry in chronic pain. Thus, this chapter reviews the various cell types and paradigms, which range from primary tissue fragments to engineered stem cell lines, that have been taken in models of chronic pain and TBI to identify breakthrough approaches in the treatment of both debilitating conditions.

MOTIVATION FOR CELL TRANSPLANTATION THERAPIES IN CHRONIC PAIN AND TBI

Chronic Pain

Despite improvements *(1)* in surgical management, physical therapy, and the availability of pharmacological agents with a variety of delivery systems, many patients following peripheral and central neural injuries continue to suffer from intractable chronic pain *(2)*. Although opioids are the most commonly used agent to control pain, only about 32% of patients receive any significant relief with long-term use *(3)*. This often leads to untoward effects associated with tolerance, tolerability, drug diversion, and other side effects *(4)*, including opioid-induced neurotoxicity. Nonopioid medications can attenuate some types of neuropathic pain but seldom remove the painful sensation completely *(5)*. Recent attempts at classification of neuropathic, nociceptive, and other pain, aided by an IASP Taskforce *(6)*, has helped the understanding of mechanisms and improvement of better treatments for chronic pain. Yet, with the frequency of inadequate or failed clinical trials, especially for chronic neuropathic pain *(5)*, the development of translational cell therapies and use of newer animal models is driving interest to more sophisticated techniques for these problems *(7–13)*.

TBI

Although the clinical outcomes following traumatic injuries to the central nervous system (CNS) can range in severity, nearly all injuries result in some degree of permanent functional deficit. Findings in recent years have characterized ischemic, excitotoxic, and inflammatory processes that occur in the early stages following primary insult; these observations have led to the exploration of acute interventive neuroprotective strategies to reduce subsequent neuropathology. Nevertheless, irreversible neuronal loss and disruption of connectivity occur and will likely require reparative strategies for more complete recovery of function. Furthermore, recent studies have revealed that progressive neuronal and white matter damage continues during prolonged periods after the initial injury. Thus, in addition to acute intervention, long-term strategies to replace lost cellular populations are being explored.

POTENTIAL STRATEGIES
FOR CELL-BASED INTERVENTIVE THERAPIES
Cellular Minipumps for the Treatment of Pain

Chronic neuropathic pain following damage to the peripheral or CNS has been difficult to treat clinically *(14)*. As an illustration of the severity of pain following spinal cord injury (SCI), patients often report pain, rather than immobility, as the major deterrent to good quality of life *(15)*. Pharmacological pain management is based on nonopioid and opioid analgesics, including nonsteroidal anti-inflammatory drugs (NSAIDs), cyclooxygenase 2 (COX-2) inhibitors *(16)*, calcium channel blockers *(17)*, capsaicin *(18)*, nicotine receptor agonists *(19)*, and opioids *(20)* (e.g., morphine and its derivatives). Adjunct drugs, such as antidepressants and anticonvulsants, often accompany more antinociceptive agents in certain types of pain, like diabetic neuropathy *(21)*. Gabapentin, an anticonvulsant, has become the most common medication for SCI pain *(22)*, probably owing to its calcium channel–blocking functions *(23)*. Combination of medications, such as NSAIDs with opioids, seems to be more effective in malignant pain that is associated with cancer *(24)*. However, the potential abuse of many of these agents, especially opioids, remains a major societal problem *(25)*. Oral administration is the favored route for all these drugs, but with the development of interfering side effects, intrathecal administration *(26)* (often by implanted mechanical pumps for long-term delivery; *27*) can be used. The intrathecal route has often been used for drugs in animal studies *(28,29)* to limit the area of application and dosage for optimum antinociception without side effects. From the initial use of cell grafts for pain *(30)*, the intrathecal placement of transplants was preferred, as cells in the intrathecal space can act as "cellular minipumps," which are able to release neuroactive antinociceptive molecules to affect spinal pain-processing centers in the dorsal horn.

Other potential interventive agents for the treatment of pain are based on current and developing strategies elucidated from recent research, especially concerning "central spinal sensitization" and the spinal mechanisms thought to be the origins and ongoing causes of chronic pain *(31)*, even when the injury is peripheral in location *(32)*. For example, persistent small afferent input, as generated by tissue or nerve damage, results in a hyperalgesia at the site of injury and a tactile allodynia in areas adjacent to the site. Hyperalgesia is the result of sensitization of the peripheral terminal and a central (or spinal) facilitation evoked by persistent small afferent input. The allodynia reflects a central sensitization, with excitatory neurotransmitter (e.g., glutamate and substance P) release, initiating a cascade of downstream events, such as release of nitric oxide (NO), various COX products, and activation of several key kinase enzymes. Specific receptors mediate the initial events, i.e., through the *N*-methyl-D-aspartate (NMDA) and non-NMDA glutamate receptors and neurokinin 1 substance P receptors. Specific activation of these receptors enhances prostaglandin E2 release, which then facilitates further release of spinal amino acids and peptides. Activation of specific receptors (μ/Δ opioid, α2 adrenergic, and neuropeptide Y) on spinal C fiber terminals prevents release of primary afferent peptides and spinal amino acids and blocks acute and facilitated pain states. In contrast, glutamate receptor antagonists, COX-2, and NO synthase inhibitors only act to diminish hyperalgesia. Spinal delivery of some of these agents diminishes human injury pain states, suggesting that such preclinical mechanisms may reflect the induction of some types of neuropathic pain.

Cellular Replacement Strategies in Pain Management

Thus far, the vast majority of cellular transplantation approaches for chronic pain management have utilized the cellular minipump method. However, it is possible to envision a cellular replacement strategy for more severe cases of chronic pain consequent to injury of the spinal cord. For example, a likely candidate for this type of strategy may be the particularly vulnerable dorsal horn inhibitory interneurons, which are thought to restrict ongoing pain under normal circumstances. The barrage of activity in damaged primary afferents and excessive excitatory amino acid release may result in excitotoxic insult to these small inhibitory interneurons in the spinal cord *(33)*. In support of this theory, an increased incidence of hyperchromatic "dark neurons" in the superficial spinal or medullary dorsal horn is found following peripheral nerve injury, and this can be further exacerbated by pharmacologic blockade of inhibitory neurotransmission *(34,35)*. Coinciding with the rise of dark neurons in these areas are spontaneously active neurons, as well as neurons with expanded receptive fields—those that respond to nonnoxious stimulation of adjacent dermatomes *(36,37)*. Dark neurons may be indicative of trans-synaptic degeneration or atrophy and are likely to include functionally impaired inhibitory interneurons *(35,38)*.

γ-Aminobutyric acid (GABA) is a major inhibitory neurotransmitter found in the spinal cord that is concentrated in the superficial laminae of the dorsal horn *(39,40)*, where sensory, particularly nociceptive, processing predominates. An important role for GABA in sensory processing is suggested by physiological and behavioral studies, which indicate primarily an inhibitory function in the transmission of noxious stimulation *(41,42)*. Thus, it is conceivable that a loss of GABAergic inhibitory mechanisms in the spinal dorsal horn leads to sustained hyperexcitability in persistent pain states. A significant decline in laminae I-III GABA immunoreactivity (GABA-IR) was found after sciatic nerve transection (43), and reductions in both GABA-IR and GAD have been observed following chronic constriction injury (CCI) of the sciatic nerve (44–47). In addition, using TUNEL labeling, cell death in the superficial dorsal horn has been observed following CCI and sciatic neurectomy; this could be prevented by NMDA antagonists (48,49). Using stereological estimates from EM sections, excitotoxic neuronal cell death in the superficial dorsal horn was also observed in sciatic nerve-lesioned animals following stimulation of A fibers (50). The magnitude of neuronal loss and spinal reorganization is likely even further exaggerated in SCI from the severe necrosis and neuronal loss as a result of the mechanical trauma, as well as secondary neuropathology. A dramatic loss in spinal GABAergic neurons occurs after ischemic spinal injury, and $GABA_B$ agonists can reverse mechanical allodynia in the early postinjury phase (51). Preliminary findings in our laboratory have also indicated a selective loss in GABA-IR in the superficial dorsal horn following excitotoxic SCI, and transplantation of GABAergic neural stem cells into the injured dorsal horn can reverse some chronic pain symptoms following SCI (52).

Cellular Replacement Strategies in TBI

In contrast with chronic pain, the majority of transplantation studies for TBI have attempted to utilize neural transplantation to replace neural elements that have been lost or damaged owing to initial trauma or secondary degeneration. Meeting this goal will be quite challenging, as it will likely require not only grafting of the appropriate complement of cell types (or potential to differentiate to appropriate cell types), but also reestablishment of proper connectivity with the intact host CNS. Nevertheless, it has been suggested that even a small (<10%) replacement of lost neurons may result in substantial functional improvement following CNS injury (53). Indeed, although overall survival of grafted cells in CNS injury models appears low, significant improvements in somatomotor and cognitive performances have been reported with a variety of transplantation interventions in the injured brain (see below for details).

Cellular Pump Strategies in TBI

In addition to transplantation approaches to replace lost cellular populations after brain trauma, several groups have refocused efforts on utilizing these grafted cells as delivery vehicles for providing therapeutic trophic molecules. Indeed, numerous studies with the initial goal of cellular replacement have revealed limited cellular differentiation to desired neuronal phenotypes and have thus attributed beneficial behavioral outcomes to the provision of local trophic factors by the grafted cells or promotion of host trophic factor upregulation. More recent attempts at taking advantage of this potential have utilized combination strategies by engineering possible replacement cells to produce increased levels of beneficial trophic support (see below for details).

CELL SOURCES FOR TRANSPLANTATION IN CHRONIC PAIN AND TBI

Cell studies that have addressed their use in animal and human pain are summarized in Table 1. Primary tissues and cells were the earliest examples and are still being studied. However, cell lines can also offer a renewable and possibly safe-to-use source of cells. Grafts of either primary or immortalized sources should reduce or eliminate side effects associated with the large doses of pharmacological agents required for centrally acting pain-reducing agents, such as opioids.

Primary Cells

Adrenal Chromaffin Cells

The earliest studies using cell transplants for pain were developed from the concept of descending inhibitory modulation of sensory information *(54–56)*, and the belief that the same agents released by cell grafts after injury could provide antinociception *(57)*. Projections from the midbrain, locus ceruleus, and ventromedial and ventrolateral medulla directly or indirectly terminate at the spinal level to modulate incoming nociceptive signals. In addition, dorsal horn interneurons provide inhibitory influences at the same termini. A variety of neurotransmitters, peptides, opioids, and (more lately) neurotrophins (e.g., brain-derived neurotrophic factor [BDNF]) have been implicated in descending inhibition. These include the endogenous neurotransmitters serotonin (5HT), noradrenaline, and GABA; the endogenous opioids β-endorphin, enkephalins, and dynorphin; such endogenous peptides as galanin; and such neurotrophins as BDNF. Many commonly used pharmacologic therapies target these agents' receptors and reuptake mechanisms to increase or imitate their presence in acute and chronic pain. Yet, it was recognized, as early as in the 1980s *(58)*, that these agents could be supplied by grafts of adrenal medullary chromaffin cells after nerve injury. Chromaffin cells contain a cocktail of antinociceptive agents, peptides, and neurotrophins *(59,60)*. These chromaffin cell grafts could be placed either

Table 1
Cell Studies for Pain

Source	Pain model
Primary	Midbrain *(223)*
Adrenal	Formalin *(62,222)*
Rat *(62)*	Nerve injury *(61)*
Bovine *(58,70,71,77,220)*	SCI *(78–81)*
Human *(82,213,221)*	Cancer *(82,83,224)*
Porcine *(222)*	Arthritis *(67,86,225)*
Cell lines	Tail-flick or chemical induction
Tumor	*(116,118–120)*
Rat PC12 *(112)*	Partial nerve injury (CCI) *(117)*
Rat B16 *(116)*	Formalin *(120)*
Human NB69 *(117)*	
AtT-20 *(118)*, Neuro2A *(119)* P19 *(120)*,	
AtT-20/hENK *(118,226)*	
Conditionally immortalized embryonic rat raphe,	Peripheral nerve injury (CCI)
SV40tsTag *(155,156,227–229)*	*(163,166,230)*
	SCI *(167,231–233)*
Embryonic rat and bovine chromaffin *(190)*	Peripheral nerve injury (CCI) *(191)*
Reversibly immortalized	Peripheral nerve injury (CCI) *(200)*
Embryonic rat chromaffin *(200)*	Formalin/c-fos induction
Embryonic rat chromaffin	*(234,235)*
overexpressing met-enkephalin *(234,235)*	
Human neuronal	
Human NT2 cell lines	Excitotoxic SCI pain *(236,237)*
Stem/precursor	Partial nerve injury (CCI) *(141)*
Rat spinal (embryonic) progenitor cells *(141)*	
Adrenal progenitors	
Bovine *(140)*; human *(87,88)*	

in the lumbar subarachnoid space after partial injury to the sciatic nerve *(61)* or after injection of formalin in the rat's hindpaw *(62)* for the antinonciceptive effect.

Many studies has sought to elucidate the agents released by these chromaffin grafts that might serve an antinociceptive role. These chromaffin cells grafts raise the levels of cerebrospinal fluid (CSF) met-enkephalin *(63)*, increase CSF levels of catecholamines *(64)*, and reduce cross-tolerance *(65)* when used with morphine for pain. More recently, changes in the spinal cord induced by nerve injury have been attenuated by chromaffin grafts, such the induction of spinal NADPH-diaphorase *(34)* and cGMP *(66)*, spinal c-fos induction *(67)*, NMDA-induced hypersensitivity *(68)*, and the loss of endogenous inhibitory GABA synthesis in the dorsal horn *(45)* that accompanies nerve injury. Likely, the adrenal transplants also block short-term spinal nociceptive facilitation, probably by stimulating some persistent cellular process that may be an important determinant, but not the only one, of their analgesic effect *(69)*. With an eye toward chromaffin cell therapy in humans, adrenal chromaffin cell grafts were prepared from xenogenic bovine sources and tested for antinociception after nerve injury *(58,70,71)*. Such sources of primary bovine chromaffin cells have been successfully and safely used in initial trials with humans with intractable cancer pain *(72–74)*. Interestingly, chromaffin cell grafts from xenogenic sources do not require extensive immunosuppressive support *(75)*, as these cells do not stimulate a significant immune response in vitro *(76)* or in vivo *(77)*.

Various animal studies have used primary medullary tissue or chromaffin cultures as a subarchnoid graft source to reduce behavioral hypersensitivity in models of SCI-induced pain *(78–80)*, including the use of polymer-encapsulated xenogenic chromaffin cells, to test removing the need for any immunosuppression *(81)*.

Similarly, human chromaffin tissue has been transplanted in humans for cancer pain *(82,83)*, because human chromaffin cells contain many of the same antinociceptive molecules *(84)*. When the immune response in the human host is examined after human chromaffin grafts, it is clear that further work on the purification and/or the immunoisolation of tissues grafted in the CNS will be necessary, particularly when the possibility of long-term and repeated grafting is considered *(85)*. In addition, there is a recent report *(86)* of human fetal adrenal transplant to treat pain associated with rheumatoid arthritis, certainly suggesting that fetal or precursor chromaffin tissue could be an antinociceptive source *(87–89)*.

Chromaffin cells are also a potential grafting source for TBI, as they can provide both pharmacologic (catecholamines) and trophic support (fibroblast growth factor 2 [FGF-2] and neurotrophins). In particular, beneficial effects of catecholamine agonists, or enhancement of noradrenergic activity, have suggested a central role for norepinephrine in improving functional recovery following TBI (including the fluid percussion model) or ischemic injury *(90,91)*. An important aspect of these findings is that treatment intervention to enhance recovery can be initiated in the subacute stage, days to weeks following injury *(90,91)*. The transplantation of catecholamine-secreting adrenal tissue in the wound cavity produced enduring restoration (at least 7–10 mo) of tactile placing in a limited frontal cortex ablation study in cats *(90)*. In addition to catecholamines, such trophic factors as FGF-2, which are also produced by chromaffin cells, have been shown to improve histopathological and behavioral outcomes following TBI. FGF-2 treatment can reduce contusion volume and cortical neuronal damage *(92)* and attenuate cognitive dysfunction *(93)* following fluid percussion injury. FGF-2 is also neuroprotective or proregenerative following SCI or focal ischemia *(94,95)*.

Primary Fetal Tissue

Neural transplantation of fetal CNS tissue has been explored in TBI models *(96,97)*. Potentially important considerations include the donor age, origin of donor tissue (homotopic vs heterotopic), maintenance of structural tissue integrity (cell suspensions vs solid tissue), interval between injury and transplantation, and graft site (at lesion, penumbra, or distant sites). Early studies used cortical ablation injuries as models for TBI, whereas models more recently used are the fluid percussion and controlled cortical impact models—both thought to more accurately mimic trauma in the clinical setting.

Fluid percussion produces irreversible neuronal atrophy in cortical areas overlying the evolving contusion, as well as areas more remote from the primary injury site, notably pyramidal cell loss in hippocampal CA3 and dentate hilar (CA4) regions and focal injury in selective thalamic nuclei *(98–100)*. Recent findings have revealed progressive axon demyelination and oligodendrocyte cell death in underlying white matter structures *(101)*. This pattern of neuropathology leads to cognitive deficits reminiscent of learning and memory dysfunction commonly observed clinically, even after mild-to-moderate TBI in humans *(100,102,103)*. The fluid percussion injury also results in sensorimotor impairments similar to those reported clinically in head injury patients *(104,105)*. Enduring cognitive and neurologic motor dysfunction may persist in parallel with progressive neurodegeneration *(98,101,106)*. Thus, successful intervention strategies may be evaluated for structural improvement in specific gray and white matter structures and behaviorally for both sensorimotor and cognitive improvements. In addition, a concern will be to establish an opportunity for these interventions amidst possible ongoing and progressive degeneration.

Results of early studies using surgical cortical ablation suggested that the CNS source of donor material was an important factor, because the fetal homotopic frontal cortex, but not cerebellar tissue transplants, could partially restore learning in a maze task *(107)*. Similarly, using the fluid percussion injury model, homotopic implants of fetal cortical cells within the contused parietal/temporal cortex produced greater neuroprotection and attenuation of cell death in the adjacent CA3 hippocampal region, compared to similar implants of fetal hippocampal cells *(108)*. Timing of transplantation postinjury may also be critical, as fetal cortical tissue grafts survived, incorporated with the host brain, and attenuated glial scarring if transplanted 2 d to 2 wk postinjury but not if transplants were delayed for 4 wk *(97)*. Nevertheless, no significant improvement in motor function was reported.

In contrast, in another study, fetal cortical tissue transplanted into the injury cavity improved motor function and attenuated cognitive dysfunction *(96)*. In this case, fetal cortical grafts were placed in the injury cavity 24 h following fluid percussion injury, and neurological motor function improved by 72-h postgrafting. However, improvement in cognitive function was only seen if animals with fetal cortical grafts also received nerve growth factor (NGF) infusion into the graft region, suggesting that combination strategies may offer the most promising approach using these transplantation paradigms. Interestingly, similar positive outcomes were not observed when cell suspensions, rather than whole-tissue pieces, were utilized. This implies that the maintenance of tissue integrity and relationships may also be important. Finally, generally, reestablishment of normal adult neurocircuitry has not been indicated with fetal tissue grafts, thus suggesting that a more likely mechanism for improved function may be neuroprotection by release of trophic substances from the grafts *(108)*.

Considering that beneficial effects of these neural grafts are most likely a result of the provision of trophic support, an alternate approach may be to provide trophic factor–rich cells or tissues. The adrenal medulla may represent one such source, as described above. Alternatively, kidney cells produce trophic factors, notably the glia-derived neurotrophic factor (GDNF). Fetal kidney tissue has been shown to reduce cortical infarction and behavioral deficits in stroke models *(109)*. Cell lines that are conditionally immortalized were also produced from fetal rat kidney cells for GDNF delivery in stroke models *(110,111)*. (Conditional immortalization approaches are discussed below.) Similar methods may be useful in reducing deficits following TBI as well. Since much of the therapeutic benefit derived from fetal tissue transplants or neural stem cell transplants (see below) in TBI models is likely owing to trophic support of local CNS tissue, rather than replacement of lost cells, which has led to the use of transplanted cells as vehicles for local delivery of trophic molecules, e.g. by genetically manipulating cells to produce desired trophic factors (see below). Thus, the grafted cells may be functioning as cellular pumps instead of integrating within the host neurocircuitry, blurring the distinction between these two theoretical strategies for repair following TBI.

Tumor Cell Lines

A cell line has the ability to be expanded in vitro; it is adequately stable in its phenotype to be characterized in vitro and after grafting; and it can be used for in vivo transplant. The archetypal adrenal medullary cell line is the rat PC12, first established from a transplantable rat adrenal pheochromocytoma *(112)*, which was shown to respond to NGF with reversible loss of mitotic activity and differentiation to a neuronal phenotype. Although originally reported to lack phenylethanolamine *N*-methyltransferase (PNMT) and epinephrine synthetic capability *(112)*, further characterization *(113)* suggest both PNMT activity and epinephrine synthesis in PC12 cells. This cell line has often been examined for its response to manipulation of agents, such as morphine analogs in pain modulation *(114)*, but it has also been tested as a grafted catecholamine source to test cell therapy for pain relief *(115)*. However, they tend to form tumors, rather than integrate and release antinociceptive agents. The mouse B16 F1C29 melanoma cell line, which also releases catecholamines, was able to reduce pain behaviors in the tail-flick model when accompanied by morphine *(116)*, but again, such grafts are tumorigenic.

The monoaminergic human NB69 neuroblastoma cell line was able to decrease neuropathic pain in a nerve injury model *(117)*, presumably related to serotonin release from the grafts, but the tumorigenic potential can be considered with a nondifferentiated tumor line. Other studies with implantation of tumor-derived cell lines (e.g., AtT-20 *[118]*, Neuro2A *[119]*, or P19 *[120]*) that overexpress opioid peptides have been attempted, but these grafts also bring the risk of tumor formation.

Although the ability of opioids to provide pain relief with SCI remains controversial, the release of enkephalin-like molecules from cell grafts genetically modified to be therapeutic was an early strategy *(118,121)*. One goal of such an opioid-based strategy would be to reduce the side effect of tolerance that develops with morphine and its analogs *(118)*.

Stem Cells and Progenitors

Neural Stem Cells/Progenitors for TBI

An exciting breakthrough in recent years was the discovery that stem cells exist in the CNS. This discovery has challenged the long-held notion that the CNS is postmitotic and suggests the possibility that the brain may possess the capacity for replacement of neural cells lost to injury or disease *(122,123)*. Cells derived from embryonic, and some areas of the adult CNS, can be harvested and grown under proliferative, self-renewing conditions and undergo differentiation to neurons, astrocytes, and oligodendrocytes in vitro. The term *neural stem cell* is used to loosely describe these multipotent self-renewing cells. Stem cells in the adult CNS are thought to be present in the subventricular zone and provide new neurons to the olfactory bulb and in the hippocampal dentate gyrus. The function of these cells in the adult CNS is unclear but may be indicative of a residual capacity for self-renewal or generation of new neurons during learning and memory processes. However, the renewal capacity of the CNS is clearly limited, as evidenced by the inability to sufficiently restore function following traumatic injury or disease.

Studies using unaltered neural stem cells in TBI models are sparse, with most focused on immortalized or engineered stem cell line derivatives (see below). Human fetal forebrain neural progenitors were shown to improve the survival of host cells when grafted immediately postinjury in a rat cortical contusion injury model *(124)*. In a cold lesion cortical injury model, producing a localized lesion and motor deficit, rat neural stem cells survived when grafted at 6 d into the penumbra but expressed few lineage markers, indicative of poor differentiation *(125)*. In our laboratory, rat cortical neural stem cells were transplanted 1 wk following a moderate fluid percussion injury *(126)*. Our preliminary findings suggested that sensorimotor deficits may be moderately improved by neural stem cell transplants, but no improvements in cognitive deficits were observed. Immunocytochemical evaluation revealed robust survival of the transplanted cells and extensive migration, particularly through white matter tracts (external capsule) and into the granular layer of the dentate gyrus, where neuronal differentiation was observed. Thus, further exogenous manipulation may be required to improve neuronal and oligodendrocyte differentiation and behavioral (particularly cognitive) outcomes.

An interesting approach that may improve survival and integration of transplanted neural stem cells in the injured brain is the use of connective tissue matrices or scaffolds to provide structural support *(127)*. Using an injectable fibronectin-based scaffold, mouse neural stem cells showed increased survival and migration when implanted in a cortical contusion injury cavity.

Embryonic stem (ES) cells have also been recently investigated for grafting in TBI *(128,129)*. Mouse ES cells predifferentiated to neural precursors were transplanted into rat cortex 1 wk following controlled cortical impact *(128)*. Rats were tested on a battery of behavioral tests, including bilateral tactile removal, locomotor placing, and reference memory in the Morris water maze. The transplanted ES cells were found to improve sensorimotor function but not cognitive function. In addition, the ES cell transplants significantly reduced lesion size.

Adult stem cell sources have also been explored for potential use in treating TBI. A potential advantage is its ability to use autologous cells, thereby avoiding the possible consequences of immunologic mismatch and graft rejection. Adult bone marrow stromal cells have been reported to contain stem-like cells with multiple differentiation potential, which may be induced to differentiate to neuronal-like phenotypes. An additional advantage of these cells may be their ability to home to sites of injury, including the CNS, even when administered peripherally. Thus, they may be useful as a delivery vehicle for therapeutic and neurotrophic molecules whether they differentiate into neural cells or not. Rat or human bone marrow stromal cells injected intravenously (IV) via the tail vein have been reported to reduce motor and neurological deficits and lower lesion volumes by 2 wk following cortical impact injury *(130–133)*. Transplanted cells migrated into the rat brain and appeared to preferentially localize around the injury site, with a subset of cells expressing neuronal and astrocytic markers. Similar findings were reported using human umbilical cord blood cells administered IV *(134)*. Bone marrow stromal cells may also be transplanted intracerebrally following culture in vitro with such neurotrophic factors as BDNF and NGF to improve graft viability and motor function *(135)*.

Stem Cells/Progenitors for Pain

To date, there are no published successful methods to treat human SCI pain with stem cells or precursors. Their promise lies in the future *(136,137)* and will likely require a degree of genetic or laboratory manipulation. For example, adrenal chromaffin progenitors *(138)* can be kept proliferating by growth factors in vitro *(139,140)*, suggesting that they might provide an alternative source for cell therapy, which is different from bioengineered, immortalized chromaffin cell lines. In a recent report *(141)*, utilizing the peripheral nerve injury with sciatic CCI to induce neuropathic pain, rat spinal embryonic progenitor cells (SPCs) that had used FGF-2 for proliferation in vitro were able to reduce thermal hyperalgesia after intrathecal transplant. Presumably, grafted cells had been induced to a GABAergic phenotype by FGF-2 in vitro and survived in its absence after transplant, maintaining their phenotype to modulate the neuropathic pain. The authors indicate that the grafts also increased the glycine content in the CSF of grafted animals, suggesting that if precursors could be induced to a phenotype that provides nociceptive inhibition, they would function much like cell lines, described earlier as minipumps, surviving in the intrathecal space. Preliminary findings in our laboratory have also explored the use of intraparenchymally placed GABAergic neural progenitors in an excitotoxic SCI pain model, and marked reductions in dysthesthetic pain behavior have been observed *(52)*.

Immortalized/Conditionally Immortalized Stem Cells and Cell Lines

Immortalized Neural Stem Cells in TBI

The mouse neural stem cell clone C17.2 has been examined as a potential neural stem cell graft in a mouse lateral–controlled cortical impact model *(142)*. These cells were originally derived from neonatal mouse cerebellum and immortalized by the retrovirus-mediated transduction of avian myc (*v-myc*) *(143)*. Cells transplanted into the cortex–hippocampus interface at 3 d postinjury resulted in significantly improved motor function via the rotating pole for coordination and integration of movement and rotorod test for vestibulomotor function *(142)*. Interestingly, when placed in the contralateral hemisphere, some improvement in motor function was also observed. However, cognitive function, as assessed by the Morris water maze, did not increase, similar to findings observed using other stem cell grafts. The grafted cells showed good survival and could be found in the hippocampus and cortical parenchyma adjacent to the injury cavity, expressing neuronal and astrocytic markers. As some neuronal differentiation appears to have occurred, the mechanisms underlying the beneficial effects of these grafts could include cell replacement as well as provision of neuroprotective and neurotrophic factors.

The MHP36 cell line is a conditionally immortalized mouse cell line derived from the hippocampus of a transgenic mouse, harboring a temperature-sensitive allele of the SV40 large T antigen. MHP36 cells have also been used as graft cells in brain injury and ischemia models *(144)*. Implantation into the hippocampus of rats with CA1 damage produced recovery of cognitive function and reconstituted a normal laminated appearance in this region. Whether these or other immortalized neural precursor lines primarily produce their effects via replacement of lost or damaged neurons and integration within the host circuitry, or more likely via provision of local trophic or structural support, they may be useful as a vehicle for delivery of neuroprotective or therapeutic molecules. Immortalized neural progenitor cells derived from embryonic rat hippocampus (HiB5) have been used to deliver NGF in injured brains *(145)*. The HiB5 cells transduced to secrete NGF were transplanted into the cerebral cortex, adjacent to fluid percussion injury sites. Both NGF-secreting and control HiB5 grafts were found to improve neuromotor function and spatial learning behavior, but significantly reduced hippocampal CA3 cell death was observed only in animals receiving the NGF grafts.

A human embryonal carcinoma cell line that exhibits stem cell-like properties and can be differentiated into postmitotic neuron-like cells has also been explored for use in the traumatically injured brain *(146,147)*. Ntera2

(NT2) cells, originally isolated from a human teratocarcinoma, terminally differentiate to postmitotic neurons when treated in vitro with retinoic acid and maintain their neuronal characteristics. Owing to their apparent safety and preclinical evaluations in ischemia models, these NT2N neural cells have been utilized in clinical transplantation trials in stroke victims *(148,149)*. For peripheral TBI evaluations cells were transplanted into the periinjured cerebral cortex of rats 24 h following moderate lateral fluid percussion injury. Viable grafted cells were observed extending processes into the surrounding host parenchyma at 2–4 wk postgrafting. However, no significant improvements in cognitive or motor function were observed *(146,147)*. Nevertheless, these cells may offer a platform for stable ex vivo gene delivery to the CNS. A recent report has illustrated this potential by transducing these cells to express NGF using a lentiviral vector *(150)*. Transplantation of the NGF-transduced NT2N cells were observed to attenuate cognitive dysfunction following TBI in mice.

Immortalized Cells for Chronic Pain

NEURONAL RAT CELL LINES FOR PAIN

Although describing engineered cell grafts as "biological minipumps" for secretion of neurotrophic or antinociceptive agents has only been recently discussed *(151)*, the isolation and use of cells genetically engineered for regeneration, or to ameliorate neurological disease *(152,153)*, has been examined for the last 15–20 yr. The same strategy, using engineered cells that might secrete potentially antinociceptive molecules when placed in the lumbar subarachnoid space after nerve injury (much like the primary adrenal chromaffin cells and opioid cell lines described above) has seen few applications for use in chronic pain *(116,118)*. However, the potential application of such cell grafts for the diverse problems with neuropathic pain in human therapy is great *(154)*. Unlike primary or immortalized chromaffin cells, the engineered cells being tested in models of acute and chronic pain are usually neuronal epithelial precursor cell lines derived from the rat medullary raphe. Two lines that have been engineered, known as RN46A and RN33B, were isolated from embryonic day-12.5 (E12.5) rat brainstem after immortalization with the SV40 tsTag sequence *(155,156)*. Although they were derived from the same primary cultured neuronal precursors, there are significant differences in their phenotypes. RN46A cells are an early serotonergic precursor neuronal cell line, with the potential to switch developmental phenotype *(155)*, depending on the timing and exposure to neurotrophic and other factors, including BDNF *(157)*, ciliary neurotrophic factor *(158)*, GDNF *(159)*, and ACTH *(160)*. This cell line synthesized and secreted the neurotrophin BDNF with the addition of the sequence for rat BDNF to its genome, causing the cells to have improved survival in vitro and in vivo and to develop a permanent serotonergic phenotype *(161)*.

Because additional 5HT might have a beneficial antinociceptive effect on neuropathic pain *(162)* if cells were placed in a lumbar subarachnoid location after sciatic nerve constriction, transplants of the serotonergic cell line 46A-B14 were tested after nerve injury. Grafts of 46A-B14 cells placed 1 wk after nerve injury, and the development of severe hypersensitivity to thermal and tactile stimuli, were able to potently and permanently reverse the symptoms of neuropathic pain *(163)*, compared to grafts of the same cells. These control cells did not receive the BDNF gene and did not synthesize 5HT in vitro or in vivo. Transplants of other cell lines genetically engineered to synthesize and secrete potentially antinociceptive molecules, such as the inhibitory peptide galanin *(164)*, neurotrophin BDNF *(165)*, and the inhibitory neurotransmitter GABA *(166)*, have been tested after nerve injury. Transplant of these cell lines has reversed the thermal and tactile allodynia and hyperalgesia that develops after nerve injury. Each engineered cell line is characterized for its particular gene expression under permissive and nonpermissive temperature conditions, as the cell lines are usually transplanted immediately after proliferation at 33°C. Following placement of the differentiating cells in the subarchnoid space, especially in pain models, both cell graft survival and continued expression of the antinociceptive phenotype can be examined in vivo. An example of such an engineered rat neuronal cell line, RN33-GAD67, synthesizes and secretes GABA after differentiation in vitro and transplant in vivo *(166)* in the CCI pain model, where GABA is synthesized after differentiation in these cells. When these cells are grafted after excitoxic SCI and the induction of neuropathic pain, they are able to reduce both thermal hyperalgesia and tactile allodynia *(167)*. But this effect for SCI pain seems to require an early transplant time, because grafts of these rat neuronal GABA cells are ineffective when placed late after nerve injury or SCI *(168)*. With both thermal and tactile hyperalgesia, rat GAD67 grafts cause a reversal of the behaviors when the behaviors are measured 1 wk later. Potent reversal is common with each engineered cell line used for therapy after peripheral nerve injury models and is more recently typical with SCI models, especially thermal hyperalgesia.

HUMAN NT2 CELL LINES FOR PAIN

The potential application of NT2 neurons in cell transplantation therapy for CNS disorders, and their use as vehicles for delivering exogenous proteins into the human brain for gene therapy, has been recently demonstrated *(169)* and is discussed above.

Two phenotypes abundantly present within the NT2 population are those that synthesize the inhibitory neurotransmitters, GABA and 5HT *(170)*. Our laboratory has recently subcloned two human neural NT2 cell lines, including one that synthesizes and secretes GABA, as well as a distinctly different cell line that synthesizes and secretes 5HT in vitro. An example of these two cell lines and their synthesis of GABA or 5HT after 2 wk of differentiation in vitro is illustrated in Fig. 1.

Characterization of these cells in vitro includes high-performance liquid chromatography for neurotransmitter analysis to determine the ability of the individual cell lines to secrete basal levels of neurotransmitters, as might be expected after grafting in vivo. These human cell lines are able to synthesize and release potential antinociceptive neurotransmitters (e.g., GABA) after differentiation. These cells are able to survive on the pial membrane and maintain their neurotransmitter phenotype for the long term in the excitotoxic SCI pain model. When testing potential subarachnoid grafts of the human GABAergic and 5HT cells to reduce chronic neuropathic pain in this rat model of SCI-associated pain, grafts are able to reduce both thermal hyperalgesia and tactile allodynia, as illustrated in Fig. 2.

Such human neuronal GABA and 5HT cells can be developed for use in clinical trials as treatment for chronic pain associated with SCI.

CONDITIONALLY IMMORTALIZED CHROMAFFIN CELL LINES FOR PAIN

Retroviral infection of neural precursors in vitro, especially with the wild-type SV40 Tag oncogene, can result in cell lines capable of undergoing differentiation under appropriate conditions *(171)*. However, infection of precursors with the temperature-sensitive allele of Tag (tsTag) in vitro *(172)* and in vivo *(173)* has allowed cells to undergo growth arrest and continue differentiation under nonpermissive temperature conditions, such as that with transplant conditions or in vitro at 39°C.

These differentiating temperatures are possible both in vitro, allowing transformed cells to revert to a near-normal primary cell phenotype, as well as in vivo, where CNS transplant temperatures are near 39°C *(174)*, and tumors are not formed by the grafts. Thus, conditional immortalization with the tsTag construct incorporates the advantages of cell lines. Advantages include the convenience of growing large quantities that can be characterized and safety tested and the ability to genetically engineer additional therapeutic molecules. The disadvantages of tumor cell lines, such as PC12 or neuroblastoma cells, are reduced.

Mitotic cells found in adrenal tissue can be conditionally immortalized with the temperature-sensitive Tag oncogene so that the differentiated cell type keeps many phenotypic features of primary chromaffin cells. Confer-

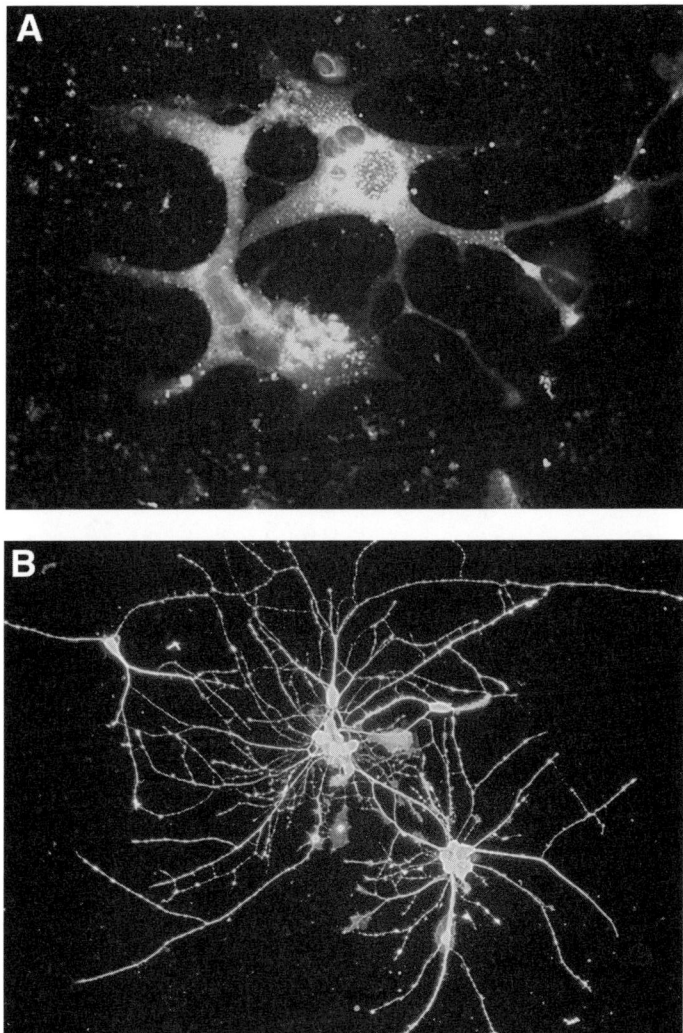

Fig. 1. Human neuronal 5HT and GABA cell lines in vitro. The NT2.19 serotonergic cell line (**A**) was subcloned by serial dilution and treated for 2 wk with retinoic acid and mitotic inhibitors. They were further differentiated for 2 wk before an antibody stain for 5HT. The 5HT NT2.19 cells have very large nuclei, are generally bipolar, with short neurites, and stain brightly for 5HT. Another cloned NT2 cell line, NT2.17 (**B**), which is positive for GABA, was subcloned by serial dilution and treated for 2 wk with retinoic acid and mitotic inhibitors. They were further differentiated for 2 wk before an antibody stain for GABA. The GABA NT2.17 cell line has small nuclei, extensive multipolar neurites, and stains brightly for GABA. Magnification: bar = 50 μm.

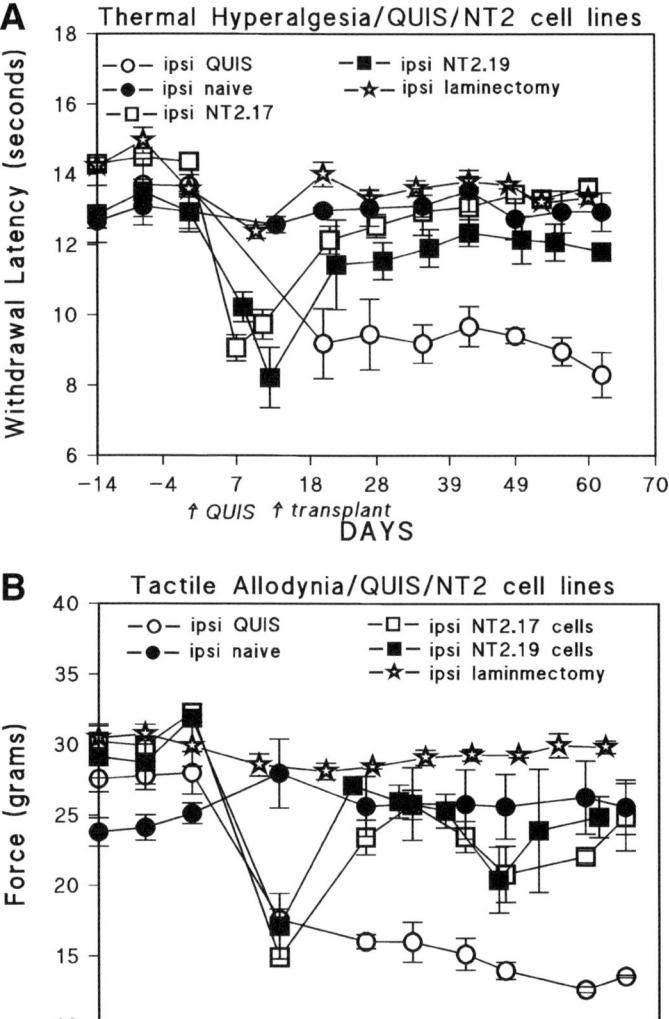

Fig. 2. Recovery of thermal hyperalgesia and tactile allodynia after subarchnoid grafts of human neuronal cell lines in the excitotoxic model of SCI pain. Adult male Wistar Furth rats were spinally injected with quisqualic acid (QUIS), an excitoxic agent, in a rat model of SCI and chronic pain *(239–242)*. Animals were left untreated or were injected with either NT2.17 GABA or NT2.19 5HT cells $(1 \times 10^6$ cells/injection) into the subarchnoid space at 2 wk (14 d) after QUIS. Animals were tested before the SCI (baseline) and weekly following QUIS and cell grafts for hypersensitivity to thermal **(A)** or tactile **(B)** stimuli in hindpaws below the SCI. The transplant was done at 2 wk after the original excitotoxic SCI.

218

ring immortalization with the SV40 large T-antigen expression has various effects on cells when the wild-type large T protein is present, including the ability of large T antigen to block the differentiation process *(175)*. However, after immortalization with the temperature-sensitive allele tsTag *(172,176)*, immortalized cells resume the stage of life span and function of an uninfected cell when they are shifted to nonpermissive temperature conditions *(177)*. These cells at the nonpermissive temperature have lost the ability to drive cell proliferation, as the large T antigen is labile at the higher temperature conditions *(178)*, and the T antigen is not able to drive mitosis in cells immortalized with the construct, and differentiation is favored *(172,176)*. In general, SV40 large T-antigen–immortalized cell lines retain the phenotype of the differentiated lineage of the parent. Cell lines generated with the SV40 large T antigen retain contact inhibition in vitro *(179,180)* and do not produce tumors or induce immune rejection, even when injected into nude mice *(181)* or rats *(182–185)*. A number of functional cell lines have been immortalized with Tag and retain their specific catecholaminergic phenotype *(186)* and efficacy to reverse neurological deficits after CNS transplant *(186,187)*.

The cell biology and developmental responsiveness during differentiation of chromaffin cells *(188)* reveals clues to the differentiation program of conditionally immortalized chromaffin cell lines in vitro. The enzyme tyrosine hydroxylase (TH; EC1.14.3.x) catalyzes the rate-limiting step in the biosynthesis of catecholamines in chromaffin cells in the adrenal medulla *(189)*. TH has been used as an antigenic marker for the mature chromaffin phenotype of primary rat and bovine chromaffin cells in vitro *(76)*, as well as dopamine-β-hydroxylase and PNMT. However, even with upregulation of enzyme expression that accompanies differentiation, these conditionally immortalized chromaffin rat and bovine cells do not synthesize catecholamines under in vitro conditions *(190)*. These conditionally immortalized cells express very low levels of TH. A more likely explanation for the

---

(Figure 2 continued) All animals were examined for chronic pain behaviors in both the contralateral and ipsilateral hindpaws. Both ipsilateral and contralateral hindpaws recovered near-normal sensory responses to thermal stimuli after grafting either the GABAergic NT2.17 or serotonergic NT2.19 cells, compared to the QUIS injury alone. QUIS injury negatively affects hindpaw responses bilaterally, but the ipsilateral hindpaw is most affected by the injection of QUIS (and for clarity of presentation is illustrated here). Neither hindpaw recovers normal tactile or thermal responses after QUIS alone by 60 d after the injection. The data reported are the mean ± standard error of the mean (SEM) of the scores for the ipsilateral-affected paw of four animals in each group.

absence of catecholamine synthesis is a continued low level of Tag expression, even though it is greatly reduced after 3 wk of differentiation at 39°C. Likely, even a low level of Tag suppresses some normal cellular functions, such as neurotransmitter synthesis. Low levels of Tag during differentiation may subvert the normal pathways of enzyme regulation, at least in vitro. However, utilizing the temperature-sensitive allele of SV40 large T antigen (tsTag) to conditionally immortalize chromaffin cells allows for cells that are able to reverse neuropathic pain following transplantation in the spinal subarachnoid space after CCI of the sciatic nerve *(191)*.

DISIMMORTALIZATION STRATEGIES FOR CHROMAFFIN CELL LINES

Transplanted primary adrenal chromaffin cells have been used for the delivery of bioactive molecules *(192)*. However, a serious limitation of using primary tissue is the necessity of harvesting fresh cells from donors, requiring safety screening for each batch of cells and a resultant mixture of cell types, which are incompletely characterized and nonhomogeneous. Even with nearly a 100% disappearance of an oncogene like Tag in the grafts within a few weeks after transplantation *(191)*, oncogene expression in vivo remains a possibility, and such oncogene-containing cells would not be an appropriate strategy for safe clinical use in humans.

Studies exploiting site-specific DNA recombination and Cre recombinase/ loxP (Cre/lox) excision have suggested that cells can be targeted in vitro *(193)* and in vivo *(194)* for removal of deleterious genes, including Tag *(195)*. Reversible immortalization with Tag and Cre/lox technology was first reported with human fibroblasts by Westerman and Leblouch *(238)* and more recently with human myogenic cells and hepatocytes *(196)*. In these latter studies, *Cre* was introduced by transfection or infection—inefficient methods that may lead to a lack of disimmortalization and the loss, via the subsequent selection of disimmortalized cells, of a significant part of the population. Moreover, in vivo excision is not possible. Use of a vector that allows a silent, but inducible, form of *Cre* is preferred for the timed excision of the oncogene.

A number of chimeric Cre-containing fusion proteins, especially fusions with the ligand-binding domains of steroid receptors, have been created to utilize the binding by synthetic ligands to activate Cre *(197)*. CrePR1 is a fusion protein *(198)*, consisting of the fusion of Cre and the ligand-binding domain of a mutant human progesterone receptor (hPRB891). Cre activity in the cells is activated by the binding of the steroid RU486, which then induces the translocation of CrePR1 to the nucleus, where Cre is active to excise the floxed sequences. The requirement for RU486, and the use of a mutated steroid receptor for disimmortalization, would assure that, if nondisimmortalized cells were transplanted, Cre would not be activated by circulating endogenous progesterone, a strategy used for inducible recombination with in vivo CNS studies *(199)*.

We recently demonstrated *(200)* that embryonic rat adrenal chromaffin cells could be immortalized with a oncogenic tsTag construct, utilizing retroviral infection of these early chromaffin precursors, where the tsTag construct (tsA-TN) was flanked by loxP sequences. Following isolation of immortalized cells using positive neomycin selection, the cells were further infected with a retrovirus expressing the CrePR1 gene, which encodes for a fusion protein that combines Cre activity plus the mutant human steroid receptor hPRB891. Cultures of embryonic rat adrenal cells were immortalized with the tsA-TN retroviral vector encoding the loxP-flanked temperature-sensitive allele of tsA-TN, which included a positive/negative *neo/HSV-TK* sequence for selection with either G418 or gancyclovir, respectively.

When immortalized chromaffin cells are disimmortalized with Cre-lox technology to disimmortalize the chromaffin cells in vitro, complete removal of the Tag sequence before differentiation seems to allow neurotransmitter synthesis and a more normal phenotype *(200)*. The disimmortalized rat chromaffin cells had increased TH expression in vitro. This was accompanied by a fivefold increase in norepinephrine synthesis in vitro *(200)*. These disimmortalizable rat chromaffin cells not only synthesize epinephrine after Tag excision, they also apparently make increased catecholamine enzymes aside from TH, as seen by qualitative immunohistochemistry for the enzymes versus both nonexcised and those conditionally immortalized with only tsTag *(190)*.

Compared to the downregulation of the Tag protein in conditionally immortalized rat chromaffin cells, disimmortalization in vitro in these disimmortalizable rat chromaffin cells, known as the loxtsTag/CrePR1/RAD chromaffin cell line, the Tag protein was completely and efficiently removed after 10 d of treatment with RU486, followed by incubation with the antibiotic gancyclovir *(200)*. Cells that were not disimmortalized were removed by their continued expression of the TK, which is toxic in the presence of gancyclovir.

Irreversible removal of a potentially subverting oncogene by its excision using the Cre/Lox system might thus be a clinically useful strategy, especially because the core temperature of humans is lower than that of rodents, and the expression of a temperature-sensitive antigen might not be completely blocked in a clinical context *(201–206)*. Note that, in this respect, use of modulatable Cre activity that can be activated by the synthetic steroid RU486 *(198,199)* has added a method to select the timing of disimmortalization and renders the overall procedure more flexible and efficient.

Importantly, transplantation of disimmortalized rat chromaffin cells was able to nearly eliminate neuropathic pain in the CCI model of peripheral nerve injury, compared to the injury alone or transplantation of immortalized cells *(200)*.

Instead of suggesting that antinociception is the result of catecholamine synthesis, release, or secretion, the existence of an equivalent functional effect by nondisimmortalized cells suggests that another agent or mechanism is responsible for the reduction of neuropathic pain by these genetically manipulated chromaffin cells. Even if chromaffin grafts do not make significant levels of catecholamines in vivo, the antinociception that the grafts provide might be a result of other antinociceptive molecules synthesized and released by the cells, such as GABA or met-enkephalin. Presumably, the increased norepinephrine phenotype recovered, following excision of the oncogene by disimmortalized cells, would function as an advantage in cell therapy. Yet, with disimmortalized rat chromaffin cell grafts, no such advantageous effect could be demonstrated. Rather, the value of disimmortalization before transplantation is to provide a measure of safety, with the complete absence of the oncogene and prevention of a remote possibility of viral transfer of the large T antigen in the host after grafting such cells.

TRANSGENIC OPIOID EXPRESSION IN IMMORTALIZED CHROMAFFIN CELLS

Differentiated primary chromaffin cells are known to synthesize such opioids as met-enkephalin *(207,208)*, as do the conditionally immortalized rat and bovine chromaffin cell lines *(190)*, but at low levels. As both tsTag-immortalized chomaffin cell types upregulate enkephalin expression with differentiation alone, enkephalin expression might be a signal of increasing maturity, independent of Tag-influenced dysregulation, in these cells in vitro. But a further advance to model genetically modified and disimmortalizable chromaffin cell lines is the recent work by Duplan and colleagues *(209)*, who infected our disimmortalizable loxtsTag/CrePR1/RAD chromaffin cell line with constructs for the synthesis and secretion of the opioid met-enkephalin (met-enk). These transgenic rat chromaffin cell lines expressed easily detectible met-enk in vitro, compared to the cell line that expressed the empty vector alone or a cell line that contained the vector for the pro-enkephalin (pro-enk) sequence. Cells containing the met-enk construct contained high levels of met-enk. The transgene also contained a NGF sequence for secretion of synthesized nascent protein. The value of opioids from chromaffin grafts in cellular therapy, especially for pain *(210)*, has seen precedents in both animal *(63,211,212)* and (more recently) human clinical work *(7,213,214)*, when primary chromaffin tissue was used as a graft source. When these disimmortalizable loxtsTag/CrePR1/RAD chromaffin cells were grafted 2 wk before injection of formalin into the hindpaw in the formalin model of tonic pain, those met-enk–overexpressing chromaffin cells significantly reduced the molecular response of rat spinal cord neurons to noxious stimulation, as illustrated with formalin-evoked c-fos protein expression *(209)*. C-fos may represent a molecular marker of nociceptive activity, and its reduction may represent a reduced response to formalin stimulation.

Although it is not yet known how disimmortalization may influence transgene expression, such as the opioid met-enk gene used here, irreversible removal of a potentially subverting oncogene by its excision using the Cre-lox system might be a clinically useful strategy. Of course, immortalization of human chromaffin tissue with an oncogene (e.g., SV40Tag) is not likely with any potential deleterious expression of SV40 proteins *(215)*, but disimmortalization utilizing Cre-lox site–directed removal of oncogenes is a growing technology to create useful graft sources for cell therapy for a variety of conditions *(201,202,216)*. There are several possible oncogeneic sequences that could be used for the reversible immortalization of human chromaffin cell lines, including v-myc *(217)*.

However, the creation of clinically safe and reversibly immortalizable human chromaffin cell lines, perhaps from embryonic precursors *(218)*, is in the near future *(219)*. Such a homogeneous source will enable the manipulation of the chromaffin cell's genome to investigate the mechanisms of action responsible for cell grafts to repair the injured CNS environment. Similar immortalization of human chromaffin precursors, and creation of human chromaffin lines, presage the advent of cellular therapy as a practical therapeutic strategy for difficult clinical problems, including intractable pain after CNS trauma.

CONCLUSIONS

During the past two decades, cell therapy as an approach to treat pain has progressed from a hypothesis for modulating pain processing, to the development of the first human cell sources that are being tested in clinical pain treatment. The near future will likely provide new challenges for the implementation in a wider audience of those who suffer chronic pain, considering problems common to all forms of cell transplantation. Such issues include immune rejection vs long-term survival and efficacy in the human host; dependable, well-characterized cell sources for grafts; cells that can safely integrate into or near the CNS, without danger of tumors or significant deleterious effects; the ability to control the antinociceptive output of cell grafts, ideally increasing with the cyclic episodes of pain; and efficacy in a wide variety of pain causalities. However, cell therapy for pain offers much promise as a replacement or adjunct to current clinical methodologies, once the mechanisms of pain are well understood, so that bioengineered cellular tools can be used appropriately. Cell transplantation for TBI is relatively less well understood and developed, and most studies are suggestive of cell survival, some modest differentiation to appropriate neural phenotypes, and little indication of direct reintegration within the host neurocircuitry. Nevertheless, behavioral improvement has been reported, particularly for sensorimotor function and less so for cognitive function. Thus, it is likely that the majority of the cell types grafted thus far are functioning like a cellular minipump, providing neuroprotective and neurotrophic agents in the damaged CNS. Future studies will likely better define these processes and mechanisms and lead to improved selection of cell types and trophic agents, which can be utilized in combination to provide improved therapeutic outcomes following traumatic brain injuries.

REFERENCES

1. MacPherson, R. D. (2002) New directions in pain management. *Drugs Today* **38,** 135–145.
2. Schwartzman, R. J, Grothusen, J., Kiefer, T. R., and Rohr, P. (2002) Neuropathic central pain: epidemiology, etiology, and treatment options. *Arch. Neurol.* **58,** 1547–1550.
3. Turk, D. C. and Okifuji, A. (1998) Treatment of chronic pain patients: Clinical outcome, cost effectiveness, and cost benefits. *Crit. Rev. Phys. Med. Rehabil. Med.* **10,** 181–208.
4. Breivik, H. (2001) Opioids in cancer and chronic noncancer pain therapy: Indications and controversies. *Acta Anesthesiol. Scand.* **45,** 1059–1066.
5. Woolf, C. J. and Mannion, R. J. (1999) Neuropathic pain: aetiology, symptoms, mechanisms, and management. *Lancet* **353,** 1959–1964.
6. Siddall, P. J., Yezierski, R. P., and Loeser, J. (2000) Pain following spinal cord injury: clinical features, prevalence, and taxonomy. *IASP Newsletter* **3,** 3–7.
7. Lazorthes, Y., Sallerin, B., Verdie, J. C., et al. (2000) Management of intractable cancer pain: from intrathecal morphine to cell allograft. *Neurochirurgie* **46,** 454–465.
8. Wilson, S. P. and Yeomans, D. C. (2000) Genetic therapy for pain management. *Curr. Rev. Pain* **4,** 445–450.
9. Carter, G. T. and Galer, B. S. (2001) Advances in the management of neuropathic pain. *Phys. Med. Rehabil. Clin. N. Am.* **12,** 447–459.
10. Martino, G., Furlan, R., Comi, G., and Adorini, L. (2001) The ependymal route to the CNS: an emerging gene-therapy approach for MS. *Trends Immunol.* **22,** 483–490.

11. Joung, I., Kim, H. S., Hong, J. S., et al. (2000) Effective gene transfer into regenerating sciatic nerves by adenoviral vectors: potentials for gene therapy of peripheral nerve injury. *Mol. Cells* **10,** 540–545.
12. Hottinger, A. F. and Aebischer, P. (1999) Treatment of diseases of the central nervous system using encapsulated cells. *Adv. Tech. Stand Neurosurg.* **25,** 3–20.
13. Latchman, D. S. (2001) Gene therapy with herpes simplex virus vectors: progress and prospects for clinical neuroscience. *Neuroscientist* **7,** 528–537.
14. Yezierski, R. P. (1996) Pain following spinal cord injury: the clinical problem and experimental studies. *Pain* **68,** 185–194.
15. Widerstrom-Noga, E., Duncan, E. G., Felipe-Cuervo, E., and Turk, D. C. (2002) Assessment of the impact of pain and impairments associated with spinal cord injuries. *Arch. Phys. Med. Rehabil.* **83,** 395–404.
16. Cicconetti, A., Bartoli, A., Ripari, F., and Ripari, A. (2004) COX-2 selective inhibitors: a literature review of analgesic efficacy and safety in oral-maxillofacil surgery. *Oral Surg. Oral Med. Oral. Pathol. Oral Radiol. Endod.* **97,** 139–146.
17. Elmslie, K. S. (2004) Calcium channels blockers in the treatment of disease. *J. Neurosci. Res.* **75,** 733–741.
18. Mason, L., Moore, R. A., Derry, S., et al. (2004) Systematic review of topical capsaicin for the treatment of chronic pain. *BMJ* **328,** 991.
19. Decker, M. W., Rueter, L. E., and Bitner, R. S. (2004) Nicotinic acetylcholine receptor agonists: a potential new class of analgesics. *Curr. Top Med. Chem.* **4,** 369–384.
20. Christo, P. J. (2003) Opioid effectiveness and side effects in chronic pain. *Anesthesiol Clin. North Am.* **21,** 699–713.
21. Nikolaus, T. and Zeyfang, A. (2004) Pharmacological treatments for persistent non-malignant pain in older persons. *Drugs Aging* **21,** 19–41.
22. Levendoglu, F., Ogun, C. O., Ozerbil, O., et al. (2004) Gabapentin is a first line drug for the treatment of neuropathic pain in spinal cord injury. *Spine* **29,** 743–751.
23. Bennett, M. I. and Simpson, K. H. (2004) Gabapentin in the treatment of neuropathic pain. *Palliat. Med.* **18,** 5–11.
24. McNicol, E., Strassels, S., Goudas, L., et al. (2004) Nonsteroidal anti-inflammatory drugs, alone or combined with opioids, for cancer pain: a systematic review. *J. Clin. Oncol.* **15,** 22–1975.
25. Woolf, C. J. and Hashmi, M. (2004) Use and abuse of opioid analgesics: potential methods to prevent and deter non-medical consumption of prescription opioids. *Curr. Opin. Investig. Drugs* **5,** 61–66.
26. Siddall, P. J., Molloy, A. R., Walker, S., et al. (2000) The efficacy of intrathecal morphine and clonidine in the treatment of pain after spinal cord injury. *Anesth. Analges* **91,** 1493–1498.
27. Chambers, S. and MacSullivan, R. (1994) Intrathecal morphine in the treatment of chronic intractable pain. *Ir. J. Med. Sci.* **163,** 318–321.
28. Bennett, A. D., Everhart, A. W., and Hulsebosch, C. E. (2000) Intrathecal administration of an NMDA or a non-NMDA receptor antagonist reduces

mechanical but not thermal allodynia in a rodent model of chronic central pain after spinal cord injury. *Brain Res.* **859,** 72–82.

29. Heijne, M., Hao, J. X., Sollevi, A., and Xu, X. J. (2001) Effects of intrathecal morphine, baclofen, clonidine, and R-PIA on the acute allodynia-like behaviors after spinal cord ischaemia in rats. *Eur. J. Pain* **5,** 1–10.

30. Sagen, J., Pappas, G. D., and Perlow, M. J. (1986) Adrenal medullary tissue transplants in the rat spinal cord reduce pain sensitivity. *Brain Res.* **384,** 189–194.

31. Siddall, P. and Cousins, M. J. (1997) Spine update. Spinal pain mechanisms. *Spine* **22,** 98–104.

32. Yaksh, T. L., Hua, X. Y., Kalcheva, I., et al. (1999) The spinal biology in humans and animals of pain states generated by persistent small afferent input. *Proc. Natl. Acad. Sci. USA* **96,** 7680–7686.

33. Dubner, R. (1991) Neuronal plasticity in the spinal and medullary dorsal horns: a possible role in central pain mechanisms. In *Pain and Central Nervous System Disease: The Central Pain Syndromes* (Casey, K. L., ed.), Raven, New York, pp. 143–155.

34. Hama, A. T. and Sagen, J. (1994) Induction of spinal NADPH-diaphorase by nerve injury is attenuated by adrenal medullary transplants. *Brain Res.* **640,** 345–351.

35. Sugimoto, T., Bennett, G. J., and Kajander, K. C. (1990) Transsynaptic degeneration in the superficial dorsal horn after sciatic nerve injury: effects of a chronic constriction injury, transection, and strychnine. *Pain* **30,** 385–393.

36. Aldskogius, H., Arvidsson, J., and Grant, G. (1985) The reaction of primary sensory neurons to peripheral nerve injury with particular emphasis on transganglionic changes. *Brain Res. Rev.* **10,** 27–46.

37. Tal, M. and Bennett, G. J. (1994) Extra-territorial pain in rats with a peripheral mononeuropathy: mechano-hyperalgesia and mechano-allodynia in the territory of an uninjured nerve. *Pain* **57,** 375–382.

38. Sugimoto, T., Takemura, M., Sakai, A., and Ishimaru, M. (1987) Rapid transneuronal destruction following peripheral nerve transection in the medullary dorsal horn is enhanced by strychnine, picrotoxin, and bicuculline. *Pain* **30,** 385–393.

39. Magoul, R., Onteniente, B., Geffard, M., and Calas, A. (1987) Anatomical distribution and ultrastructural organization of the GABAergic system in the rat spinal cord: an immunocytochemical study using anti-GABA antibodies. *Neurosci* **20,** 1001–1009.

40. Todd, A. J. and McKenzie, J. (1989) GABA-immunoreactive neurons in the dorsal horn of the rat spinal cord. *Neurosci* **31,** 799–806.

41. Kangraga, I., Jiang, M., and Randic, M. (1991) Actions of (-)-baclofen on rat dorsal horn neurons. *Brain Res.* **562,** 265–275.

42. Levy, R. A. and Prodfit, H. K. (1977) The analgesic action of baclofen [(4-chlorophenyl)-aminobutyric acid]. *J. Pharmacol. Exp. Ther.* **202,** 437–445.

43. Castro-Lopes, J. M., Tavares, I., and Coimbra, A. (1993) GABA decreases in the spinal cord dorsal horn after peripheral neurectomy. *Brain Res.* **620,** 287–291.

44. Eaton, M. J., Plunkett, J. A., Martinez, M. A., et al. (1998) Changes in GAD and GABA immunoreactivity in the spinal dorsal horn after peripheral nerve

injury and promotion of recovery by lumbar transplant of immortalized serotonergic neurons. *J. Chem. Neuroanat.* **16,** 57–72.

45. Ibuki, T., Hama, A. T., Wang, X.-T., et al. (1997) Loss of GABA immunoreactivity in the spinal dorsal horn of rats with peripheral nerve injury and promotion of recovery by adrenal medullary grafts. *Neurosci* **76,** 845–858.

46. Meng, X.-W., Canchola, S., Ralston, D. D., and Ralston, H. J. (1997) Changes in expression for glutamic acid decarboxylase (GAD) mRNA and nitric oxide synthase (NOS) in rat spinal cord dorsal horn following sciatic nerve injury. *Soc. Neurosci. Abstr.* **23,** 1532.

47. Ralston, D. D., Coyle, D. E., Tedesco, C. S., et al. (1997) Decreased GABA immunoreactivity in rat dorsal horn is correlated with pain behavior: A light and electron microscopic study. In *Proceedings 8th World Congress on Pain* (Jensen, T. S., Turner, J. A., Wiesenfeld-Hallin, Z., eds.), IASP Press, Seattle, WA, pp. 547–560.

48. Azkue, J. J, Zimmermann, M., Hxieh, T.-F., and Herdegen, T. (1997) Peripheral nerve insult induces NMDA receptor-mediated, delayed degeneration in spinal neurons. *Eur. J. Neurosci.* **10,** 2204–2206.

49. Whiteside, G. T. and Munglani, R. (2001) Cell death in the superficial dorsal horn in a model of neuropathic pain. *J. Neurosci. Res.* **64,** 168–173.

50. Coggeshall, R. E., Lekan, H. A., White, F. A., and Woolf, C. J. (2001) A-fiber sensory input induces neuronal cell death in the dorsal horn of the adult rat spinal cord. *J. Comp. Neurol.* **435,** 276–282.

51. Hao, J.-X., Xu, X.-J., Yu, Y.-X., et al. (1992) Baclofen reverses the hypersensitivity of dorsal horn wide dynamic range neurons to mechanical stimulation after transient spinal cord ischemia: implications for a tonic GABAergic inhibitory control of myelinated fiber input. *J. Neurophysiol.* **68,** 392–396.

52. Lee, J. W., Yezierski, R. P., and Sagen, J. (2001) Transplantation of embryonic progenitor cells into excitotoxically lesioned adult spinal cord: in vivo survival and differentiation in quisqualic acid-treated spinal cord. *Soc. Neurosci. Abstr.* **27,** 369.

53. Park, K. I., Liu, S., Flax, J. D., et al. (1999) Transplantation of neural progenitor and stem cells: developmental insights may suggest new therapies for spinal cord and other CNS dysfunction. *J. Neurotrauma* **16,** 675–687.

54. Sandkuhler, J. (1996) The organization and function of endogenous antinociceptive systems. *Prog. Neurobiol.* **50,** 49–81.

55. Stamford, J. A. (1993) Descending control of pain. *Br. J. Anaesth.* **75,** 217–227.

56. Jones, S. L. (1991) Descending noradrenergic influences on pain. *Prog. Brain Res.* **88,** 381–394.

57. Czech, K. A. and Sagen, J. (1995) Update on cellular transplantation into the CNS as a novel therapy for chronic pain. *Prog. Neurobiol.* **46,** 507–529.

58. Sagen, J., Pappas, G. D., and Pollard, H. B. (1986) Analgesia induced by isolated bovine chromaffin cells implanted in rat spinal cord. *Proc. Natl. Acad. Sci. USA* **83,** 7522–7526.

59. Unsicker, K. (1993) The trophic cocktail made by adrenal chromaffin cells. *Exp. Neurol.* **123,** 167–173.

60. Unsicker, K. and Krieglstein, K. (1996) Growth factors in chromaffin cells. *Prog. Neurobiol.* **48,** 307–324.
61. Hama, A. T. and Sagen, J. (1994) Reduced pain-related behavior by adrenal medullary transplants in rats with experimental painful peripheral neuropathy. *Pain* **52,** 223–231.
62. Siegan, J. B. and Sagen, J. (1997) Attenuation of formalin pain responses in the rat by adrenal medullary transplants in the spinal subarachnoid space. *Pain* **70,** 279–285.
63. Sagen, J. and Kemmler, J. E. (1989) Increased levels of met-enkephalin like-immunoreactivity in the spinal cord CSF of rats with adrenal medullary transplants. *Brain Res.* **502,** 1–10.
64. Sagen, J., Kemmler, J. E., and Wang, H. (1991) Adrenal medullary transplants increase spinal cord cerebral spinal fluid catecholamine levels and reduce pain sensitivity. *J. Neurochem.* **56,** 623–627.
65. Wang, H. and Sagen, J. (1994) Absence of appreciable tolerance and morphine cross-tolerance in rats with adrenal medullary transplants in the spinal cord. *Neuropharm* **33,** 681–692.
66. Siegan, J. B., Hama, A. T., and Sagen, J. (1996) Alterations in rat spinal cord cGMP by peripheral nerve injury and adrenal medullary transplantation. *Neurosci. Lett.* **215,** 49–52.
67. Sagen, J. and Wang, H. (1995) Adrenal medullary grafts suppress c-fos induction in spinal neurons of arthritic rats. *Neurosci. Lett.* **192,** 181–184.
68. Siegan, J. B. and Sagen, J. (1995) Attenuation of NMDA-induced spinal hypersensitivity by adrenal medullary transplants. *Brain Res.* **680,** 88–98.
69. Hentall, I. D., Noga, B. R., and Sagen, J. (2001) Spinal allografts of adrenal medulla block nociceptive facilitation in the dorsal horn. *J. Neurophysiol.* **85,** 1788–1792.
70. Ortega, J. D., Sagen, J., and Pappas, G. D. (1992) Survival and integration of bovine chromaffin cells transplanted into rat central nervous system with exogenous trophic factors. *J. Comp. Neurol.* **323,** 13–24.
71. Sagen, J., Wang, H., Tresco, P. A., and Aebischer, P. (1993) Transplants of immunologically isolated xenogenic chromaffin cells provide a long-term source of pain-reducing neuroactive substances. *J. Neurosci.* **13,** 2415–2423.
72. Buchser, E., Goddard, M., Heyd, B., et al. (1996) Immunoisolated xenogenic chromaffin cell therapy for chronic pain. Initial clinical experience. *Anesthesiology* **85,** 1005–1012.
73. Lazorthes, Y., Bes, J. C., Sagen, J., et al. (1995) Transplantation of human chromaffin cells for control of intractable cancer pain. *Acta Neurochir. Supp* **64,** 97–100.
74. Sagen, J. (1996) Chromaffin cell transplants in the CNS: basic and clinical update. In *Yearbook of Cell and Tissue Transplantation* (Lanza, R. P. and Chick, W. L., eds.), Kluwer Academic, Netherlands, pp. 71–89.
75. Ortega, J. D., Sagen, J., and Pappas, G. D. (1992) Short-term immunosuppression enhances long-term graft survival of bovine chromaffin cell xenografts in rat CNS. *Cell Transplant.* **1,** 33–41.

76. Czech, K. A., Pollak, R., Pappas, G. D., and Sagen, J. (1996) Bovine chromaffin cells for CNS transplantation do not elicit xenogenic T cell proliferative responses in vitro. *Cell Transplant.* **5,** 257–267.
77. Czech, K. A., Ryan, J. W., Sagen, J., and Pappas, G. D. (1997) The influence of xenotransplant immunogenicity and immunosuppression on host MHC expression in the rat CNS. *Exp. Neurol.* **147,** 66–83.
78. Brewer, K. L. and Yezierski, R. P. (1998) Effects of adrenal medullary transplants on pain-related behaviors following excitotoxic spinal cord injury. *Pain* **798,** 83–92.
79. Yu, W., Hao, X.-J., Xu, X.-J., et al. (1998) Long-term alleviation of allodynia-like behaviors by intrathecal implantation of bovine chromaffin cells in rats with spinal cord injury. *Pain* **74,** 115–122.
80. Hains, B. C., Chastain, K. M., Everhart, A. W., et al. (2000) Transplants of adrenal medullary chromaffin cells reduce forelimb and hindlimb allodynia in a rodent model of chronic central pain after spinal hemisection injury. *Exp. Neurol.* **164,** 426–437.
81. Yu, W., Hao, J. X., Xu, X.-J., et al. (1998) Immunoisolating encapsulation of intrathecally implanted bovine chromaffin cells prolongs their survival and produces anti-allodynic effect in spinally injured rats. *Eur. J. Pain* **2,** 143–151.
82. Pappas, G. D., Lazorthes, Y., Bes, J. C., et al. (1997) Relief of intractable cancer pain by human chromaffin cell transplants: experience at two medical centers. *Neurol. Res.* **19,** 71–77.
83. Tkaczuk, J., Bes, J. C., deBouet du Portal, H., et al. (1997) Intrathecal allograft of chromaffin cells for intractable pain treatment: a model for understanding CNS tolerance mechanisms in humans. *Transplant Proc.* **29,** 2356–2357.
84. Wilson, S., Chang, K., and Viveros, O. (1981) Opioid peptide synthesis in bovine and human adrenal chromaffin cells. *Peptides* **2(Suppl),** 83–88.
85. Tkaczuk, J., Bes, J. C., Duplan, H., et al. (2000) Intrathecal grafting of unencapsulated adrenal medullary tissue can bring CD4 T lymphocytes into CSF: a potentially deleterious event for the graft. *Cell Transplant.* **9,** 79–91.
86. Bhattacharya, N., Chhetri, M. K., Mukherjee, K. L., et al. (2002) Human fetal adrenal transplant: a possible role in relieving intractable pain in advanced rheumatoid arthritis. *Clin. Exp. Obstet. Gynecol.* **29,** 197–206.
87. Bes, J. C. and Sagen, J. (2002) Dissociated human embryonic and fetal adrenal glands in neural stem cell culture system: open fate for neuronal, nonneuronal, and chromaffin lineages? *Ann. NY Acad. Sci.* **971,** 563–572.
88. Bes, J. C., Frydel, B. R., Walters, W. M., et al. (2000) Generation of chromaffin cell precursors from human embryonic adrenal glands for potential use in transplantation. *Soc. Neurosci. Abstr.* **26,** 829.
89. Bes, J. C., Frydel, B. R., Potter, E. D., et al. (2001) Human embryonic and fetal adrenal glands as sources of neural precursors for possible transplantation strategies. *Am. Soc. Neuro. Transplant Repair* **8,** 61.
90. Feeney, D. M., Weisend, M. P., and Kline, A. E. (1993) Noradrenergic pharmacotherapy, intracerebral infusion and adrenal transplantation promote functional recovery after cortical damage. *J. Neural Transplant. Plast.* **4,** 199–213.

91. Gladstone, D. J. and Black, S. E. (2000) Enhancing recovery after stroke with noradrenergic pharmacotherapy: a new frontier? *Can. J. Neurol.* **27,** 97–105.
92. Dietrich, W. D., Alonso, O., Busto, R., and Finklestein, S. P. (1996) Posttreatment with intravenous basic fibroblast growth factor reduces histopathological damage following fluid-percussion brain injury in rats. *J. Neurotrauma* **13,** 309–316.
93. McDermott, K. L., Raghupathi, R., Fernandez, S. C., et al. (1997) Delayed administration of basic fibroblast growth factor (bFGF) attenuates cognitive dysfunction following parasagittal fluid percussion brain injury in the rat. *J. Neurotrauma* **4,** 191–200.
94. Bethel, A. M., Kirsch, J. R., Koehler, R. C., et al. (1997) Intravenous basic fibroblast growth factor decreases brain injury resulting from focal ischemia in cats. *Stroke* **28,** 609–616.
95. Lee, T. T., Green, B. A., Dietrich, W. D., and Yezierski, R. P. (1999) Neuroprotective effects of basic fibroblast growth factor following spinal cord contusion injury in the rat. *J. Neurotrauma* **16,** 347–356.
96. Sinson, G., Voddi, M., and McIntosh, T. K. (1996) Combined fetal neural transplantation and nerve growth factor infusion: effects on neurological outcome following fluid-percussion brain injury in the rat. *J. Neurosurg.* **84,** 655–662.
97. Soares, H. and McIntosh, T. K. (1991) Fetal cortical transplants in adult rats subjected to experimental brain injury. *J. Neural Transplant. Plast.* **2,** 207–220.
98. Bramlett, H. M., Kraydieh, S., Green, E. J., and Dietrich, W. D. (1997) Temporal and regional patterns of axonal damage following traumatic brain injury: a beta-amyloid precursor protein immunocytochemical study in rats. *J. Neuropathol. Exp. Neurol.* **56,** 1132–1141.
99. Dietrich, W. D., Alonso, O., and Halley, M. (1994) Early microvascular and neuronal consequences of traumatic brain injury: A light and electron microscopic study in rats. *J. Neurotrauma* **11,** 289–301.
100. Hicks, R. R., Smith, D. H., Lowenstein, D. H., et al. (1993) Mild experimental brain injury in the rat induces cognitive deficits associated with regional neuronal loss in the hippocampus. *J. Neurotrauma* **10,** 405–414.
101. Bramlett, H. M. and Dietrich, W. D. (2002) Characterization of gray and white matter structural changes one year after traumatic brain injury in the rat. *Acta Neuropathol.* **103,** 607–614.
102. Gorman, L. K., Shook, B. L., and Becker, D. P. (1993) Traumatic brain injury produces impairments in long-term and recent memory. *Brain Res.* **614,** 29–36.
103. Yamaki, T., Murakami, N., Iwamoto, Y., et al. (1997) Evaluation of learning and memory dysfunction and histological findings in rats with chronic stage contusion and diffuse axonal injury. *Brain Res.* **28,** 151–160.
104. Bramlett, H. M., Dietrich, W. D., and Green, E. J. (1999) Secondary hypoxia following moderate fluid percussion brain injury in rats exacerbates sensorimotor and cognitive deficits. *J. Neurotrauma* **16,** 1035–1047.
105. Pierce, J. E., Smith, D. H., Trojanowski, J. Q., and McIntosh, T. K. Enduring cognitive, neurobehavioral, and histopathological changes persist for up to

one year following severe experimental brain injury in rats. *Neurosci* **87,** 359–369.

106. Smith, D. H., Chen, X. H., Pierce, J. E., et al. (1997) Progressive atrophy and neuron death for one year following brain trauma in the rat. *J. Neurotrauma* **14,** 715–727.

107. Labbe, R., Firl, A., Mufson, E., and Stein, D. (1983) Fetal brain transplants: Reduction of cognitive deficits in rats with frontal cortex lesions. *Science* **2,** 470–472.

108. Soares, H., Sinson, G., and McIntosh, T. K. (1995) Fetal hippocampal transplants attenuate CA3 pyramidal cell death resulting from fluid percussion brain injury in the rat. *J. Neurotrauma* **12,** 1059–1067.

109. Chiang, Y. H., Lin, S. Z., Borlogan, C. V., et al. (1999) Transplantation of fetal kidney tissue reduces cerebral infarct induced by middle cerebral artery ligation. *J. Cereb. Blood Flow Metab.* **19,** 1329–1335.

110. Dillon-Carter, O., Kohnston, R. E., Borlongan, C. V., et al. (2002) T155g-immortalized kidney cells produce growth factors and reduce sequelae of cerebral ischemia. *Cell Transplant.* **11,** 251–259.

111. Johnston, R. E., Dillon-Carter, O., Freed, W. J., and Borlongan, C. V. (2001) Trophic factor secreting kidney cell lines: in vitro characterization and functional effects following transplantation in ischemic rats. *Brain Res.* **900,** 268–276.

112. Greene, L. A. and Tischler, A. S. (1976) Establishment of a noradrenergic clonal cell line of rat adrenal pheochromocytoma cells which respond to nerve growth factor. *Proc. Natl. Acad. Sci. USA* **73,** 2424–2428.

113. Byrd, J. C., Hadjicònstantinou, M., and Cavalla, D. (1986) Epinephrine synthesis in the PC12 pheochromocytoma cell line. *Eur. J. Pharamacol.* **127,** 139–142.

114. Yoshikawa, M., Nakayama, H., Ueno, S., et al. (2000) Chronic fentanyl treatments induce the up-regulation of mu opioid receptor mRNA in rat pheochromocytoma cells. *Brain Res.* **859,** 217–223.

115. Stockley, T. L. and Chang, P. L. (1997) Non-autologous transplantation with immuno-isolation in large animals—a review. *Ann. NY Acad. Sci.* **831,** 408–426.

116. Wu, H. H., Lester, B. R., Sun, Z., and Wilcox, G. L. (1994) Antinociception following implantation of mouse B16 melanoma cells in mouse and rat spinal cord. *Pain* **56,** 203–210.

117. De la Calle, J. L., Mena, M. A., Gonzalez-Escalada, J. R., and Paino, C. L. (2002) Intrathecal transplantation of neuroblastoma cells decreases heat hyperalgesia and cold allodynia in a rat model of neuropathic pain. *Brain Res. Bull.* **59,** 205–211.

118. Wu, H. H., Wilcox, G. L., and McLoon, S. C. (1994) Implantation of AtT-20 or genetically modified AtT-20/hENK cells in mouse spinal cord induced antinociception and opioid tolerance. *J. Neurosci.* **14,** 4806–4814.

119. Saitoh, Y., Taki, T., Arita, N., et al. (1995) Analgesia induced by transplantation of encapsulated tumor cells secreting beta-endorphin. *J. Neurosurg.* **82,** 630–634.

120. Ishii, K., Isono, M., Inoue, R., and Hori, S. (2000) Attempted gene therapy for intractable cancer pain: dexamethasone-mediated exogenous control of

beta-endorphin secretion in genetically modified cells and intrathecal transplantation. *Exp. Neurol.* **166,** 90–98.

121. Saitoh, Y., Taki, T., Arita, N., et al. (1995) Cell therapy with encapsulated xenogenic tumor cells secreting B-endorphin for treatment of peripheral pain. *Cell Transplant.* **4,** S13–S17.
122. Gage, F. H., Ray, J., and Fisher, L. J. (1995) Isolation, characterization, and use of stem cells from the CNS. *Ann. Rev. Neurosci.* **18,** 159–192.
123. McKay, R. (1997) Stem cells in the central nervous system. *Science* **276,** 66–71.
124. Hagan, M., Wennersten, A., Meijer, X., et al. (2003) Neuroprotection by human neural progenitor cells after experimental contusion in rats. *Neurosci. Lett.* **351,** 149–152.
125. Lacza, Z., Horvath, E., and Busija, D. W. (2003) Neural stem cell transplantation in cold lesion: a novel approach for the investigation of brain trauma and repair. *Brain Res. Protoc.* **11,** 145–154.
126. Castellanos, D. A., Daniels, L. D., Farahvar, A., et al. (2003) Cellular and functional outcome of neuronal stem cell transplants after fluid-percussive brain injury in rats. *Soc. Neurosci. Abstr.* **33,** 4141.
127. Tate, M. C., Shear, D. A., Hoffman, S. W., et al. (2002) Fibronectin promotes survival and migration of primary neural stem cells transplanted into the traumatically injured mouse brain. *Cell Transplant.* **11,** 283–295.
128. Hoane, M. R., Becerra, D., Shank, J. E., et al. (2004) Transplantation of neuronal and glial precursors dramatically improves sensorimotor function but not cognitive function in the traumatically injured brain. *J. Neurotrauma* **21,** 163–174.
129. Chiba, S., Ikeda, R., Kurokawa, M. S., et al. (2004) Anatomical and functional recovery by embryonic stem cell-derived neural tissue of a mouse model of brain damage. *J. Neurotrauma* **219,** 107–117.
130. Mahmood, A., Lu, D., Wang, L., et al. (2001) Treatment of traumatic brain injury in female rats with intravenous administration of bone marrow stromal cells. *Neurosurg* **49,** 1196–1203.
131. Mahmood, A., Lu, D., Lu, M., and Chop, M. (2003) Treatment of traumatic brain injury in adult rats with intravenous administration of human bone marrow stromal cells. *Neurosurg* **53,** 697–702.
132. Lu, D., Li, Y., Mahmood, A., et al. (2002) Neural and marrow-derived stromal cell sphere transplantation in a rat model of traumatic brain injury. *J. Neurosurg.* **97,** 935–940.
133. Lu, M., Chen, J., Lu, D., et al. (2003) Global test statistics for treatment effect of stroke and traumatic brain injury in rats with administration of bone marrow stromal cells. *J. Neurosci. Methods* **128,** 183–190.
134. Lu, D., Sanberg, P. R., Mahmood, A., et al. (2002) Intravenous administration of human umbilical cord blood reduces neurological deficit in the rat after traumatic brain injury. *Cell Transplant.* **11,** 275–281.
135. Mahmood, A., Lu, D., Wang, L., and Choop, M. (2002) Intracerebral transplantation of marrow stromal cells cultured with neurotrophic factors

promotes functional recovery in adult rats subjected to traumatic brain injury. *J. Neurotrauma* **19,** 1609–1617.

136. Alessandri, G., Emanueli, C., and Madeddu, P. (2004) Genetically engineered stem cell therapy for tissue regeneration. *Ann. NY Acad. Sci.* **1015,** 271–284.
137. Myckatyn, T. M., Mackinnon, S. E., and McDonald, J. W. (2004) Stem cell transplantation and other novel techniques for promting recovery from spinal cord injury. *Transpl. Immunol.* **12,** 343–358.
138. Bes, J. C., Walters, W. M., Potter, E. D., et al. (2000) Ultrastructural characterization of chromaffin cell progenitor cultures. *Amer. Soc. Neuro. Transplant Repair* **7,** 42.
139. Potter, E. D., Walters, W. M., Frydel, B. R., et al. (1999) Serum-free media and FGF-2 induce proliferation of neonatal and adult bovine chromaffin cells/. *Am. Soc. Neuro. Transplant Repair* **6,** 54.
140. Potter, E. D., Walters, W. M., Frydel, B. R., et al. (1999) Serum-free N2 media and basic fibroblast growth factor induce proliferation in bovine neonatal chromaffin cells. *Soc. Neurosci. Abstr.* **25,** 2045.
141. Lin, C. R., Wu, P. G., Shih, H. C., et al. (2002) Intrathecal spinal progenitor cell transplantation for the treatment of neuropathic pain. *Cell Transplant.* **11,** 17–24.
142. Reiss, P., Zhang, C., Saatman, K. E., et al. (2002) Transplanted neural stem cells survive, differentiate, and improve neurological motor function after experimental traumatic brain injury. *Neurosurg* **51,** 1043–1054.
143. Snyder, E. Y., Deitcher, D. L., Walsh, C., et al. (1992) Multipotent neural cell lines can engraft and participate in development of mouse cerebellum. *Cell* **66,** 33–51.
144. Gray, J. A., Grigoryan, G., Virley, D., et al. (2000) Prospects for the clinical application of neural transplants with the use of conditionally immortalized neuroepithelial cells. *Philos. Trans. R. Soc. Lond. Biol. Sci.* **9,** 153–168.
145. Philips, M. F., Mattiasson, G., Wielock, T., et al. (2001) Neuroprotective and behavioral efficacy of nerve growth factor-transfected hippocampal progenitor cell transplants after experimental traumatic brain injury. *J. Neurosurg.* **94,** 765–774.
146. Muir, J. K., Raghupathi, R., Saatman, K. E., et al. (1999) Terminally differentiated human neurons survive and integrate following transplantation into the traumatically injured rat brain. *J. Neurotrauma* **16,** 403–414.
147. Philips, M. F., Muir, J. K., Saatman, K. E., et al. (1999) Survival and integration of transplanted postmitotic human neurons following experimental brain injury in immunocompetent rats. *J. Neurosurg.* **90,** 116–124.
148. Kondziolka, D., Wechsler, L., Goldstein, S., et al. (2000) Transplantation of cultured human neuronal cells for patients with stroke. *Neurology* **55,** 565–569.
149. Meltzer, C. C., Kondziolka, D., Villermagne, V. L., et al. (2001) Serial [18F] fluorodeoxyglucose positron emission tomography after human neuronal implantation for stroke. *Neurosurg.* **49,** 586–592.
150. Watson, D. J., Longhi, L., Lee, E. B., et al. (2003) Genetically modified NT2N human neuronal cells mediate long-term gene expression as CNS grafts in

vivo and improve functional cognitive outcome following experimental traumatic brain injury. *J. Neuropathol. Exp. Neurol.* **62,** 368–380.

151. Chang, T. M. and Prakash, S. (1998) Therapeutic uses of microencapsulated genetically engineered cells. *Mol. Med. Today* **4,** 221–227.

152. Zurn, A. D., Tseng, J., and Aebischer, P. (1996) Treatment of Parkinson's disease. Symptomatic cell therapies: cells as biologic minipumps. *Eur. Neurol.* **36,** 405–408.

153. Gage, F. H., Wolff, J. A., Rosenberg, M. D., et al. (1987) Grafting genetically modified cells to the brain: possibilities for the future. *Neurosci* **23,** 795–807.

154. Sagen, J. and Eaton, M. J. (2003) Cellular implantation for the treatment of chronic pain. In *Pain: Current Understanding, Emerging Therapies, and Novel Approaches to Drug Discovery* (Schmidt, W. K., ed.), Marcel Dekker, New York, pp. 815–833.

155. White, L. A., Eaton, M. J., Castro, M. C., et al. (1994) Distinct regulatory pathways control neurofilament expression and neurotransmitter synthesis in immortalized serotonergic neurons. *J. Neurosci.* **14,** 6744–6753.

156. Whittemore, S. R. and White, L. A. (1993) Target regulation of neuronal differentiation in a temperature-sensitive cell line derived from medullary raphe. *Brain Res.* **615,** 27–40.

157. Eaton, M. J., Staley, J. K., Globus, M. Y. T., and Whittemore, S. R. (1995) Developmental regulation of early serotonergic neuronal differentiation: the role of brain-derived neurotrophic factor and membrane depolarization. *Dev. Biol.* **170,** 169–182.

158. Rudge, J. S., Eaton, M. J., Mather, P., et al. (1996) CNTF induces raphe neuronal precursors to switch from a serotonergic to a cholinergic phenotype in vitro. *Mol. Cell Biol.* **7,** 204–221.

159. Eaton, M. J., Dancausse, H. A., and Whittemore, S. R. (1997) GDNF and neurturin affect the survival and 5HT synthesis in immortalized raphe neuronal precursors. *Summ. Neuropep. Conf.* 1997.

160. Eaton, M. J. and Whittemore, S. R. (1995) ACTH activation of adenylate cyclase in raphe neurons: multiple regulatory pathways control serotonergic neuronal differentiation. *J. Neurobiol.* **28,** 465–481.

161. Eaton, M. J. and Whittemore, S. R. (1996) Autocrine BDNF secretion enhances the survival and serotonergic differentiation of raphe neuronal precursor cells grafted into the adult rat CNS. *Exp. Neurol.* **140,** 105–114.

162. Farakash, A. E. and Portney, R. K. (1986) The pharmacological management of chronic pain in the paraplegic patient. *J. Amer. Paraplegia Soc.* **9,** 41–50.

163. Eaton, M. J., Dancausse, H. R., Santiago, D. I., and Whittemore, S. R. (1997) Lumbar transplants of immortalized serotonergic neurons alleviates chronic neuropathic pain. *Pain* **72,** 59–69.

164. Eaton, M. J., Karmally, S., Martinez, M. A., et al. (1999) Lumbar transplants of neurons genetically modified to secrete galanin reverse pain-like behaviors after partial sciatic nerve injury. *J. Peripher. Nerv. Sys.* **4,** 245–257.

165. Cejas, P., Plunkett, J. A., Martinez, M. A., et al. (1999) A BDNF-synthesizing cell line ameliorates chronic neuropathic pain. *Soc. Neurosci. Abstr.* **25,** 1945.

166. Eaton, M. J., Plunkett, J. A., Martinez, M. A., et al. (1999) Transplants of neuronal cells bio-engineered to synthesize GABA alleviate chronic neuropathic pain. *Cell Transplant.* **8,** 87–101.

167. Eaton, M. J., Martinez, M. A., Karmally, S., et al. (2001) Transplants of GABA-secreting cells for chronic pain after excitotoxic spinal cord injury. *Am. Soc. Neuro. Transplant Repair* **8,** 76.

168. Eaton, M. J. (2004) SCI pain is a GABA-sensitive phenomenon amendable to GABA cell therapy. *7th Intnl Conf Mechs Treatment Neuropathic Pain* 2004.

169. Trojanowski, J. Q., Kleppner, S. R., Hartley, R. S., et al. (1997) Transfectable and transplantable postmitotic human neurons: potential "platform" for gene therapy of nervous system diseases. *Exp. Neurol.* **144,** 92–97.

170. Yoshioka, A., Yudkoff, M., and Pleasure, D. (1997) Expression of glutamic acid decarboxylase during human neuronal differentiation: studies using the NTera-2 culture system. *Brain Res.* **767,** 333–339.

171. Cepko, C. L. (1989) Immortalization of neural cells via retrovirus-mediated oncogene transduction. *Ann. Rev. Neurosci.* **12,** 47–65.

172. Jat, P. S. and Sharp, P. A. (1989) Cell lines established by a temperature-sensitive simian virus 40 large T antigen gene are growth restricted at nonpermissive temperature. *Mol. Cell Biol.* **9,** 1672–1681.

173. Jat, P. S., Noble, M. D., Ataliotis, P., et al. (1991) Direct derivation of conditionally immortal cell lines from an H-2k^b-tsA58 transgenic mouse. *Proc. Natl. Acad. Sci. USA* **88,** 5096–5100.

174. Jiang, J. Y., Lyeth, B. G., Clifton, G. L., et al. (1991) Relationship between body and brain temperature in traumatically brain-injured rodents. *J. Neurosurg.* **74,** 492–496.

175. Cherington, V., Brown, M., Paucha, E., et al. (1988) Separation of simian virus 40 large-T-antigen-transforming and origin-binding functions from the ability to block differentiation. *Mol. Cell Biol.* **8,** 1380–1384.

176. Frederiksen, K., Jat, P. S., Valtz, N., et al. (1988) Immortalization of precursor cells from the mammalian CNS. *Neuron* **1,** 439–448.

177. Ikram, Z., Norton, T., and Jat, P. S. (1994) The biological clock that measures the mitotic life-span of mouse embryo fibroblasts continues to function in the presence of simian virus 40 large tumor antigen. *Proc. Natl. Acad. Sci. USA* **91,** 6448–6452.

178. Reynisdottir, I., O'Reilly, D., Miller, L., and Prives, C. (1990) Thermally inactivated simian virus 40 tsA58 mutant T antigen cannot initiate viral DNA replication in vitro. *J. Virol.* **64,** 6234–6245.

179. Frisa, P., Goodman, M., Smith, G., et al. (1994) Immortalization of immature and mature astrocytes by SV40 T antigen. *J. Neurosci. Res.* **39,** 47–56.

180. Goodman, G., Eckert, R., Utian, W., and Rorke, E. (1993) Establishment of and neurite outgrowth properties of neonatal and adult rat olfactory bulb glial cell lines. *Brain Res.* **619,** 199–213.

181. Jat, P. S. and Sharp, P. (1986) Large t antigen of simian virus 40 and polyomavirus efficiently establish primary fibroblasts. *J. Virol.* **59,** 746–750.

182. Renfranz, P. J., Cunningham, M. G., and McKay, R. D. G. (1991) Region-specific differentiation of the hippocampal stem cell line HiB5 upon implantation into the developing mammalian brain. *Cell* **66,** 713–729.
183. Onifer, S. M., Whittemore, S. R., and Holets, V. R. (1993) Variable morphological differentiation of a raphe-derived neuronal cell line following transplantation into the adult rat CNS. *Exp. Neurol.* **122,** 130–142.
184. Shihabuddin, L. S., Hertz, J. A., Holets, V. R., and Whittemore, S. R. (1995) The adult CNS retains the potential to direct region-specific differentiation of a transplanted neuronal precursor cell line. *J. Neurosci.* **15,** 6666–6678.
185. Shihabuddin, L. S., Holets, V. R., and Whittemore, S. R. (1996) Selective hippocampal lesions differentially affect the phenotypic fate of transplanted neuronal precursor cells. *Exp. Neurol.* **139,** 61–72.
186. Tornatore, C., Baker-Cairns, B., Yadid, G., et al. (1996) Expression of tyrosine hydroxylase in an immortalized human fetal astrocyte cell line in vitro characterization and engraftment into the rodent striatum. *Cell Transplant.* **5,** 145–163.
187. Clarkson, E., Rosa, F., Edwards-Prasad, J., et al. (1998) Improvement of neurological deficits in 6-hydroxydopamine-lesioned rats after transplantation with allogeneic simian virus 40 large tumor antigen gene-induced immortalized dopamine cells. *Proc. Natl. Acad. Sci. USA* **95,** 1265–1270.
188. Unsicker, K. (1993) The chromaffin cell: paradigm in cell, developmental and growth factor biology. *J. Anat.* **183,** 207–221.
189. Totzauer, I., Amselgruber, W., Sinowatz, F., and Gratzl, M. (1995) Early expression of chromogranin A and tyrosine hydroxylase during prenatal development of the bovine adrenal gland. *Anat. Embryol. (Berl.)* **191,** 139–143.
190. Eaton, M. J., Frydel, B., Lopez, T., et al. (1999) Generation and initial characterization of conditionally immortalized chromaffin cells. *J. Cell Biochem.* **79,** 38–57.
191. Eaton, M. J., Martinez, M., Frydel, B., et al. (2000) Initial characterization of the transplant of immortalized chromaffin cells for the attenuation of chronic neuropathic pain. *Cell Transplant.* **9,** 637–656.
192. Sortwell, C. and Sagen, J. (1995) Induction of antidepressive activity by monoaminergic transplants in rat neocortex. *Pharmacol. Biochem. Behav.* **46,** 225–230.
193. Sauer, B. and Henderson, N. (1988) Site-specific DNA recombination in mammalian cells by the Cre recombinase of bacteriophage P1. *Proc. Natl. Acad. Sci. USA* **85,** 5166–5170.
194. Sauer, B. (1998) Inducible gene targeting in mice using the cre/lox system. *Methods* **14,** 381–392.
195. Paillard, F. (1999) Reversible cell immortalization with the cre-lox system. *Hum. Gene Ther.* **10,** 1597–1598.
196. Kobayashi, N., Miyazaki, M., Fukaya, K., et al. (2000) Treatment of surgically induced acute liver failure with transplantation of highly differentiated immortalized human hepatocytes. *Cell Transplant.* **9,** 733–735.
197. Metzger, D. and Feil, R. (1999) Engineering the mouse genome by site-specific recombination. *Curr. Opin. Biotechnol.* **10,** 470–476.

198. Kellendonk, C., Tronche, F., Monaghan, A.-P., et al. (1996) Regulation of Cre recombinase activity by the synthetic steroid RU 486. *Nuc. Acids Res.* **24,** 1404–1411.

199. Kellendonk, C., Tronche, F., Casanova, E., et al. (1999) Inducible site-specific recombination in the brain. *J. Mol. Biol.* **285,** 175–182.

200. Eaton, M. J., Herman, J. P., Jullien, N., et al. (2002) Immortalized chromaffin cells disimmortalized with Cre/lox site-directed recombination for use in cell therapy for pain. *Exp. Neurol.* **175,** 49–60.

201. Cai, J., Ito, M., Westerman, K. A., et al. (2000) Construction of a non-tumorigenic rat hepatocyte cell line for transplantation: reversal of hepatocyte immortalization by site-specific excision of the SV40 T antigen. *J. Hepatol.* **33,** 701–708.

202. Kobayashi, N., Fujiwara, T., Westerman, K. A., et al. (2000) Prevention of acute liver failure in rats with reversibly immortalized human hepatocytes. *Science* **287,** 1258–1262.

203. Herman, J. P., Becq, H., and Enjalbert, A. (1997) A reversible immortalization procedure to obtain neural cell lines. *Soc. Neurosci. Abstr.* **23,** 319.

204. Berghella, L., De Angelis, L., Coletta, M., et al. (1999) Reversible immortalization of human myogenic cells by site-specific excision of a retrovirally transferred oncogene. *Hum. Gene Ther.* **10,** 1607–1617.

205. Kobayashi, N., Noguchi, H., Westerman, K. A., et al. (2000) Efficient Cre/loxP site-specific recombination in a HepG2 human liver cell line. *Cell Transplant.* **9,** 737–742.

206. Westerman, K. A. and Leboulch, P. (1996) Reversible immortalization of mammalian cells mediated by retroviral transfer and site-specific recombination. *Proc. Natl. Acad. Sci. USA* **93,** 8971–8976.

207. Eiden, L. E., Giraud, P., Affolter, H., et al. (1984) Alternative modes of enkephalin biosynthesis regulation by reserpine and cyclic AMP in cultured chromaffin cells. *Proc. Natl. Acad. Sci. USA* **81,** 3949–3953.

208. Eiden, L. E. and Hotchkiss, A. (1983) Cyclic adenosine monophosphate regulates vasoactive intestinal polypeptide and enkephalin biosynthesis in cultured chromaffin cells. *Neuropep.* **4,** 1–9.

209. Duplan, H., Li, R. Y., Vue, C., et al. (2004) Grafts of immortalized chromaffin cells bio-engineered to improve Met-enkephalin release also reduce formalin-evoked c-fos expression in rat spinal cord. *Neurosci. Lett.* **370,** 1–6.

210. Wang, H. and Sagen, J. (1994) Optimization of adrenal medullary allograft conditions for pain alleviation. *J. Neural. Transpl. Plastic* **5,** 49–64.

211. Sagen, J., Wang, H., Tresco, P. A., and Aebischer, P. (1993) Transplants of immunologically isolated xenogeneic chromaffin cells provide a long-term source of pain-reducing neuroactive substances. *J. Neurosci.* **13,** 2415–2423.

212. Ortega, J. D. and Sagen, J. (1993) Pharmacologic characterization of opioid peptide release from chromaffin cell transplants using a brain slice superfusion method. *Exp. Brain Res.* **95,** 381–387.

213. Lazorthes, Y., Sagen, J., Sallerin, B., et al. (2000) Human chromaffin cell graft into the CSF for cancer pain management: a prospective phase II clinical study. *Pain* **87,** 19–32.

214. Duplan, H., Bes, J. C., Tafani, M., et al. (2000) Adrenal medullary explants as an efficient tool for pain control: adhesive biomolecular components are involved in graft function ex vivo. *Exp. Neurol.* **163,** 331–347.
215. Ferber, D. (2002) Public Health. Creeping consensus on SV40 and polio vaccine. **725,** 727.
216. Liu, J., Pan, J., Naik, S., et al. (1999) Characterization and evaluation of detoxification functions of a nontumorigenic immortalized porcine hepatocyte cell line (HepLiu). *Cell Transplant.* **8,** 219–232.
217. Villa, A., Snyder, E. Y., Vescovi, A., and Martinez-Serrano, A. (2000) Establishment and properties of a growth factor-dependent, perpetual neural stem cell line from the human CNS. *Exp. Neurol.* **161,** 67–84.
218. Deimling, F., Finotto, S., Lindner, K., et al. (1998) Characterization of adrenal chromaffin progenitor cells in mice. *Adv. Pharmacol.* **42,** 932–935.
219. Schober, A., Krieglstein, K., and Unsicker, K. (2000) Molecular cues for the development of adrenal chromaffin cells and their preganglionic innervation. *Eur. J. Clin. Invest.* **30(Suppl 3),** 87–90.
220. Decosterd, I., Buchser, E., Gilliard, N., et al. (1998) Intrathecal implants of bovine chromaffin cells alleviate mechanical allodynia in a rat model of neuropathic pain. *Pain* **76,** 159–166.
221. Hansen, J. T., Notter, M. F. D., Okawara, S.-H., and Gash, D. M. (1988) Organization, fine structure, and viability of the human adrenal medulla: considerations for neural transplantation. *Ann. Neurol.* **24,** 599–609.
222. Sol, J. C., Larrue, S., Li, R. Y., et al. (2002) Intrathecal grafting of porcine chromaffin cells in the rat reduce nociception in a tonic pain model. *Exp. Neurol.* **175,** 433.
223. Sagen, J., Pappas, G. D., and Perlow, M. J. (1987) Alterations in nociception following adrenal medullary transplants into the periaqueductal gray. *Exp. Brain Res.* **67,** 373–379.
224. Joseph, J. M., Goddard, M. B., Mills, J., et al. (1994) Transplantation of encapsulated bovine chromaffin cells in the sheep subarachnoid space: a preclinical study for the treatment of cancer pain. *Cell Transplant.* **3,** 355–364.
225. Wang, H. and Sagen, J. (1995) Attenuation of pain-related hyperventilation in adjuvant arthritic rats with adrenal medullary transplants in the spinal subarachnoid space. *Pain* **63,** 313–320.
226. Wu, H. H., McLoon, S. C., and Wilcox, G. L. (1993) Antinociception following implantation of ATT-20 and genetically modified AtT-20/hENK cells in rat spinal cord. *J. Neur. Trans. Plastic* **4,** 15–26.
227. White, L. A. and Whittemore, S. R. (1992) Immortalization of raphe neurons: an approach to neuronal function *in vitro* and *in vivo*. *J. Chem. Neuroanat.* **5,** 327–330.
228. White, L. A., Keane, R. W., and Whittemore, S. R. (1994) Differentiation of an immortalized CNS neuronal cell line decreases their susceptibility to cytotoxic T lymphocyte cell lysis in vitro. *J. Neuroimmunol.* **49,** 135–143.
229. Whittemore, S. R., White, L. A., Shihabuddin, L. S., and Eaton, M. J. (1995) Phenotypic diversity in neuronal cell lines derived from raphe nucleus by retroviral transduction. *Methods* **7,** 285–296.

230. Cejas, P. J., Martinez, M., Karmally, S., et al. (2000) Lumbar transplant of neurons genetically modified to secrete brain-derived neurotrophic factor attenuate allodynia and hyperalgesia after sciatic nerve constriction. *Pain* **86,** 195–210.
231. Eaton, M. J., Martinez, M., Karmally, S., and Lopez, T. (2000) GABA cell therapy reverses neurogenic pain after excitotoxic spinal cord injury. 18th Natl Soc Neurotrauma Abstr.
232. Hains, B. C., Johnson, K. M., McAdoo, D. J., et al. (2001) Engraftment of immortalized serotonergic neurons enhances locomotor function and attenuates pain-like behavior following spinal hemisection injury in the rat. *Exp. Neurol.* **171,** 361–378.
233. Hains, B. C., Fullwood, S. D., Eaton, M. J., and Hulsebosch, C. E. (2001) Subdural engraftment of serotonergic neurons following spinal hemisection restores spinal serotonin, downregulates the serotonin transporter, and increases BDNF tissue content in the rat. *Brain Res.* **913,** 35–46.
234. Duplan, H., Li, R. Y., Eaton, M. J., et al. (2002) Generation of conditionally immortalized cell lines genetically modified to over-express opioid peptides for pain control. *Exp. Neurol.* **175,** 433.
235. Duplan, H., Li, R. Y., Vue, C., et al. (2002) Anti-nociceptive effect of a conditionally immortalized chromaffin cell line genetically modified to express met-enkephalin in a rodent tonic pain model. *Acta. Neurochir.* **144,** 1106.
236. Eaton, M. J. (2004) Development of human cell therapy for functional recovery following SCI. *J. Spinal Cord Med.* **27,** 155.
237. Eaton, M. J. (2003) Human neuronal cell lines for the treatment of pain and spasticity following SCI. *JJRD* **40(Suppl. 3),** 40.
238. Westerman, K. A. and Lebouleh, P. (1996) Reversible immortalization of mammalian cells mediated by retroviral transfer and site-specific recombination. *Proc. Natl. Acad. Sci. USA* **93,** 8971–8976.
239. Yezierski, R. P., Santana, M., Park, S. H., and Madsen, P. W. (1993) Neuronal degeneration and spinal cavitation following intraspinal injections of quisqualic acid in the rat. *J. Neurotrauma* **10,** 445–456.
240. Yezierski, R. P., Liu, S., Ruenes, G. L., and Kajander, K. J. (1996) Behavioral and pathological characteristics of a central pain model following spinal injury. VIIIth World Congress on Pain.
241. Yezierski, R. P., Liu, S., Ruenes, G. L., et al. (1998) Excitotoxic spinal cord injury: behavioral and morphological characteristics of a central pain model. *Pain* **75,** 141–155.
242. Yezierski, R. P. and Park, S. H. (1998) The mechanosensitivity of spinal sensory neurons following intraspinal injections of quisqualic acid in the rat. *Neurosci. Lett.* **157,** 115–119.

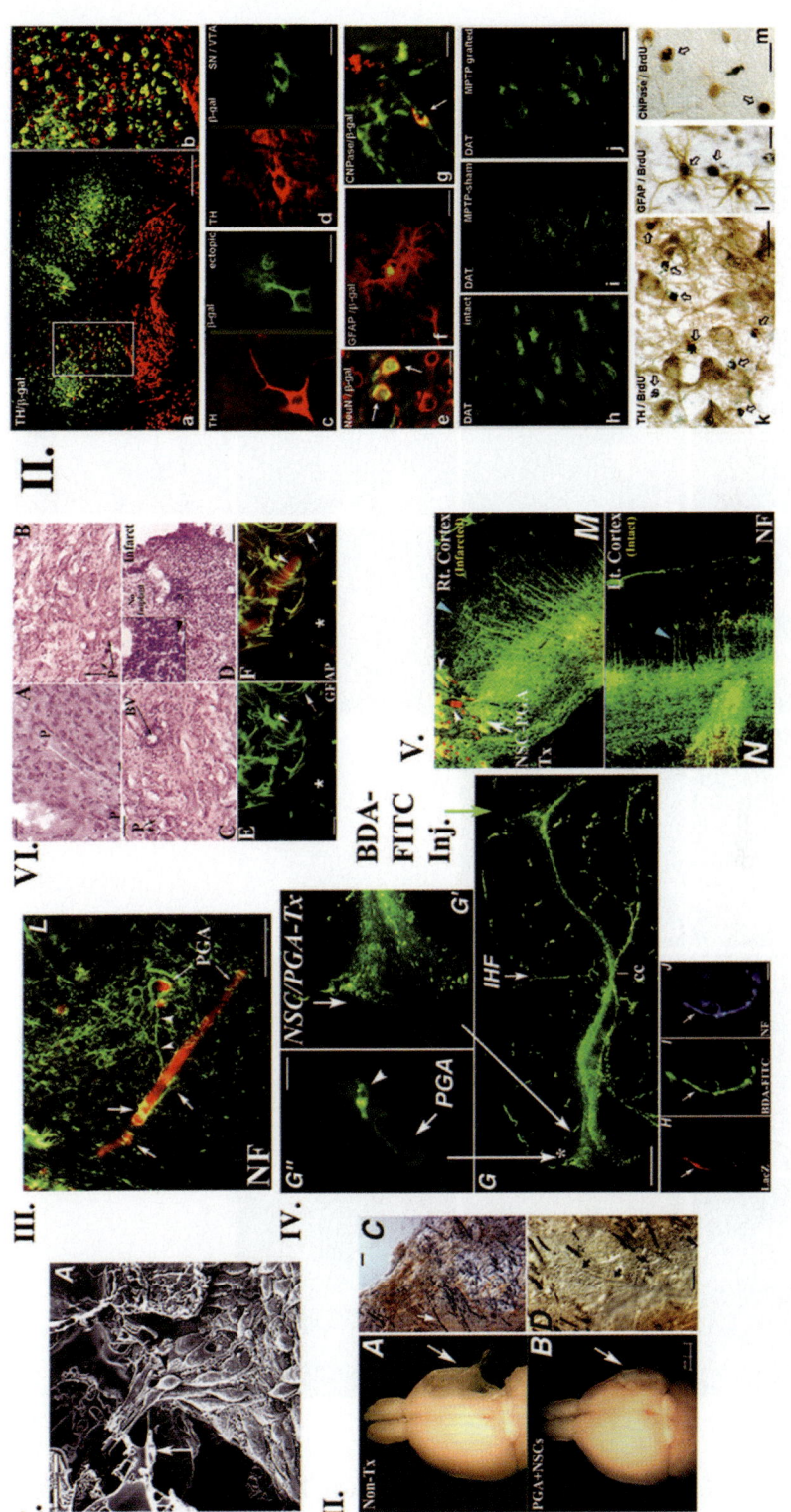

Color Plate 2, Ch. 1, Fig. 3 (right side). (see discussion and caption in Ch 1, on pp. 15–16.)

Color Plate 1, Ch. 1, Fig. 2. (see discussion in and caption in Ch. 1 on pp. 9–11.)

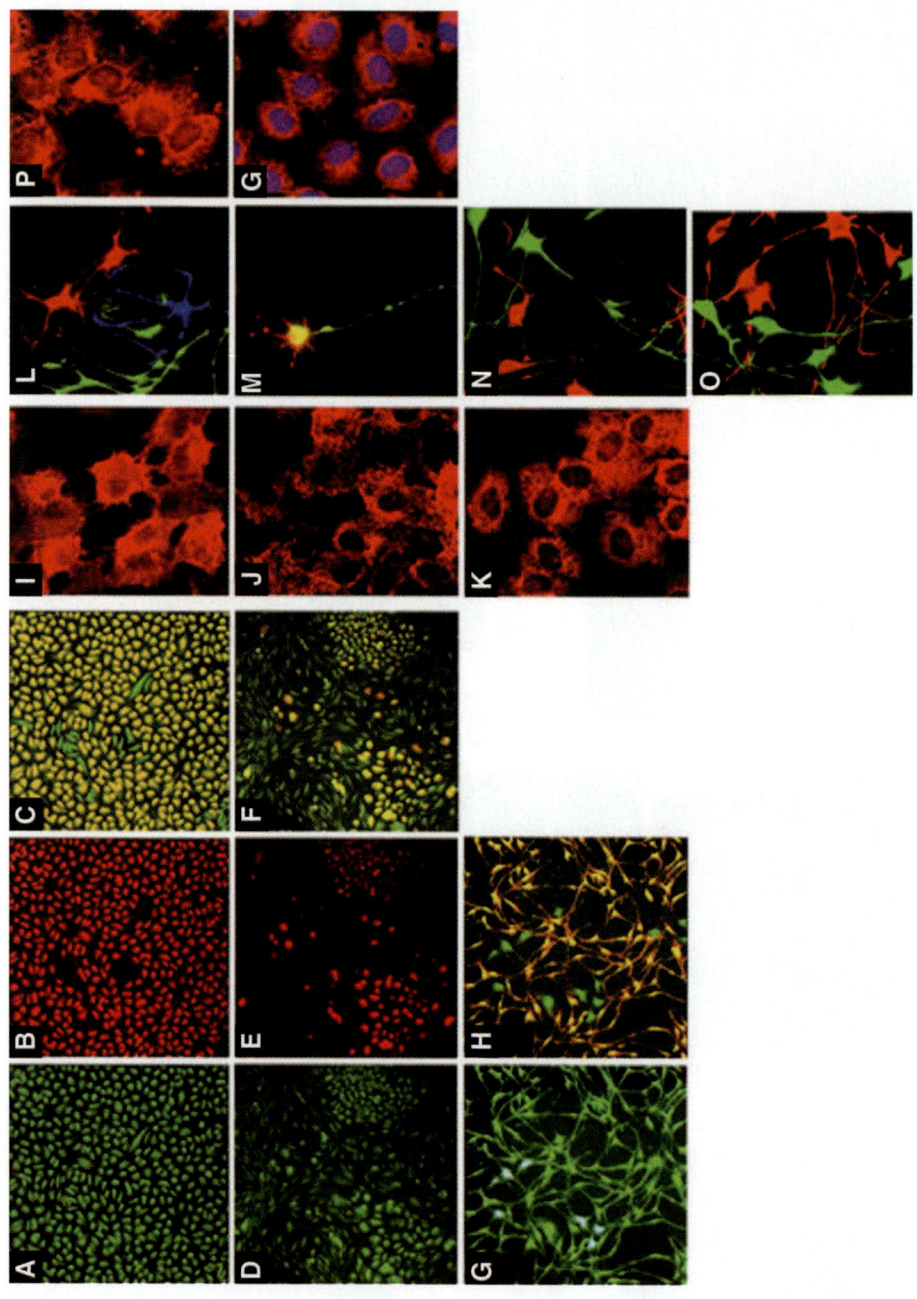

Color Plate 3, Ch. 7, Fig. 1. (see discussion in Ch. 7 and caption on p. 179.)

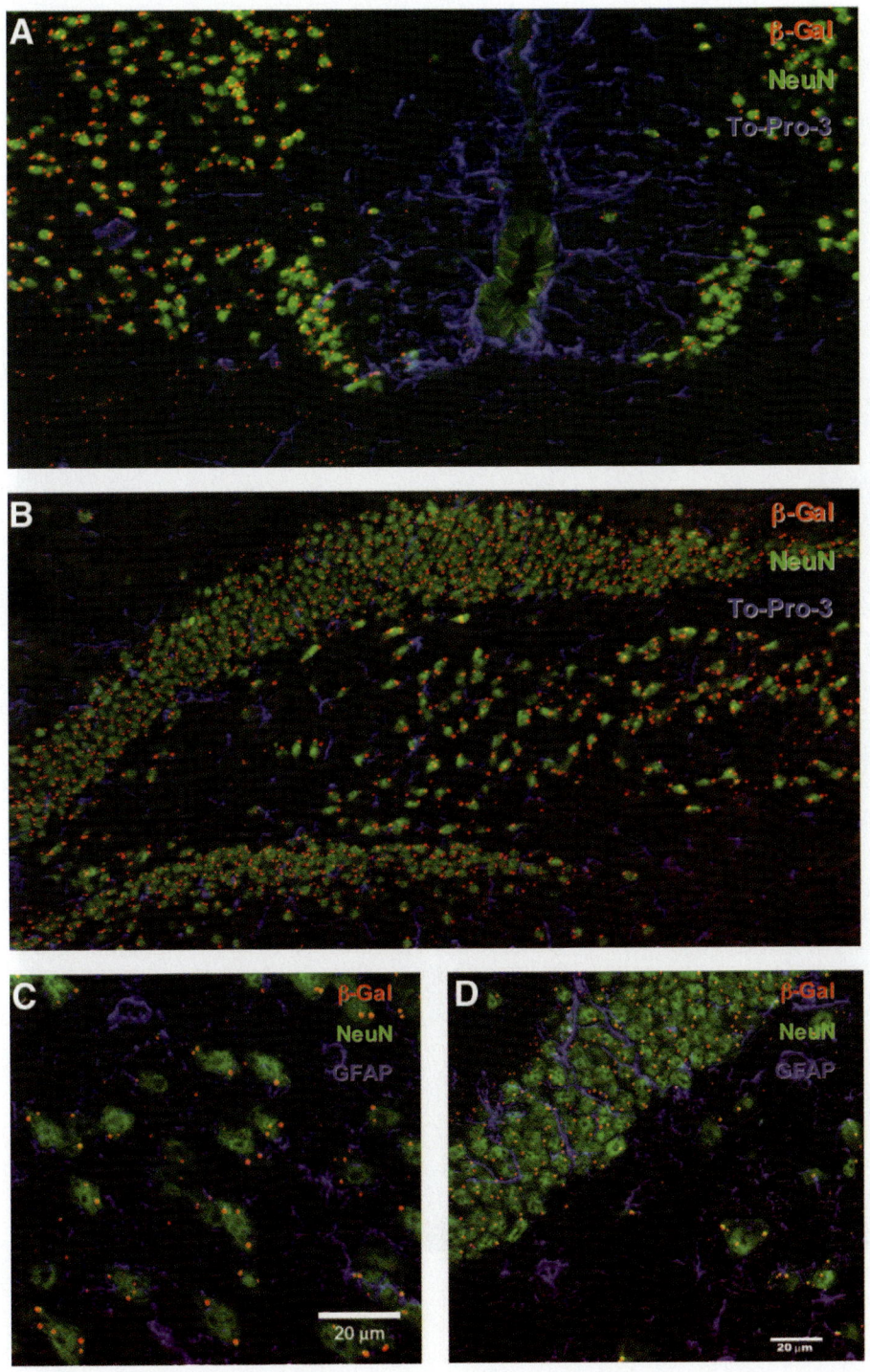

Color Plate 4, Ch. 7, Fig. 5. (see discussion in Ch. 7 and caption on p. 187.)

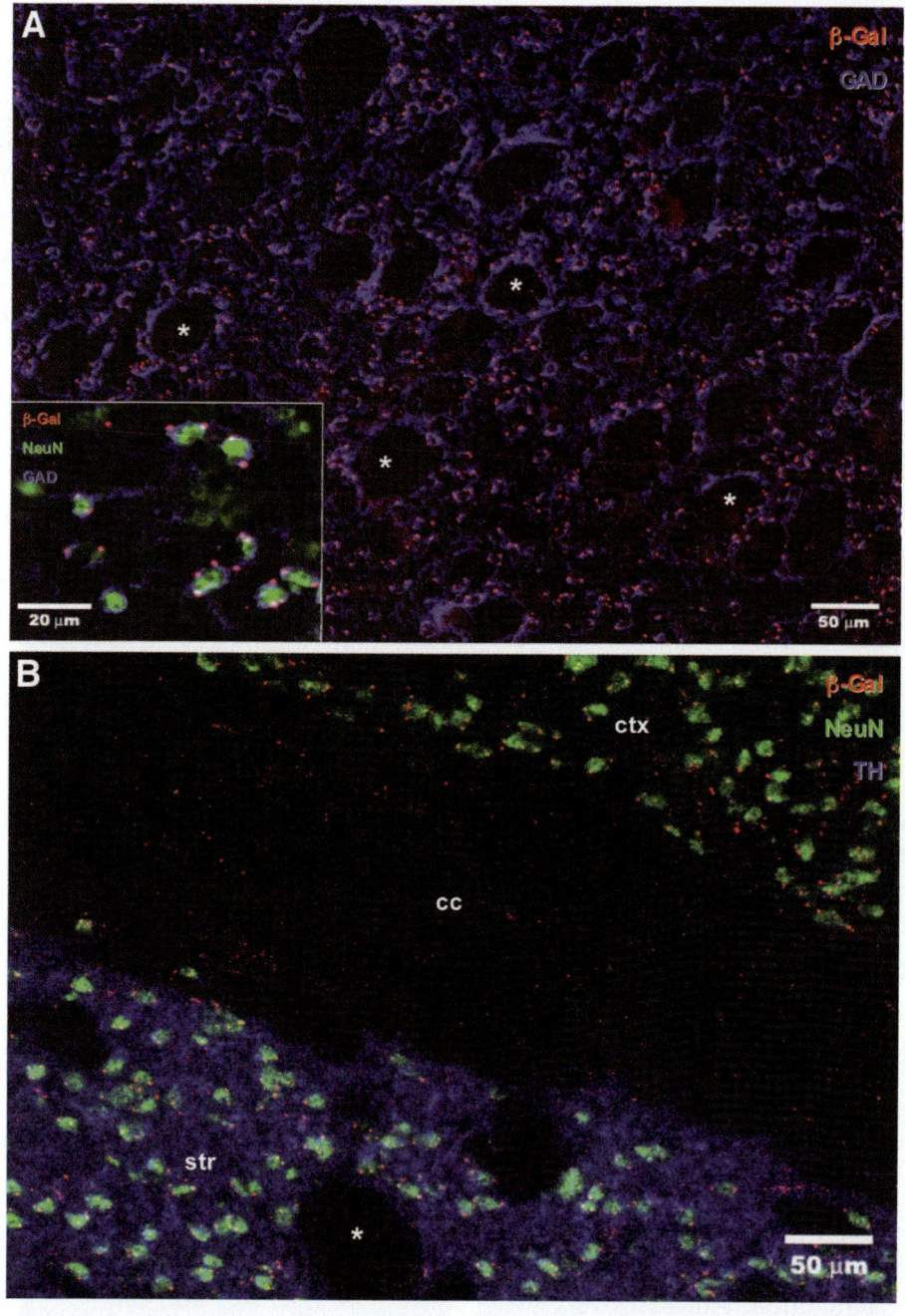

Color Plate 5, Ch. 7, Fig. 6. (see discussion in Ch. 1 and caption on p 189.)

9

The Use of Sertoli Cells in Neural Transplantation

Dwaine F. Emerich, Cesario V. Borlongan, and Craig R. Halberstadt

ABSTRACT

Neural transplantation is emerging as a viable long-term method to treat degenerative diseases of the nervous system. Unfortunately, the shortage of suitable donor tissue necessitates the development of novel approaches to enable transplantation on a widespread basis. Transplantable cells obtained from xenogeneic sources have shown considerable promise in preclinical animal models, as well as in a limited number of human trials conducted to date. Although animal-sourced tissue overcomes the shortage of available cells, transplant survival remains poor, and immunosuppressive regimens are suboptimal in improving graft viability, with deleterious side effects. One potential way of enabling xenotransplantation takes advantage of the immunoprotective capabilities of Sertoli cells, which are normally found in the testes, where they provide local immunologic protection to developing germ cells. In animal models, allogeneic and xenogeneic Sertoli cells survive and function for extended periods of time when grafted into the brain. Moreover, isolated Sertoli cells protect cografted dopaminergic neurons from immune destruction and reverse lesion-induced deficits in Parkinsonian rodents. This chapter discusses these benefits within the context of several potential underlying biological mechanisms.

Key Words: Sertoli cells; Parkinson's disease; transplant.

INTRODUCTION

Transplanting living cells to replace damaged or lost tissues has emerged as a viable long-term treatment for diseases characterized by secretory cell dysfunction. Animal studies and recent clinical research have demonstrated that transplanted cells can survive and promote functional recovery.

From: *Contemporary Neuroscience: Cell Therapy, Stem Cells, and Brain Repair*
Edited by: C. D. Sanberg and P. R. Sanberg © Humana Press Inc., Totowa, NJ

Despite this evidence, limitations in tissue availability, and the need for chronic systemic immunosuppression, to minimize graft rejection are major stumbling blocks to widespread clinical evaluation. Poor availability, ethical concerns, and the lack of usable cell lines complicate the utility of human tissue. Irregular, unpredictable delivery of the desired proteins or modulators hampers the use of cell lines, as well as the possibility that the cell lines used may become neoplastic owing to the incorporation of transforming elements into the cellular genome.

Given these concerns, animals have been proposed as a possible source of transplantable cells. Using xenogeneic tissue can help overcome the limitations in tissue availability. However, even the most successful examples of clinical cell transplantation using allogeneic islets for diabetes require life-long immunosuppressive drugs to prevent the rejection of the transplanted cells *(1,2)*. Although immunosuppressive drugs are considered a breakthrough, their chronic use is burdened with problems. Many immunosuppressive regimens have such severe side effects that their use is precluded in individuals who do not suffer from life-threatening organ failure or other serious disease complications *(3–6)*. Even if these side effects could be eliminated, reducing the incidence of graft rejection of xenogeneic tissue transplants remains a challenge for current immunosuppressive drugs.

Enabling transplantation of xenogeneic cells involves the cotransplantation of Sertoli cells to provide immunologic and trophic support to cografted cells. Sertoli cells are normal constituents of the testes and act as "nurse" cells to support the developing germ cells. Recent data suggest that Sertoli cells allow the survival of cotransplanted allogeneic or xenogeneic cells by secreting potent immunosuppressive and survival-enhancing molecules. For central nervous system (CNS) applications, the most research accumulated to date indicates that Sertoli cells provide a locally protective environment for the successful transplantation of dopaminergic cells as a possible treatment for Parkinson's disease (PD). This chapter describes the development of Sertoli cells as a unique and potentially important cell type to facilitate and enable the transplantation of xenogeneic tissue for PD and discusses several mechanisms by which Sertoli cells may be immunoprotective.

SURVIVAL AND FUNCTIONAL RECOVERY OF XENOGRAFTS IN PRECLINICAL MODELS

Likely, neural xenografts could function anatomically and behaviorally if their survival could be enhanced. Xenogeneic tissues mediate behavioral recovery in many animal models of neurodegeneration *(7)*. In rodent models of PD, transplantation of fetal brain tissue obtained from mouse, hamster, rabbit, pig, and humans results in structural and functional recovery. Similarly, xenografts of fetal preoptic nuclei, superchiasmatic nuclei, and striatal neurons reverse genetic hypergonadism in mice, as well as lesion-induced disturbances of circadian rhythm in the Siberian chipmunk and motor deficits in an excitotoxic primate model of Huntington's disease *(8–12)*.

Although these data are generally encouraging, xenotransplants are nearly completely killed following transplantation *(7)*. Even with extensive immunosuppression, at least 90–95% of the implanted allo- or xenogeneic cells die because of intrinsic developmental signals, damage incurred during cell preparation, inadequate trophic support, and oxidative stress *(13–15)*. Increasing the numbers of cells implanted, or protecting the cells with trophic factors and antioxidants, can partially overcome nonimmunological cell death, but cell- and antibody-mediated rejection processes still actively kill the transplanted cells. This happens, despite that the inflammatory response occurring to CNS grafts is relatively minor, compared to rejection of peripheral tissue grafts. The relatively low immunological reaction in the brain results partly because neurons express virtually no major histocompatibility complex (MHC)-I, and the blood–brain barrier restricts the entry of complement and immune cells into the graft site *(7,16)*. Accordingly, cyclosporine and other conventional immune suppressive regimes can usually prevent catastrophic rejection. However, during rejection, an up regulation of MHC class I and class II antigens occurs in donor and host tissue, further triggering immune responses from class I- and II-stimulated T cells and, by clonal expansion, fully activating the efferent arm of the immune system *(7,13,17)*.

ENCOURAGING HUMAN EXPERIENCE WITH PIG TRANSPLANTS

An Food and Drug Administration-approved US clinical phase I study is being conducted using neural xenotransplantation in 12 PD patients *(18,19)*. This safety trial is focused on the unilateral transplantation of fetal pig mesencephalic neural cells into the putamen and caudate. Six patients received continuous systemic cyclosporine. An interim report at 12 mo listed no evidence of serious adverse events. Cultures were negative for bacterial and viral contamination, and no porcine endogenous retrovirus DNA sequences were found. As assessed by magnetic resonance imaging, the cannula tracts were localized within the putamen and caudate without noticeable tissue trauma. In the medication-off state, total Unified Parkinson's Disease Rating Scale scores improved by 19% ($p = 0.01$), with three patients improving over 30% and two patients exhibiting improved gait. 18F-Levodopa positron emission tomography failed to show changes on the transplanted side. Histological analysis of the graft in a patient that died 7 mo after surgery revealed that, even though only about 0.1% of the total number of transplanted fetal pig dopaminergic cells survived the 7-mo period, extensive axonal connections had formed between the surviving pig neurons and human neurons. Nonneuronal graft cells (likely glial) outnumbered neuronal cells by a 10:1 ratio. Similar results have been obtained in patients with Huntington's disease (HD) who were transplanted with up to 24 million fetal porcine striatal cells (20). Reportedly, these patients have shown a favorable safety profile, but no change was seen in total functional capacity score after 1-yr posttransplant. The combined results of these studies have led to the conclusion that one major challenge to the successful use of xenogeneic cells in neurodegenerative disease is minimizing the immune-mediated rejection of the grafted cells. The sections below outline the potential of Sertoli cell transplants to accomplish this goal.

SERTOLI CELLS: THE BASICS

Sertoli cells are one of four main cell types in the testes; the others are Leydig, peritubular myoid, and germ cells. The architecture of the Sertoli cell is complex and can be readily visualized using reconstructive electron microscopy. Sertoli cells are large (75–100 µm in length), irregularly columnar, and stellate, with a base that abuts neighboring Sertoli cells (21). They are attached to the basement membrane of the seminiferous tubules and extend radially from the basal lamina toward the center of the seminiferous tubule. Sertoli cells are marked by numerous processes that envelope individual germinal cells. While the sperm develop in the testis, Sertoli cells regulate the expression of many proteins. This diverse protein expression is related to the Sertoli cells' primary function—to serve as a "nurse cell" and construct a favorable environment for the sperm to grow and mature. The process of sperm differentiation is too complicated to describe here but deserves some mention (21). The pathway of the developing sperm starts with the spermatogonium, located at the basal end of the Sertoli cell/peritubular myoid cell juncture. Via secretory signals from the Sertoli cell, the spermatogonium actively divides, forming the spermatocytes that will differentiate into spermatids. During meiosis, junctions between the Sertoli cells continuously break and reform to allow the maturing spermatocytes to move. As the spermatocyte differentiates, it migrates between the adjacent Sertoli cells to move into the apical point of the Sertoli cell to be eventually released into the seminiferous tubule. During this process of development and migration, the head of the developing spermatocytes is impregnated into the cytoplasm of the Sertoli cell. Machinery is lacking in the spermatocytes to break down complex sugars; thus, the Sertoli cell actively processes glucose and transports lactic acid to the spermatocytes. The role of the Sertoli cells in spermatogenesis involves the following:

1. Localizing and maintaining developing spermatocytes and spermatids within the seminiferous endothelium.
2. Internalization or phagocytosis of cytoplasmic bodies that have become detached from spermatids.
3. Formation of tight junctional complexes or the blood–testis barrier by adjacent Sertoli cells anchored into the basal lamina of the seminiferous tubule.
4. Lifting early meiotic spermatocytes from the basal to the adluminal portion of the cell.
5. Spermiation, or the release of spermatozoa, from the seminiferous endothelium.
6. A nutritional role characterized by the breakdown of complex sugars to lactic acid and subsequent secretion and delivery of lactic acid, amino acids, growth factors, and so on, to the germinal cells.
7. A secretory role characterized by the release of proteins, including transforming growth factor-β (TGFβ), platelet-derived growth factor, vascular endothelial growth factor, stem cell factor, and desert hedgehog protein.
8. Extracellular matrix protein secretion to create the tubular structures of the testis and provide a substrate for the spermatogonia.
9. Recognition and removal of apoptotic germ cells.

During spermatogenesis, Sertoli cells actively modulate their protein and nutrient-secretory patterns to differentiate the developing sperm. This dynamic pattern of growth factor and cytokine secretion has a prominent role in regulating the immunologic privilege of the testes, which is critical for protecting the developing germ cells. The immune system becomes competent during the perinatal phase of development and, at that time, recognizes all present antigens as "self." However, germ cells develop after puberty; and their new genomic program produces autoantigens that are recognized as unique and not part of the family considered "self" by the immune system. Without the ability of the immune system to tolerate the germ cells, they would normally be recognized as foreign and would be immediately rejected.

The immune privilege of the testes has been traditionally explained on the basis that the blood–testes barrier physically segregates germ cell autoantigens *(22)*. Although the structural integrity formed by the continuous layer of Sertoli cell–Sertoli cell tight junctions dampens the passage of substances, it is not an absolute barrier and does not restrict circulating antibodies and T cells *(23,24)*. Given the limited protection provided by the physical barrier to developing germ cells, it has been suggested and confirmed that Sertoli cells secrete immunosuppressive molecules into the testicular fluids and interstitium. These molecules inhibit the proliferation of activated B and T lymphocytes (see below for a detailed discussion of this and other mechanistic possibilities), strongly implying a natural role in contributing to the maintenance of the local immune privilege of the testes. The following sections describe recent efforts to harness these immunomodulatory capacities to facilitate cell grafting in PD.

SERTOLI CELLS: IS THERE A ROLE IN TRANSPLANTATION?

A relatively long history of research identifies the testes as an immunologically privileged transplant site (25). As early as the 1940s, intratesticular grafts of human breast sarcoma were shown to remain viable in guinea pig recipients. Discordant tissues, including skin, parathyroid fragments, pancreatic islets, and insulinoma tissue, were also shown (in varying degrees) to survive and function in the testes. More recent studies in large animals have demonstrated that xenogeneic islets survive well when grafted into the testes of dogs. Combined with more recent studies showing that Sertoli cells could be isolated from the testes and transplanted into an ectopic site to facilitate the survival of cografted cells, these data provide the foundation for the feasibility of using Sertoli cells to enable allogeneic and xenogeneic transplantation for CNS diseases.

PD is a progressive neurodegenerative disease characterized by a selective and severe loss of dopaminergic neurons in the substantia nigra. The most common therapy for PD is the administration of levodopa (L-dopa). Although quite effective in the early disease stages, L-dopa does not stop disease progression, and debilitating side effects manifest over several years. Transplantation of fetal dopaminergic neurons into the striatum of PD patients has been under clinical investigation for nearly two decades and has been met with generally encouraging results (26). As already mentioned, however, the limited availability and ethical issues concerning the use of human fetal tissue precludes this approach from widespread practice. Therefore, there is a great deal of interest in developing the use of fetal dopaminergic tissue transplants from other species. Based largely on the successful demonstration that Sertoli cells promote the survival of discordant islet cell grafts to reverse diabetes in rodents, numerous studies have been performed to determine whether Sertoli cells enhance the viability and function of dopaminergic cells grafted to the brain. To date, these efforts have focused on the transplantation of Sertoli cells to augment the survival and efficacy of cotransplanted dopaminergic cells in rodent models of PD and to promote functional and anatomical recovery when transplanted alone into parkinsonian rodent models.

In Vitro Studies: Benefits of Factors Secreted by Sertoli Cells

Sertoli cells exert a trophic effect on cocultured fetal dopaminergic neurons *(27–29)*. Porcine Sertoli cells cultured with rat or human ventral mesencephalic neurons for 7 d significantly increased the number of surviving tyrosine hydroxylase–positive dopaminergic neurons. This effect was dose-dependent and optimal (two- to threefold increases) when the ratio of Sertoli cells to dopaminergic neurons was between 1:1 and 1:5. A substantially blunted effect occurred at both lower and higher Sertoli/neuron ratios. Quantitative histology revealed significant increases in soma size, neurite outgrowth, and number of neuritic branching points of neurons cocultured with porcine Sertoli cells. The benefits produced by freshly harvested Sertoli cells were also comparable to those produced by previously cryopreserved Sertoli cells. Together, these data support the concept that Sertoli cells secrete proteins or other bioactive molecules that are capable of increasing survival and enhancing the maturation of dopaminergic neurons in vitro.

Protection of Sertoli Cell Transplants

Based on the successful in vitro demonstration that Sertoli cells exert a pronounced trophic effect on dopaminergic neurons, in vivo studies determined if discordant Sertoli cells can self-protect and survive following transplantation into the brain. Rat and porcine Sertoli cells were prelabeled with DiI, then transplanted into the striatum of rats *(30)*. After 2 mo, histological analysis revealed the presence of viable DiI-labeled Sertoli cells within, and surrounding, the injection tract. Interestingly, in these and other studies, significant decreases are seen in the microglial response surrounding grafts of Sertoli cells *(31)*. It remains to be determined whether this represents a diminished response to the initial surgical trauma via the secretion of protective growth factors from the Sertoli cells, a diminished immunologic response, or a combination of the two.

The data described above indicates that both allografts and xenografts of Sertoli cells self-protect and survive following transplantation into the CNS. However, these studies do not necessarily indicate that the cells remain functionally viable. Two different approaches have been used to define the continued function of grafted Sertoli cells. In the first approach, Sertoli cells were cotransplanted with dopaminergic cells into the rat striatum. The extent to which transplanted Sertoli cells retained their native immunosuppressant and survival-enhancing properties was then determined by examining the survival of the cografted cells (which would normally survive poorly when transplanted alone). In this model, fetal dopaminergic allografts, as well as human hNT neurons, survived better in the rat striatum when cografted with allogeneic Sertoli cells *(31–33)*. In a separate study, cografts of bovine chromaffin cells were transplanted together with rat Sertoli cells (previously labeled using latex microbeads) into the rat striatum *(32)*. Chromaffin cells did not survive when transplanted alone. However, when cografted with Sertoli cells, numerous viable chromaffin cells were found within the injection tract. Other studies demonstrated that hNT neurons and porcine adrenal chromaffin cells survive for extended periods of time (up to 10 mo) when cografted with Sertoli cells, opposed to when grafted alone *(33)*.

The second way of verifying that Sertoli cells continue to function postgrafting is to determine their trophic effects on the lesioned striatum, and if these effects are associated with behavioral recovery. Using this general paradigm, Sertoli cells were transplanted into the 6-hydroxydopamine–lesioned rat striatum. Sertoli cells from both allogeneic and xenogeneic sources survived and exerted a pronounced regenerative effect on the host tissue surrounding the transplant site *(29,34,35)*. Histological analysis revealed that the striatum of control animals was virtually absent of dopaminergic terminals. In contrast, a dense plexus of dopaminergic fibers was observed surrounding the site of surviving Sertoli cell aggregates in the transplanted animals, suggesting that locally secreted Sertoli cell products produced a trophic/regenerative effect on residual dopaminergic fibers in the striatum. The increase in striatal dopaminergic fibers, produced by the Sertoli cell grafts, was associated with significantly reduced (approx 60%) apomorphine- and amphetamine-induced rotational behavior, as well as spontaneous motor asymmetries (biased swing behavior). Together, the in vitro and in vivo data suggest that Sertoli cells provide a unique, and perhaps safe *(36)*, means of facilitating xenotransplantation for PD without the use of systemic immunosuppression.

Recently, the use of Sertoli cell grafts has been extended beyond PD to include HD *(37)*. Sertoli cells were transplanted into the striatum of rats following systemic treatment with 3-nitropropionic acid (3-NP). Then, the animals were monitored to determine whether the Sertoli cells could alleviate the motor abnormalities in this HD model. 3-NP produced a characteristic locomotor hyperactivity that was lowered substantially by Sertoli cell transplants. The grafted Sertoli cells survived in the striatum and reportedly formed tubule-like structures. These results show that Sertoli cell transplants ameliorate locomotor abnormalities in a 3-NP model of HD, adding to the possibility that Sertoli cells might be generally useful to treat a wide range of CNS diseases.

MECHANISM OF IMMUNOMODULATION
BY SERTOLI CELL TRANSPLANTS

The Role of Secreted Factors

The mechanism(s) underlying the ability of Sertoli cells to protect themselves and cografted cells is not well understood. Unlike the brain and the anterior chamber of the eye, the native testes have excellent lymphatic drainage *(38)*. Antigens grafted into the testes stimulate humoral and cellular immune response, indicating an intact afferent limb of the immune system *(39)*. The efferent limb also appears to be intact, because intratesticular islet allografts can be destroyed in presensitized recipients *(40)*. These observations have led to the hypothesis that local factors produced within the testes are responsible for the inhibition of the immune response *(41)*. Indeed, tissue culture experiments support this notion *(42–45)*. As determined using cytofluorimetric analysis, cultured Sertoli cells secrete molecules that inhibit the proliferation of B and T lymphocytes, arresting those same lymphocytes in the G_1 phase of the cell cycle *(42)*. Conditioned media from Sertoli cell cultures also significantly blunted the proliferation of T cells following the addition of concanavalin A *(44)*. These effects were greatest when the conditioned media was obtained from cells cultured at 37°C, and when stimulated by follicle-stimulating hormone. The blockade of lymphocyte proliferation was associated with the inhibition of interleukin 2 (IL-2) and could not be overcome with the further addition of IL-2 to the cultures. The isolation and identification of the inhibitor of IL-2 produced by Sertoli cells deserves attention. Unfortunately, the studies of Sertoli cell grafts within the CNS have not focused on the mechanism of action by which they exert their immunological benefits. Instead, studies of peripheral cell grafts need to be examined to understand the putative underlying biology of Sertoli cell grafts. However, caution is advised, as the immunological effects of Sertoli cell grafts may vary widely between sites as disparate as the CNS and the periphery.

Fas and TGF

The Fas/FasL system has been considered critical to maintain the immune privilege of the testes *(46)*. Fas is a cell surface receptor in the tumor necrosis factor (TNF) receptor family and mediates apoptosis upon binding with its ligand FasL *(47–49)*. Fas is inducibly expressed on T cells, B cells, and natural killer cells and has a role in immune modulation in numerous organ systems under both normal and pathological circumstances *(50–54)*. Based on its effect in immune function, and because the testes are a major nonlymphoid source of FasL in the body, it was logical to investigate Fas/FasL in testicular immune privilege *(55)*. Based on immunocytochemical staining of FasL in grafted Sertoli cells, and the use of FasL knockout mice in transplantation experiments, FasL may help protect cografted cells. However, the role of FasL in grafted Sertoli cells is not well understood, and other groups have questioned its potential in protecting tissue grafts *(56–62)*.

In vivo studies have suggested that Sertoli cells do not necessarily prevent T-cell proliferation and infiltration into the transplant site but rather are capable of changing the phenotype of the infiltrating T cells in a manner dependent on the secretion of TGF-β. Using the NOD mouse model of diabetes—an autoimmune model where syngeneic islet transplantation is followed by rejection—Suarez-Pinzon et al. *(62)* showed that implantation of syngeneic Sertoli cells prolongs the survival of cografted syngeneic islets and maintains normoglycemia. In contrast to the loss of Fas immunoreactivity in grafted Sertoli cells, immunocytochemical analysis revealed that Sertoli cell grafts maintained a high expression of TGF-β throughout the experiment (60 d). Plasma levels of TGF-β were elevated approximately twofold in mice that received Sertoli cell and islet grafts, and the administration of anti-TGF-β antibody blocked the protective effects of Sertoli cells on the islet grafts. The destruction of the islet grafts in the antibody-treated animals was associated with increases in interferon-α (IFN-α)–producing cells and decreases in IL-4–producing cells in the islet grafts. No changes in the distribution of CD4+ T cells, CD8+ T cells, B cells, or macrophages were seen. Anti-TGF-β treatment increasing IFN-α–producing cells and lowering IL-4–producing cells suggest that TGF-β from Sertoli cells modulates cell activity and favors Th2 over Th1 cell differentiation.

The Potential Impact of TNF-α and Cell Adhesion

Most of the data obtained thus far indicates that Sertoli cells exert their effects on the local immune environment via a paracrine interaction with immune cells. An underexplored possibility is the contribution of a more direct physical interaction between Sertoli cells and immune cells. TNF-α is a cytokine with diverse functions, including modulating inflammation and immune function. Characteristically found at the site of infiltrating mononuclear cells, when stressed, TNF-α is released by testicular macrophages *(63)*. Interestingly, Sertoli cells express the p55 TNF-α receptor, which enables them to be responsive to TNF-α *(64)*. TNF-α also increases the expression of intracellular adhesion and vascular cell adhesion molecules that are specific binding substrates for lymphocytes *(65)*. Electron microscopy revealed an enhanced binding between Sertoli cells and radiolabeled lymphocytes under conditions of inflammation. An intriguing avenue of future research will be the individual and potentially synergistic interactions of the paracrine and direct cell-binding effects of Sertoli cells on immune function.

ZOONOSES AND XENOTRANSPLANTATION

The overriding limitations of neural transplantation are the severe shortages of donor tissue and the need for lifelong immunosuppression in those individuals fortunate enough to receive a transplant. The use of Sertoli cells could, in theory, help avoid these limitations by protecting themselves and protecting cografted cells from immune destruction. The issue then becomes if an animal source should be used and which source is appropriate. Pigs can be bred and raised under clean and controlled conditions, providing a well-characterized source of cells and tissue for transplantation. The most commonly recognized problem associated with using pig cells is the risk that the donor tissue may transmit an infectious agent to the recipient *(66)*. The risk of infection with an exogenous agent can be largely controlled by eliminating those agents from the herd. Nonetheless, zoonotic agents that are benign in the donor could become opportunistic and aggressive within the human host. The risk to an individual would be clearly outweighed by the benefit of a life-saving transplant, but a potentially contagious zoonotic infection spread to the general public brings significantly greater risk. Much of this concern has been generated from the examples of viruses transmitted from nonhuman primates to humans, including Marburg, which is benign in primates but has had catastrophic results when transmitted to humans *(67)*. Despite these dramatic cases, the likelihood of porcine tissue is greatly reduced relative to primate tissue. In addition, the danger of infections known to be transmitted from pigs to humans (e.g., *Salmonella, Campylobacter*, influenza) is made more theoretical than real with high standards of animal husbandry *(68)*. Reports of thorough necroscopic analysis of pigs have revealed the presence of stool-borne parasites but not any pathogenic agents for humans *(69)*.

The risk of infecting recipients with porcine endogenous retrovirus (PERV) from pig grafts represents a more difficult problem than simple infection from exogenous sources, because the virus is coded within the genome of the animal *(70,71)*. First, mammalian genomes contain numerous retrovirally derived sequences, including those capable of producing replication-competent virus. Second, mutational and recombinational events are not uncommon within the retroviral genome. Third, retroviral infections can cause a variety of diseases, such as malignant and degenerative conditions. Retroviral sequences of one species are believed to have produced diseases in another species. Finally, there can be a long latent period between initial exposure and manifestation of the virus. Fortunately, there are a number of natural biological limitations relating to the spread of retroviral infection. Some limitations are short half-lives, infections that are limited to dividing cells that express the appropriate receptor for the particular virion envelope protein, cellular resistance to subsequent infections by virions of the same envelope type, and transfer among individuals that is limited to tissue exchange or sexual contact. Along with these built-in defenses, the normal adult brain may be particularly safe to transplant xenogenic cells, as it normally has relatively few dividing cells.

The concern that a virus could infect humans posttransplantation, acting more aggressively than it did in its original host, is still only theoretical. It is encouraging to note that an examination of 160 patients who had previously received porcine tissue grafts found no infection with PERV *(72,73)*, despite that some patients formed a microchimerism for up to 8 yr after transplantation. Moreover, unilateral transplantation of porcine cells into PD patients has been well tolerated, with no evidence of PERV transmission *(19)*. These studies help to alleviate some concerns about PERV, because some patients were exposed to various porcine tissues for long periods of time (up to 12 yr), the specimens were analyzed in several laboratories (e.g., the Center for Disease Control), and a highly sensitive polymerase chain reaction assay was used to distinguish the presence of PERV was a result of chimerism or true infection *(74)*. Also, the virus was not transmitted to baboons following transplantation with porcine endothelial cells, to patients after extracorporeal connection to pig kidneys, or following injection of pig islets into either the portal vein or under the kidney capsule *(75–78)*.

CONCLUSIONS

The recent success of grafting allogeneic islets into diabetic patients demonstrates that cell therapy is feasible, and the future treatment of many degenerative diseases may benefit from some form of cellular transplant. Despite the astounding success of such studies as the "Edmonton Protocol," limited tissue availability and the need for lifelong immunosuppression remain major challenges that must be resolved before widespread benefits from cell transplantation can be achieved. Continued refinements in creating newer, gentler immunosuppressive regimens, gene therapy, stem cell biology, and tolerization protocols are needed to ultimately result in sufficiently available, safe, and efficacious cell therapies. However, at present, these possibilities are in their infancy and are far from clinical evaluation. Using Sertoli cells to provide local immunoprotection of either themselves or cografted cells might represent a method to overcome the limitations in donor tissue availability and the requirement for chronic systemic immuno-suppression. Data generated in several animal models clearly show that allogeneic and xenogeneic Sertoli cells survive following transplantation. Moreover, initial (but not yet peer-reviewed) clinical research has demonstrated the safety and possible efficacy of grafting porcine Sertoli cells and islets for the treatment of juvenile-onset diabetes *(79)*.

Clearly, sufficient proof of principle exists for the notion that grafted Sertoli cells can uniquely provide an immunoprotective environment into which other cell types can survive and flourish. Yet, as a practical matter, improvements are warranted in the ability to selectively identify Sertoli cells postgrafting. Although numerous questions must be asked prior to clinical evaluation, many cannot be adequately addressed until a selective marker for Sertoli cells is identified and routinely used. Conventional techniques (e.g., the incorporation of DiI or latex beads and the use of antibodies to P-cadherin, vimentin, and Fas ligand) have been used to identify grafted Sertoli cells, but each approach suffers from practical concerns and lack of specificity. Recent reports have indicated that immunocytochemistry for the Mullerian-inhibiting substance and the nuclear transcription factor *Sox9* (activator of the Mullerian-inhibiting substance gene) selectively identifies rat and porcine Sertoli cells when grafted in the periphery *(80–82)*. These findings, providing the first use of a specific marker to identify Sertoli cells within a transplant site, will allow subsequent immunocytochemical protocols to be developed with enough resolution to permit quantitative evaluation of cell numbers and morphology. Subsequent experiments should be conducted to expand these basic findings to ensure that protein expression remains adequate across species, donor age, and transplant site (i.e., periphery vs. brain).

Long-term in vivo studies may help validate that grafted Sertoli cells continue to express these proteins at levels for sufficient resolution using immunocytochemistry, which may enable quantitative analysis of cell count and morphology. Recently, Sertoli cells that were harvested from transgenic mice producing green fluorescent protein (GFP) were transplanted into the kidney capsule. Immunohistochemistry of the grafts revealed the presence of GFP-expressing Sertoli cells at both 30 and 60 d posttransplantation *(83)*. This study demonstrates that GFP labeling remains a specific marker of Sertoli cells; yet, it also raises the possibility that Sertoli cells can be engineered using transgenic technology to produce genes of interest that are therapeutically beneficial for CNS disorders.

One of the more fertile areas of future research efforts will be the mechanism(s) by which Sertoli cells provide immunoprotection. Sertoli cells secrete several cytokines and proteins that act on multiple cell types of the immune system. Despite that the phenomenon of immune protection in the testes has long been recognized, most efforts to uncover the mechanism of this protection have focused on a single cause, such as the anatomical wall provided by the blood–testes barrier, the killing of lymphocytes by FasL, or the production of individual factors by Sertoli cells. Realistically, the blood–testes barrier, paracrine functions of the Sertoli cell, and its potential direct interaction with immune cells each likely contribute (in varying degrees) and might even interact in synergistic ways within the native testes. Indeed, multiple actions of the Sertoli cell seem to be probable, given its natural role in a function as complex and critical as spermatogenesis. An understanding of how this immunoprivilege arises will likely require a more complete knowledge of the factors produced by Sertoli cells under specific conditions of immunologic recognition. In addition, the delicate balance achieved between the beneficial and deleterious effects of these secreted factors will have to be explored, as well as the extent to which Sertoli cells can modulate the human immune system. Thus, the possibilities of using Sertoli cells to locally modulate the immune system can be fully examined. Sertoli cells may ultimately provide a platform technology from which numerous and fundamentally different cell therapy applications will emerge, ranging from transplants of Sertoli cells alone, to cografts of Sertoli cells and other distinctive cell types useful for acute and chronic neurodegenerative conditions.

REFERENCES

1. Ryan, E. A., Lakay, J. R., Paty, B. W., et al. (2002) Successful islet transplantation: continued insulin reserve provides long-term glycemic control. *Diabetes* **51**, 2148–2157.
2. Shapiro, A. M., Lakey, J. R., Ryan, E. A., et al. (2000) Islet transplantation in seven patients with type 1 diabetes mellitus using a glucocorticoid-free immunosuppressive regimen. *N. Engl. J. Med.* **343**, 230–238.
3. Balistreri, W. F., Bucuvalas, J. C., and Ryckman, F. C. (1995) The effect of immunosuppression on growth and development. *Liver Transpl. Surg.* **1**, 64–73.
4. Gaya, S. B., Rees, A. J., Lechler, R. I., et al. (1995) Malignant disease in patients with long-term renal transplants. *Transplantation* **59**, 1705–1709.
5. Gremlich, S., Roduit, R., and Thorens, B. (1997) Dexamethasone induces post-translational degradation of GLUT2 and inhibition of insulin secretion in isolated pancreatic beta cells. Comparison with the effects of fatty acids. *J. Biol. Chem.* **272**, 3216–3222.
6. London, N. J., Farmery, S. M., Will, E. J., et al. (1995) Risk of neoplasia in renal transplant patients. *Lancet* **346**, 403–406.

7. Pakzaban, P. and Isacson, O. (1994) Neural xenotransplantation: reconstruction of neuronal circuitry across species barriers. *Neuroscience* **62,** 989–1001.
8. Klassen, H. and Lund, R. D. (1988) Anatomical and behavioral correlates of a xenograft-mediated pupillary reflex. *Exp. Neurol.* **102,** 102–108.
9. Klassen, H. and Lund, R. D. (1990) Parameters of retinal graft-mediated responses are related to underlying target innervation. *Brain Res.* **533,** 181–91.
10. Honey, C. R., Charlton, H. M., and Wood, K. J. (1991) Rat brain xenografts reverse hypogonadism in mice immunosuppressed with anti-CD4 monoclonal antibody. *Exp. Brain Res.* **85,** 149–152.
11. Saitoh, Y., Matsui, Y., Nihonmatsu, I., and Kawamura, H. (1991) Cross-species transplantation of the suprachiasmatic nuclei from rats to Siberian chipmunks (Eutamias sibiricus) with suprachiasmatic lesions. *Neurosci. Lett.* **123,** 77–81.
12. Hantraye, P., Riche, D., Maziere, M., and Isacson, O. (1992) Intrastriatal transplantation of cross-species fetal striatal cells reduces abnormal movements in a primate model of Huntington disease. *Proc. Natl. Acad. Sci. USA* **89,** 4187–4191.
13. Galpern, W. G., Burns, L. H., Deacon, T. W., et al. (1996) Xenotransplantation of porcine fetal ventral mesencephalon in a rat model of Parkinson's disease: functional recovery and graft morphology. *Exp. Neurol.* **140,** 1–13.
14. Mayer, E., Dunnett, S. B., Pellitteri, R., and Fawcett, J. W. (1993) Basic fibroblast growth factor promotes the survival of embryonic ventral mesencephalic dopaminergic neurons-I. Effects in vitro. *Neuroscience* **56,** 379–388.
15. Nakao, N. (1995) Overexpressing Cu/Zn superoxide dismutase enhances survival of transplanted neurons in a rat model of Parkinson's disease. *Nature Med.* **1,** 226–231.
16. Lampson, L. A. (1987) Molecular bases of the immune response to neural antigens. *Trends Neurosci.* **10,** 211–216.
17. Pedersen, E. B., Poulson, F. R., Zimmer, J., and Finsen, B. (1995) Prevention of mouse-rat xenograft rejection by a combination therapy of cyclosporin A, prednisolone and azathioprine. *Exper. Brain Res.* **106,** 181–186.
18. Deacon, T., Schumacher, J., Dinsmore, J., et al. (1997) Histological evidence of fetal pig neural survival after transplantation into a patient with Parkinson's disease. *Nat. Med.* **3,** 350–353.
19. Schumacher, J. M., Ellias, S. A., Palmer, E. P., et al. (2000) Transplantation of embryonic porcine mesencephalic tissue in patients with PD. *Neurology* **14,** 1042–1050.
20. Fink, J. S., Schumacher, J. M., Ellias, S. L., et al. (2000) Porcine xenografts in Parkinson's disease and Huntington's disease patients: preliminary results. *Cell Transplant.* **9,** 273–278.
21. deKretser, D. M. and Kerr, J. B. (1994) The cytology of the testis. In *The Physiology of Reproduction, 2nd ed.* (Knobil, E. and Neill, J. D., eds.), Raven Press, LTD, New York, NY, pp. 1177–1290.
22. Dym, M. and Fawcett, D. W. (1970) The blood-testes barrier in the rat and the physiological compartmentalization of the seminiferous epithelium. *Biol. Reprod.* **3,** 308–326.

23. Filipini, A., Riccioli, A., Padula, F., et al. (2001) Control and impairment of immune privilege in the testes and in semen. *Hum. Reprod. Update* **7,** 444–449.
24. Yule, T. D., Montoya, G. D., Russell, L. D., et al. (1988) Autoantigenic germ cells exist outside the blood-testes barrier. *J. Immunol.* **15,** 1161–1167.
25. Emerich, D. F., Hemendinger, R., and Halberstadt, C. (2003) The testicular-derived Sertoli cell: Cellular immunoscience to enable transplantation. *Cell Transplant.* **12,** 335–349.
26. Freeman, T. B. and Widner, H. (1998) *Cell Transplantation for Neurological Disorders*. Humana Press, Totowa, NJ.
27. Cameron, D. F., Othberg, A. J., Borlongan, C. V., et al. (1997) Post-thaw viability and function of cryopreserved rat fetal brain cell co-cultured with Sertoli cells. *Cell Transplant.* **6,** 185–189.
28. Othberg, A. I., Willing, A. E., Cameron, D. F., et al. (1988) Trophic effect of porcine Sertoli cells on rat and human ventral mesencephalic cells and hNT neurons in vitro. *Cell Transplant.* **7,** 157–164.
29. Sanberg, P. R., Borlongan, C. V., Othberg, A. I., et al. (1997) Testes-derived Sertoli cells have a trophic effect on dopaminergic neurons and alleviate hemiparkinsonism in rats. *Nat. Med.* **3,** 1129–1132.
30. Saporta, S., Cameron, D. F., Borlongan, C. V., and Sanberg, P. R. (1997) Survival of rat and porcine Sertoli cell transplants in the rat striatum without cyclosporin-A immunosuppression. *Exp. Neurol.* **146,** 299–304.
31. Willing, A. E., Sudberry, J. J., Othberg, A. I., et al. (1999) Sertoli cells decrease microglial response and increase engraftment of human hNT neurons in the hemiparkinsonian rat striatum. *Brain Res. Bull.* **48,** 441–444.
32. Sanberg, P. R., Borlongan, C. V., Saporta, S., and Cameron, D. F. (1996) Testis-derived Sertoli cells survive and provide localized immunoprotection for xenografts in rat brain. *Nat. Biotechnol.* **14,** 1692–1695.
33. Willing, A. E., Cameron, D. F., and Sanberg, P. R. (1998) Sertoli cell transplants: their use in the treatment of neurodegenerative disease. *Mol. Med. Today* **4,** 471–477.
34. Borlongan, C. V., Cameron, D. F., Saporta, A., and Sanberg, P. R. (1997) Intracerebral transplantation of testis-derived Sertoli cells promotes functional recovery in female rats with 6-hydroxydopamine-induced hemiparkinsonism. *Exp. Neurol.* **148,** 388–392.
35. Liu, H.-W., Kaung, Y.-J., Wu, J.-C., et al. (1999) Intrastriatal transplantation of Sertoli cells may improve amphetamine-induced rotation in hemiparkinsonian rats. *Brain Res.* **838,** 227–233.
36. Rodriguez, A. I., Willing, A. E., Cameron, D. F., et al. (2002) Neurobehavioral assessment of transplanted porcine Sertoli cells into the intact rat striatum. *Neurotox. Res.* **4,** 103–109.
37. Rodriguez, A. I., Willing, A. E., Saporta, S., et al. (2003) Effects of Sertoli cell transplants in a 3-nitropropionic acid model of early Huntington's disease: a preliminary study. *Neurotox. Res.* **5,** 443–450.
38. Head, J. R., Neaves, W. B., and Billingham, R. E. (1983) Reconsideration of the lymphatic drainage of the rat testes. *Transplantation* **35,** 91–95.

39. Head, J. R., Neaves, W. B., and Billingham, R. E. (1983) Immune privilege in the testes. I. Basic parameters of allograft survival. *Transplantation* **36,** 423–431.
40. Ferguson, J. and Scothorne, R. J. (1977) Further studies on the transplantation of isolated islets. *J. Anat.* **124,** 9–20.
41. Head, J. R. and Billingham, R. E. (1985) Immune privilege in the testes. II. Evaluation of potential local factors. *Transplantation* **40,** 269–275.
42. DeCesaris, P., Filippini, A., Cervelli, C., et al. (1992) Immunosuppressive molecules produced by Sertoli cells cultured in vitro: biological effects on lymphocytes. *Biochem. Biophys. Res. Commun.* **186,** 1639–1646.
43. Pollanen, P., Sodor, O., and Uksila, J. (1988) Testicular immunosuppressive protein. *J. Reprod. Immunol.* **14,** 125–138.
44. Selawry, H. P., Kotb, M., Herrod, H. G., and Lu, Z. N. (1991) Production of factor, or factors, suppressing IL-2 production and T cell proliferation by Sertoli cell-enriched preparations. *Transplantation* **52,** 846–850.
45. Wyatt, C. R., Law, L., Magnuson, J. A., et al. (1988) Suppression of lymphocyte proliferation by proteins secreted by cultured Sertoli cells. *J. Reprod. Immunol.* **14,** 27–40.
46. Lynch, D. H., Ramsdell, F., and Alderson, M. R. (1995) Fas and FasL in the homeostatic regulation of immune responses. *Immunol. Today* **16,** 569–574.
47. Dhein, J., Walczak, H., Baumler, C., et al. (1995) Autocrine T-cell suicide mediated by APO-1 (Fas/CD95). *Nature* **373,** 438–441.
48. Itoh, N., Yonehara, S., Ishii, A., et al. (1991) The polypeptide encoded by the cDNA for human cell surface antigen Fas can mediate apoptosis. *Cell* **66,** 233–243.
49. Watanbe-Fukunaga, R., Brannan, C. I., Itoh, N., et al. (1992) The cDNA structure, expression, and chromosomal assignment of the mouse Fas antigen. *J. Immunol.* **148,** 1274–1279.
50. Graves, J. D., Draves, K. E., Craxton, A., et al. (1998) A comparison of signaling requirements for apoptosis of human B lymphocytes induced by the B cell receptor and CD95/Fas. *J. Immunol.* **161,** 168–174.
51. Griffith, T. S., Brunner, T., Fletcher, S. M., et al. (1999) Fas ligand-induced apoptosis as a mechanism of immune privilege. *Science* **270,** 1189–1192.
52. Hahne, M., Renno, T., Schroeter, M., et al. (1996) Activated B cells express functional Fas ligand. *Eur. J. Immunol.* **26,** 721–724.
53. Hanabuchi, S., Koyanagi, M., Kawasaki, A., et al. (1994) Fas and Fas ligand in a general mechanism of T-cell mediated cytotoxicity. *Proc. Natl. Acad. Sci. USA* **91,** 4930–4934.
54. Katsikis, P. D., Wunderlich, E. S., Smith, C. A., and Herzenberg, L. A. (1995) Fas antigen stimulation induces marked apoptosis of T lymphocytes in human immunodeficiency virus-infected individuals. *J. Exp. Med.* **181,** 2029–2036.
55. Suda, T., Takahashi, T., Golstein, P., and Nagata, S. (1993) Molecular cloning and expression of the Fas ligand, a novel member of the tumor necrosis factor family. *Cell* **75,** 1169–1178.
56. Bellgrau, D., Gold, D., Selawry, H. P., et al. (1995) A role for CD95 ligand in preventing graft rejection. *Nature* **377,** 630–632.

57. Lau, H. T., Yu, M., Fontana, A., and Stoeckert, C. J. (1996) Prevention of islet allograft rejection with engineered myoblasts expressing FasL in mice. *Science* **273,** 109–112.
58. Kang, S., Schneider, D. B., Lin, Z., et al. (1997) Fas ligand expression in islets of Langerhans does not confer immune privilege and instead targets them for rapid destruction. *Nat. Med.* **7,** 738–743.
59. Zhan, W., Cai, S., and Wang, J. (2002) The effect of FasL expression on pancreatic allografts. *Chin. Med. J.* **115,** 1006–1009.
60. Chervonsky, A. V., Wang, Y., Wong, F. S., et al. (1997) The role of Fas in autoimmune diabetes. *Cell* **89,** 17–24.
61. Restifo, N. P. (2000) Not so Fas: Re-evaluating the mechanism of immune privilege and tumor escape. *Nat. Med.* **6,** 493–495.
62. Suarez-Pinzon, W., Korbutt, G. S., Power, R., et al. (2000) Testicular Sertoli cells protect islet B-cells from autoimmune destruction by a transforming growth factor-β1-dependent mechanism. *Diabetes* **49,** 1810–1818.
63. Moore, C. and Hutson, J. C. (1994) Physiological relevance of tumor necrosis factor in mediating macrophage-Leydig cell interactions. *Endocrinology* **134,** 63–69.
64. Mauduit, C., Besset, V., Aussanel, V., and Benahmed, M. (1996) Tumor necrosis factor alpha receptor p55 is under hormonal (follicle-stimulating hormone) control in testicular Sertoli cells. *Biochem. Biophys. Res. Comm.* **25,** 631–637.
65. Riccioli, A., Fillippini, A., De Cesaris, P., et al. (1995) Inflammatory mediators increase surface expression of integrin ligands, adhesion to lymphocytes, and secretion of interleukin 6 in mouse Sertoli cells. *Proc. Natl. Acad. Sci. USA* **92,** 5808–5812.
66. Chapman, L. E., Folks, T. M., Saloman, D. R., et al. (1995) Xenotransplantation and xenogeneic infections. *N. Engl. J. Med.* **333,** 1498–1501.
67. Martini, G. A. (1969) Marburg agent disease in man. *Trans. Roy. Soc. Trop. Med. Hyg.* **63,** 295–302.
68. Michaels, M. G. and Simmons, R. L. (1994) Xenotransplant-associated zoonoses, strategies for prevention. *Transplantation* **57,** 1–7.
69. Ye, Y., Niekrasz, M., Kosanke, S., et al. (1994) The pig as a potential organ donor for man. A study of potentially transferable disease from donor pig to recipient man. *Transplantation* **57,** 694–703.
70. Stoye, J. (1998) No clear answers on safety of pigs as tissue donor source. *Lancet* **352,** 666–667.
71. Stoye, J. P. and Coffin, J. M. (1995) The dangers of xenotransplantation. *Nat. Med.* **1,** 1100.
72. Paradis, K., Langford, G., Long, Z., et al. (1999) Search for cross-species transmission of porcine endogenous retrovirus in patients treated with living pig tissue. *Science* **285,** 1236–1241.
73. Weiss, R. A. (1999) Xenografts and retroviruses. *Science* **285,** 1221–1222.
74. Switzer, W. M., Shanmugam, V., Chapman, V., and Heneine, W. (1999) Polymerase chain reaction assays for the diagnosis of infection with the porcine

endogenous retrovirus and the detection of pig cells in human and nonhuman recipients of pig xenografts. *Transplantation* **68**, 183–188.

75. Martin, U., Steinhoff, G., Kiessig, V., et al. (1998) Porcine endogenous retrovirus (PERV) was not transmitted from transplanted porcine endothelial cells to baboons in vivo. *Transplant. Int.* **11**, 247–251.

76. Patience, C., Patton, G. S., Takeuchi, Y., et al. (1998) No evidence of pig DNA or retroviral infection in patients with short-term extracorporeal connection to pig kidneys. *Lancet* **352**, 699–701.

77. Groth, C. G., Korsgren, O., Tibell, A., et al. (1994) Transplantation of porcine fetal pancreas to diabetic patients. *Lancet* **344**, 1402–1404.

78. Heneine, W., Tibell, A., Switzer, W. M., et al. (1988) No evidence of infection with porcine endogenous retrovirus in recipients of porcine islet-cell xenografts. *Lancet* **352**, 695–699.

79. Valdes-Gomzalaez, R. A., Elliot, R. B., Dorantes, L. M., et al. (2002) Porcine islet xenografts can survive and function in type I diabetic patients in the presence of both pre-existing and elicited anti-pig antibodies. *Transplant. Suppl.* **74**, 94.

80. Cameron, D. F., Hushen, J. J., Dejarlais, T., et al. (2002) A unique cytoplasmic marker for extratesticular Sertoli cells. *Cell Transplant.* **11**, 507–512.

81. Emerich, D. F. and Sanberg, P. R. (2002) Novel means to selectively identify Sertoli cell transplants. *Cell Transplant.* **11**, 495–497.

82. Hemendinger, R. A., Gores, P., Blacksten, L., et al. (2002) Identification of a specific Sertoli cell marker, Sox9, for use in transplantation. *Cell Transplant.* **11**, 499–505.

83. Dufour, J. M., Hemedinger, R., Halberstadt, C. R., et al. (2004) Genetically engineered Sertoli cells are able to survive allogeneic transplantation. *Gene Ther.* **11**, 694–700.

10

The Choroid Plexus

A Novel Graft Source for Neural Transplantation

Cesario V. Borlongan, Stephen J. M. Skinner, Alfred Vasconcellos, Robert B. Elliott, and Dwaine F. Emerich

ABSTRACT

The choroid plexus (CP) produces cerebrospinal fluid (CSF) and forms a portion of the physical structure of the CSF–blood barrier. More recently, the CP been implicated in other basic aspects of neural functioning, such as surveying the chemical and immunological status of the brain, detoxifying the brain, secreting a nutritive cocktail of polypeptides for neuronal function and survival, and participating in repair processes following trauma. The CP also has a role in maintaining the extracellular milieu of the brain by actively modulating the chemical exchange between the CSF and brain parenchyma and by secreting numerous growth factors into the CSF. Preclinical and clinical studies in aging and neurodegeneration demonstrate anatomical and physiological changes in the CP, suggesting effects not only in normal development and pathological conditions, but also in potential endogenous repair processes following trauma. CP dysfunction in central nervous system (CNS) diseases, and the endogenous secretion of growth factors, indicates that transplantable CP might enable delivery of growth factors to the brain while avoiding the conventional molecular and genetic alterations associated with modifying cells to secrete selected products. Thus, this enables the possibility of replacing or transplanting CP as a means of treating acute and chronic brain diseases. This chapter focuses on the various functions of the CP, how these functions are altered in aging and neurodegeneration, and recent demonstrations of the therapeutic potential of transplanted CP for neural trauma.

Key Words: Choroid plexus; cerebrospinal fluid (CSF); transplantation; growth factor.

From: *Contemporary Neuroscience: Cell Therapy, Stem Cells, and Brain Repair*
Edited by: C. D. Sanberg and P. R. Sanberg © Humana Press Inc., Totowa, NJ

INTRODUCTION

The choroid plexus (CP) lies in the four ventricles within the brain as a fibrous network of tissue and vasculature. The CP forms a unique interface between peripheral blood and cerebrospinal fluid (CSF). Recent studies indicate that the CP also has important roles in the establishment and maintenance of the extracellular milieu throughout the central nervous system (CNS) under both normal and pathological conditions. Physically, the CP is ideally located and structured to survey the biochemical and cellular status of the brain. Receptors involved in transcytosis and clearance of polypeptides are diversely expressed in the CP, where the ependymal cells secrete numerous neuropeptides, growth factors, and cytokines into the CSF, permitting both local and distal endocrine-like effects on target cells in the brain *(1,2)*. The CP has a pivotal impact in brain development and maintenance of normal neuronal functioning. When there is physiological stress in the brain, the CP actively changes its secretory, metabolic, and transport profile.

The CP is a unique organ responsible for a diverse array of vital functions; therefore, even subtle anatomical and physiological changes in the CP can have wide-ranging consequences and can contribute to pathological processes. Distinct alterations in choroidal polypeptide synthesis occur during brain development and in CNS disorders, including Alzheimer's disease (AD), traumatic brain injury, and stroke. In AD, the ependymal cells of the CP atrophy, with subsequent decreases in CSF production, enzymatic activity, and polypeptide transport. The results of these changes can be widespread, leading to dysfunctional methylation, increased oxidative stress, and augmented accumulation of toxins. At the same time, the CP appears to assist in recovery processes following neuronal damage, particularly by secreting neuroprotective compounds and by acting as a site for neurogenesis, which suggests an ongoing and "as needed" role in cellular repair and replacement *(3–6)*. Based on this profile, CP may have an integral role not only in normal brain function, but also in the degenerative processes associated with aging, as well as in recovery from injury. This chapter discusses the impact of the CP and CP-derived polypeptides in development, normal brain functions, and selected CNS disorders. In addition, this chapter describes the future potential for using transplantable CP cells to deliver neurotrophic factors to the brain and spinal cord.

BASIC STRUCTURE OF THE CHOROID PLEXUS

The CP is located throughout the ventricles of the brain and weighs about 3 g in humans *(7,8)*. Within the lateral ventricles, it projects from the choroidal fissure and extends from the interventricular foramen to the end of the temporal horn. It is invaginated into the third and fourth ventricles from the ventricular roof. Grossly, the CP is lobulated with a single continuous layer of cells, which are derived from the ependymal lining of the ventricles. Despite their origins from the ependymal lining, they possess epithelial cell characteristics and are typically referred to as *choroidal epithelial cells*. The epithelial cells rest on a basal lamina and contain a large central spherical nucleus and abundant cytoplasm *(9)*. The luminal surface of the choroidal epithelial cells is covered with microvilli, and adjacent cells are bound by tight junctions that physically restrict the exchange of substances with the CSF. Ultrastructurally, the choroidal epithelial cells contain numerous mitochondria to maintain their high respiratory metabolism and energy requirements *(10)*. The Golgi apparatus contains columns of cisternae. Smooth endoplasmic reticulum and clear vesicles are distributed throughout the apical cytoplasm. A dense vascular bed, providing blood flow four to seven times greater than the rest of the brain, underlies the choroidal epithelial cells and basal lamina *(11)*. The capillaries are large with thin, fenestrated endothelial walls, and bridging diaphragms overlie the fenestrations. An extensive array of adrenergic, cholinergic, peptidergic, and serotoninergic nerve fibers innervate the blood vessels and the epithelium *(12)*.

TRADITIONAL VIEWS OF THE CP

Production of CSF

The most recognized function of the CP is CSF production *(13)*. CSF formation across species is generally proportional to the weight of the CP. In humans, CSF volume is 80–150 mL and forms at a rate of approx 500 mL per d. CSF production follows a circadian rhythm, with nighttime levels (0200 hours) doubling the levels produced during the daytime *(14)*. As much as 30% of the CSF is formed at other sites, particularly under pathological conditions, including the ependymal lining of the ventricles and the endothelium of brain capillaries.

Secretion and Composition

CSF is produced mainly by active secretion; water enters the CSF from the blood along an osmotic gradient or by specific water channels (e.g., aquaporin). The epithelial cells of the CP secrete CSF by moving $Na(+)$, $Cl(-)$, and $HCO_3(-)$ from the blood to the ventricles, creating a gradient that drives the secretion of H_2O. The CSF is clear, with few cells and little protein *(15)*. Compared to blood plasma, CSF has a lower pH and concentrations of glucose, potassium, calcium, bicarbonate, and amino acids *(16,17)*. In contrast, sodium, chloride, and magnesium contents are greater in CSF than in blood plasma. Folate levels are two to three times higher in CSF than in plasma, and transthyretin (TTR) represents 25% of all CSF proteins *(18,19)*. Interestingly, TTR is produced exclusively by the CP. Notably, a link has been described between TTR and depression. Studies in both TTR-null mice and depressed patients suggest a relationship between lowered TTR and increased exploratory behavior and increased Hamilton depression scores.

Circulation and Absorption

CSF in the subarachnoid space flows from the lateral foramin of Lushka and the cisterna pontis, anteriorly along the base of the brain through the Sylvian fissure, and along the lateral convex and medial surfaces of the hemispheres. From the midline foramen of Magendie and cisterna magna, the CSF flows over the cerebellar hemispheres toward the tentorial incisure and downward into the subarachnoid space that surrounds the spinal cord. The circulation of CSF involves pressure waves generated by pulsatile arterial blood flow and brain expansion, pressure gradients created by the production and absorption of CSF, and currents induced by ependymal cilia *(20)*.

CSF accesses the blood primarily through the arachnoid villi, which are continuous with the subarachnoid space. Because the hydrostatic pressure of CSF exceeds the venous pressure, the villi act as one-way valves that return CSF from the subarachnoid space to the dural venous sinuses. Small amounts of CSF are absorbed via pial vessels, across capillary walls, and via lymphatic channels adjacent to extensions of the subarachnoid space surrounding cranial and spinal nerves. These routes of absorption are particularly important under pathological conditions, such as hydrocephalus *(21)*.

Formation of the CSF–Blood Barrier

The typical function of the blood–brain barrier (BBB) within the CP is shifted from the vasculature to the epithelium, where tight junctions form between the epithelial cells to confer the permeability properties of the individual cells *(22)*. Briefly, the CP and arachnoid membrane act together at the barriers between blood and CSF. On the external brain surface, the ependymal cells fold over onto themselves to form a double-layered structure between the dura and pia, referred to as the *arachnoid membrane*. The passage of substances from the blood to the arachnoid membrane is prevented by the tight junctions between adjacent cells. The arachnoid membrane is generally, but not entirely, impermeable to hydrophilic substances and has a largely passive role in forming the blood–CSF barrier *(23)*.

CP: First-Line Defense for the Brain

Lying within the central ventricular system, the CP is in an ideally suited position to monitor and modulate the functional status of the brain. The CP protects the brain against acute neurotoxic insults by using a complex, multilayered detoxification system. First, the CP contains high concentrations of glutathione (GSH), cysteine, and metallothioneins that potently sequester toxic agents. Second, the CP uses protective enzymes, e.g., superoxide dismutase, GSH-S-transferase, and GSH peroxidase and reductase, to provide a barrier against free-radical oxidative stress. Third, the CP aids in the overall biodistribution of drugs and toxic compounds by using a full compliment of metabolizing enzymes, including phase I enzymes used for the functionalization of such drugs as cytochrome P-450 (CYP) isoform CYP2B1,2 and monoamine oxidase, phase II enzymes used for the conjugation of drugs (e.g., UDP-glucuronosyl transferase), and phase III activity, which provides "kidney-like" transport systems. These include indirectly coupled Na^+/dicarboxylate cotransport and dicarboxylate/organic anion exchange, such as the organic anion transporter (OAT), the organic anion transporter polypeptide 1 (Oatp1) and Oatp2, and the multidrug resistance protein Mrp1/MRP1 and *p*-glycoprotein Mdr1/MDR1. Taken together, the CP contains the machinery needed to impede the entrance of noxious compounds to the brain and to control the efflux, binding, and metabolism of toxins.

CP and the Neuroimmune System

The CP also functions within the neuroimmune system. Traditionally, the CNS has been considered an immunologically privileged site, with no inherent need for immunosurveillance. The first indication that the CP-mediated interactions, as well as signaling between the peripheral immune system and the brain, came from evidence showing that the CP contains inducible lymphoid cells. The rapid and transient induction of interleukin (IL)-1β and tumor necrosis factor (TNF)-α following systemic lipopolysaccharide or IL-6 is a clear example of this interaction. This activation in the CP, and also in circumventricular organs, leptomeninges, and surrounding blood vessels, is the initiation of a process that ultimately spreads throughout the brain. This sequence suggests that CP transfers information between the peripheral immune system and the brain through a coordinated local induction of proinflammatory cytokines. Choroidal epithelial cells also constitutively express major histocompatibility complex (MHC) class II molecules, and class I molecules can be induced with such infectious agents as the rabies virus. In vitro, epithelial cells present foreign antigens and stimulate T-lymphocyte proliferation through a MHC class II–restricted mechanism. Accessory molecules that are important for leukocyte adhesion, such as L-selectin, intracellular cell adhesion molecule-1, and vascular cell adhesion molecule-1, are found at low levels on CP epithelial cells but can be upregulated during inflammatory conditions, experimental autoimmune encephalomyelitis. Other cells, including the Kolmer cells of the CP that normally act as phagocytic scavenger cells, also display inducible MHC class I and II antigens and proliferate when challenged with endotoxins. This antigen presentation capacity implies that the CP is part of an intrinsic surveillance system that defends against blood-borne pathogens and CSF-localized antigens.

The pathogen-induced inflammation within the CP is not surprising given the tropism that bacteria, parasites, and viruses (e.g., *Neisseria meningitidis*, *Trypanosoma brucei*, Sendai virus, lymphocytic choriomeningitis virus, mumps virus, and perhaps even human immunodeficiency virus (HIV)-1 and human T-cell leukemia virus 1 [HTLV-1]), have for the CP. HIV-infected T lymphocytes and monocytes are often observed in the stroma and supraepithelial portions of the CP, suggesting that the CP may be a pathway of entry for infected cells into the brain. HIV-1–positive cells are initially found in the subarachnoid and perivascular spaces but extend into the CSF of HIV-1– and HTLV-1–infected patients during the later stages of infection.

How infected leukocytes or activated T lymphocytes cross the CP is unknown, but it is intuitively obvious that their crossing can have disastrous consequences. The CP might be involved in the CNS entry of activated, myelin-directed autoreactive T lymphocytes during multiple sclerosis (MS). Activated T-lymphocyte infiltration into the brain results in the formation of demyelination plaques that underlie the clinical symptoms of MS. Because these plaques are frequently located in the periventricular area, the CP may constitute a preferential way for T lymphocytes to reach these structures. T lymphocytes and T-lymphocyte chemo-attractants are found in the CSF from MS patients.

CP IN DEVELOPMENT AND AGING

The Development of CP

The CP, which forms in the seventh week of gestation in humans, develops early in embryogenesis, thus indicating that the CP may help provide the maturing brain with a controlled extracellular environment. Even in the immature brain, the blood–CSF barrier is functional and limits passage of unwanted substances from CSF to the parenchyma. Although the CP is functional early in development, its permeability appears to be suited for the embryonic brain by allowing low-molecular-weight compounds to enter the brain more easily during development, compared to adulthood. The CP also provides developing surrounding tissues with morphogens, mitogens, and trophic factors that guide and pattern the general and specific growth of the CNS. For instance, the embryonic CP contains highly localized levels of insulin-like growth factor (IGF)-II. Based on the localization and unusually high expression of IGF receptors in the floor plate of the hindbrain, it has been hypothesized that IGF-II secreted by the CP diffuses to the receptors on the floor plate cells and allows them to guide spinal axon growth. More direct evidence for the impact of CP in morphogenesis comes from findings that the radial migration of cerebral cortical neurons from the ventricular and subventricular zone to the cortical plate is governed by gradients of soluble factors, e.g., secreted semaphorins, netrins, or Slit proteins secreted partly by the CP. A soluble factor secreted by the fetal CP and immunologically related to Slit2 diffuses through the CSF to establish a gradient of a repulsive cue, guiding cortical neurons away from the ventricular surface. Yamamoto and colleagues showed that the CP modulates neurite outgrowth in the developing cerebellum. Using explant cocultures of cerebellum and fourth-ventricle CP from fetal and infant rats, studies confirmed that CP secretes a soluble neurite growth-promoting activity that is high, biphasic, and correlates with the major milestones of cerebellar morphogenesis. The importance of CSF distribution of soluble factors is highlighted by the work of Miyan and colleagues, which demonstrates that hydrocephalus in rats leads to significantly abnormal cortical development. Based on their studies, it has been hypothesized that the circulating CSF and its associated factors are vital for development along the entire length of the neural tube.

CP and Aging

Structural and functional changes in the CP are also associated with, and perhaps contribute to, various pathological conditions, including aging *(3,24)*. Aged rat choroid epithelial cells undergo significant atrophy and lose approx 15% of their normal height as they are reduced, from 12.5-µm high at 6 mo to 11.5- and 10.5-µm high at 18 and 30 mo, respectively *(24)*. At 18 mo, additional changes include a morphological shift to a dome-like appearance. This effect becomes more prominent at 30 mo, as the epithelial cells are flattened with an irregular and elongated nucleus, significantly shortened microvilli, and lipid vacuoles. An irregular fibrosis of stroma occurs as the epithelial basement membrane thickens, from 100 at 6 mo, to more than 200 nm thick by 30 mo. The endothelial basement membrane also thickens but not as robustly. Age-related increases in the number of dark epithelial cells and reductions in the size of the microvilli also occur in aged mice *(25)*. Similar alterations are seen in human CP, where epithelial cell height decreases by about 10%, from 15-μm high in newborns to 13.7-μm high in the elderly *(26)*. The aged epithelial cell cytoplasm becomes rich with fibrillar Biondi bodies and lipofuchsin deposits *(27)*. The nuclei appear irregular and flattened, and the basement membrane becomes irregular and thick, from 95-nm thick in newborns to 116-nm thick in aged humans *(26)*. The stroma thickens and contains collagen fibers, hyaline bodies, calcifications, and psammomas, and the local arterial walls become thicker and fragmented *(3,28)*.

Functions of the CP are all energy-dependent, and evidence indicates that aging CP is unable to maintain its normal energy output. Thus, the enzymatic activity of the CP also changes in ways that profoundly impact normal cellular metabolism. Synthesis of lactate dehydrogenase and succinate-dehydrogenase decreases by 9% and 26%, respectively, over the life span of rats *(29)*. Significant age-related increases occur in the number of epithelial cells deficient in cytochrome C oxidase, altering the respiratory mitochondrial chain and decreasing cellular production of ATP *(30)*. Reductions in Na^+K^+-ATPase and the Na^+K^+-$2Cl^-$ cotransporter also occur *(31,32)*. These anatomical and enzymatic deteriorations likely underlie the diminution of CSF secretion, which can be as great as 45% in animal models *(4)*. Owing to the decreased secretion and the simultaneously increased CSF volume caused by brain atrophy, CSF turnover takes approx 7.9 h in elderly rats, much longer than the 2.2 h needed for CSF turnover in young rats. In humans, CSF-secreted volume diminishes with age, from 0.41 mL/min at 28 yr to 0.19 mL/min at 77 yr. As a result of age-related cerebral atrophy, the turnover of CSF decreases to less than two times daily in elderly subjects.

Elevated CSF:plasma ratios have been reported for several proteins during aging *(4)*. The effects of these cumulative changes on brain functioning have not been directly tested, but it seems logical to assume that such dramatic alterations in the CP and CSF would be associated with inadequate distribution of nutritive substances, additional cellular stress, and reduced clearance of toxic compounds—all of which could be influential factors in age-related cognitive and motor decline or the development of specific neurological disorders.

CP AND NEUROLOGICAL DISORDERS: AD AND ISCHEMIA
AD

The majority of our knowledge regarding the morphology of human CP during aging is derived from control tissues in studies investigating changes in AD, where the anatomical changes that occur in the CP during AD are exacerbations of normal aging. As the impairments in CP function become greater, the accompanying medical consequences become more devastating. Epithelial cell atrophy is even greater, with cell height decreasing up to 22%, compared to age-matched controls *(26)*. More epithelial cells contain lipofuchsin vacuoles and Biondi bodies, and the intracellular distribution of these pathological entities is more extensive *(27,34)*. The epithelial basement membrane becomes very irregular and thickens an additional 28% beyond that seen in age-matched controls *(26)*. The stroma of the microvilli become fibrotic, with extensive vascular thickening *(26,35)*. Numerous hyaline bodies and calcifications are found, and deposits of immunoglobulin (Ig)G, IgM, and C1q are common along the epithelial basement membrane. These anatomical changes are accompanied by further decreases in CSF secretion, with CSF turnover requiring up to 36 h in AD patients *(36)*.

Further atrophy of the choroidal epithelial cells in AD is linked to pronounced declines in secretory activities, transport function, and enzymatic activities. Levels of TTR—a CP-synthesized molecule that associates with β-amyloid protein to form complexes—are more than 10% lower in patients with AD *(37)*. Ascorbic acid and α-tocopherol levels, the two major scavengers of free radicals in the CSF, are decreased in AD, likely adding to oxidative stress *(38,39)*. The reduction of these molecules, and the decreased CSF turnover, translates to an even greater impoverishment of the brain, potentially leading to methylation problems, increased oxidative stress and lipid peroxidation, decreased amyloid clearance, augmented tau protein polymerization, and amyloid protein oligomers and fibrillo formation *(3)*.

CP transport functions are also impaired in AD. CSF folate and vitamin B12, which are both important for the methylation of numerous molecules, are significantly lower in AD patients *(40–42)*. Homocysteine, as well as mediating lipid peroxidation and increasing the production of toxic (E)-4-hydroxy-2-nonenal, is increased in AD CSF. The impaired ability of CP to clear molecules from the CSF of AD patients has potentially profound implications *(3)*. The choroid epithelium is ideally formed to clear the CSF of peptide degradation products and noxious polypeptides, such as soluble amyloid β-protein. In rats, clearance of intraventricularly injected β-amyloid protein decreases, from 10.4 μL/min at 3 mo to 0.71 μL/min at 30 mo. Consequently, the brain content of amyloid peptides increases, from 7% at the end of CSF perfusion in young rats to 49% in old animals *(4)*. The increase of β1-40 and β1-42 amyloid protein levels in elderly humans could also be related to decreased clearance from the CNS. Lowered CSF production could enhance protein glycation and the formation of β-amyloid protein oligomers. The AD brain contains elevated levels of glycation products and deposits of amyloid protein; senile plaques and fibrillary tangles contain advanced glycation products. Glycation promotes protein aggregation, the polymerization of tau proteins, and protein β-amyloid protein fibrillo trans-formation. The decrease of CSF turnover, increase of protein glycation, and diminution of β-amyloid protein clearance could induce oligomer formation and retention.

Ischemia

Several anatomical and secretory changes occur in CP following cerebral ischemia *(5)*. The methods to induce ischemia, and the duration of ischemia, vary across studies; however, time-related changes in the CP are linked to both degenerative and repair processes. Early rodent studies using 10 min of bilateral carotid artery occlusion demonstrated that within 30 min, potassium, sodium, and water content increased by 32%, 85%, and 22%, respectively *(43)*. Although these changes were robust, they all normalized within 24 h. Transient ultrastructural changes were also noted in the CP. Increasing the severity of the ischemic episode results in more profound and long-lasting changes in the CP. Middle cerebral artery (MCA) occlusion in rats induces apoptotic cell death, with nuclear DNA breaks occurring after 6 h but not after 1.5 h of ischemia *(44)*. Enhanced immunostaining for pro-apoptotic Bax protein and Bcl-X is evident in the choroidal epithelium. DNA fragmentation and cell death in the CP can affect neighboring neuronal systems. Following cerebral ischemia, the CA1 pyramidal neurons in the hippocampus undergo delayed neuronal death, evident about 48 h after the insult. It is unclear why this delay occurs, but extrinsic mechanisms, rather than an intrinsic mechanism of the CA1 neurons, might have an important role. An analysis of the time course of cell loss in the CP vs the hippocampus supports this possibility *(45)*. DNA fragmentation occurs in the CP 24 h after ischemia but diminishes by 48 h. In contrast, at 48 h, DNA breaks are widespread in the medial CA1 region; at 72 h, DNA fragmentation spreads laterally within the CA1 region *(5)*. Subsequent studies confirmed and extended these data by showing a dramatic increase in terminal dUTP nick-end labeling (TUNEL)-positive cells in the CP at 18–36-h postischemia but not at 48 h. No TUNEL-positive cells were seen at 24 h in the CA1 hippocampal region, clearly showing that cell death in the CP preceded cell death in the CA1 region. These studies suggest that ischemia initiates a cycle of cell loss in which the death of CP cells at least precedes, and perhaps adversely affects, neighboring neurons.

Part of the CP response to an ischemic event is based on a time-dependent, increased production of factors capable of mitigating neuronal loss, indicating that the CP likely participates in endogenous repair processes following trauma. One example of a secretory change by the CP after ischemia occurs with cystatin C—a cysteine protease inhibitor produced by the CP, which is elevated in the hippocampus of animals after 10 min of forebrain ischemia. Cystatin C immunoreactivity is minimal in CA1 neurons in normal animals and during the first 24-h postischemia but becomes significantly elevated in degenerative CA1 pyramidal cells and reactive glia of the stratum radiatum and stratum oriens at 3-, 7-, and 14-d postischemia. Given the similarity in the temporal expression of cystatin C to the delayed neuronal death following forebrain ischemia, it is interesting to speculate about the effect that CP-derived cystatin C and its substrates have in the postischemic degenerative and repair processes (46).

Choroidal ependymal cells also exhibit enhanced growth hormone reactivity following hypoxic ischemia in rats (47). Growth hormone–like immunoreactivity is seen in injured neurons, axons, glial cells proximal to the infarction, and on the CP ependymal cells. The pattern of immunoreactivity implies that growth hormone is transported via the CSF to act in a neurotrophic manner specifically targeted to injured neurons and glia. This hypothesis was supported by injecting growth hormone into the lateral ventricle 2 h after injury. After 3 d, growth hormone treatment was found to reduce neuronal loss in those areas containing the neural growth hormone receptor (47).

Finally, transforming growth factor (TGF)-β (isoforms β1, β2, and β3) is expressed in CP and increases following ischemia in rats (1,48,49). Augmented synthesis of choroidal TGF-β may have a role in neuronal survival and function recovery after ischemia by regulating Ca^{++} homeostasis and the expression of the *Bcl2* proto-oncogene (1,50). The distal activity of CP-derived TGF-β is supported by the ability of intraventicularly administered TGF-$β_1$ to reduce the size of the infarct following ischemia (51,52). Because the CP also expresses the TGF-β receptor, this growth factor may also be involved in regulating the synthesis and secretion of other CP proteins. Although the response of the CP to an ischemic event is extremely complex, data do suggest that CP cell loss can impact the death of neighboring neurons adversely, while at the same time, the surviving CP cells produce factors that can beneficially affect the outcome of neurons. Extensive cell death in the CP will lead to a leaky blood–CSF barrier, allowing substances released into the CSF to gain greater access to parenchymal tissues.

CP AS A BIOLOGICAL SOURCE
OF NEUROTROPHIC FACTORS

The CP is in an ideal location for distributing molecules both locally and globally to the brain. The CP possesses numerous specific transport systems, contains an array of receptors, and also serves as a major source of biologically active compounds. Choroid epithelial cells synthesize many proteins, e.g., TTR, transferrin, ceruloplasmin, cytokines, and growth factors, including TGF-α, TGF-β, basic fibroblast growth factor, nerve growth factor, glial-derived neurotrophic factor, neurotrophin (NT)-3, NT-4, TNF, and IGF-II (Tables 1 and 2). The secretion of these proteins underlies the broad function of the CP to respond to the biochemistry of the brain and establish and maintain baseline levels of the extracellular milieu throughout the CNS. The molecules secreted by the CP can gain proximal and distal access to the brain parenchyma. The traditional view of synaptic- and gap junctional–based signaling of cell–cell communication and distribution of molecules in the brain has been modified to include volume transmission within the extracellular fluid and ventricular system *(53)*. The principle of volume transmission underlies diffusional and convective distribution of molecules within the CNS and accounts for both autocrine/paracrine– and distal/endocrine–like effects on target cells in the brain. In this broader view, the CP has an integral impact by continually producing CSF and polypeptides and distributing them from the ventricles to cells within the brain and spinal cord. Polypeptides released from the CP into the CSF may have access to distal targets in the brain owing to bulk flow of fluid and intraparenchymal diffusion/receptor–mediated retrograde transport in neurons whose endings are located near the ependyma or pia-astroglial membrane *(54,55)*.

CP AND NEUROGENESIS

In the last 10 yr, it has become apparent that the brain is not static and immutable but possesses a considerable ability to generate new CNS cells from locally discrete populations of stem cells. The ependymal and subependymal layers of the lateral ventricles have received the greatest attention as stem cell sources, but it is interesting to note that the CP is largely an extension of these structures and also possesses remarkable neurogenic capabilities. Proliferative and immunocytochemical markers show stem cells in the CP. Moreover, CP actively generates such mitogens as amphigegulin for regulating stem cell proliferation and neurogenesis.

Stem cells within the CP appear to proliferate in response to trauma. Li and colleagues *(68)* demonstrated that CP cells proliferate and differentiate after stroke in adult rats. Following MCA occlusion in rats, authors noted

Table 1
Receptors and Polypeptides in the CP

Receptors found in CP	Compounds synthesized in CP
Angiotensin II	Adrenomedullin
Anionic pesticides	Apolipoprotein J
Antipyrine	Arginine
Apolipoprotein E	β-Amyloid precursor protein
Apolipoprotein J (clusterin)	Basic fibroblast growth factor 1 and 2
Atrial natriuretic peptide	Brain-derived neurotrophic factor
Atropine	Prostaglandin D synthase
Barbitol	Cystatin C
Bradykinin	Endothelin-1
Brain-derived neurotrophic factor	GD-15
Cefodizime	Glial-derived neurotrophic factor
Cimetidine	Hepatocyte growth factor
Corticotropin-releasing factor	IGF-II
Digotoxin	IGF-binding protein 2–6
Diphenhydramine	IL-1α
Endothelin	IL-6
Gentimicine	Nerve growth factor
Insulin	NF-3 and 4
IGF	Transferrin
IL-1	TGF-α
Leptin	TGF-β1-3
Lidocaine	Transthyretin
Methadone	TNF-α
Methotrexate	Vascular endothelial growth factor
Morphine	Vasopressin
Nerve growth factor	
NT-4	
Penicillin	
Prolactin	
Proline	
Salicylic acid	
Tetrahydrocannabinol	
TGF-β	
Vascular endothelial growth factor	
Vasoactin intestinal polypeptide	
Vasopressin	

Table 2
Medical Conditions Associated With CP Pathology

Disease	Pathology
Aicardi syndrome	CP tumors/cysts common
Aging	Atrophy of CP, decreased CSF production, diminished polypeptide synthesis, and metabolic activity
Amyotrophic lateral sclerosis (ALS)	Upregulation of lysyl oxidase in CP of SAO mice and in ALS patients
	CP of ALS patients has elevated SOD activity, as determined by enhanced Cu/Zn-SOD staining
AD	Atrophy of CP, decreased CSF production, diminished polypeptide synthesis, and metabolic activity beyond that seen in age-matched controls
Amyloidotic polyneuropathy	Amyloid deposits seen in CP of humans
Autoimmune encephalomyelitis	Enhanced expression of mRNA encoding the toll-like receptor 2 in CP of mice
Chester-Erdheim disease	Choroidal lesions in subset of patients
Choroidal-cerebral calcification syndrome	Dense calcification of CP and increased CSF protein
Dandy-Walker syndrome	Hypoplasia of CP in rats following injection of 6-aminonicotinamide
Depression	CSF levels of TTR are decreased in depressed patients
Diabetes	Intraperitoneal streptozotocin in rats disrupts ion transport in CP
Hendra virus encephalitis	Choroid invasion of virus seen in symptomatic guinea pigs
Huntington's disease	Enhanced immunostaining of 3-nitrotyrosine in transgenic R6/2 mice
Leigh disease	Ubiquitous increase in mitochondria in epithelial cells of CP
Listeriosis	Upregulation of intracellular cell adhesion molecule-1 in CP associated with leptomeningitis in infected mice
Lupus	MRL-lpr mice show infiltration of lymphoid cells in CP

274

Menkes disease	Mottled gene expression highest in CP of macular mutant mice
Minamata disease	Mercury deposition on CP
Mitochondrial encephalomyopathy	Increased mitochondria, loss of microvilli, attenuated apical processes, and electron-dense bodies in epithelial cells of CP
Myotonic dystrophy	Accumulation of myotonic dystrophy protein in CP of rats and humans
Neu-Laxova syndrome	CP cysts
Schizophrenia	Calcification of CP associated and hallucinations
Sturge-Weber syndrome	Angiomatous enlargement of CP in humans
Trypanosomiasis	Trypanosomes and inflammatory cells in CP of rats infected with *Trypanosoma brucei*; Fibronectin and its receptor (VLA-4) also increased
Tuberculosis meningitis	Granulomatous lesions of CP
Unknown syndrome	Calcification of CP associated with moderate facial irregularities in a mother and twin daughters. No impact on intelligence
Von Hippel-Lindau disease	CP papilloma with chromosome 3 allele loss

increased bromodeoxyuridine immunoreactivity colocalized with neuronal nuclear antigen and glial fibrillary acidic protein (GFAP) immunoreactivity in the CP of the ischemia-affected hemisphere. Interestingly, transplanted circulating bone marrow–derived endothelial progenitor cells also incorporated to a high degree into the CP *(69,70)*.

Transplant studies also show that when grafted into damaged regions of the CNS, CP cells have the ability to differentiate. Kitada et al. *(67)* isolated CP from green fluorescent protein (GFP)–transgenic mice and grafted it into the damaged spinal cord of syngeneic mice. Interestingly, the grafted cells survived quite well and tended to differentiate into astrocytes. One week after injection, GFP-positive transplanted cells became immunohisto-chemically positive for astrocytic markers, including GFAP, but because negative for neuronal markers. Two weeks after grafting, immunoelectron microscopy showed that the GFP-positive transplanted cells that had gained GFAP immunoreactivity contained numerous bundles of intermediate filaments—a morphological characteristic similar to that of astrocytes—and were in close contact with adjacent host tissue *(67)*. Future studies should focus on isolating these cells, evaluating their multilineage potential, and exploring pharmacological methods to facilitate the native neurogenesis response that occurs following stroke to further optimize the endogenous repair mechanisms of the CP.

TRANSPLANTATION STUDIES USING CP
FOR NEUROPROTECTION AND REGENERATION

The profound age- and disease-related changes in CP have clear implications for understanding disease processes and raise an unexplored possibility that replacing added or damaged CP can be therapeutic. The diminished function of the aged CP is much like any number of diseases characterized by secretory cell dysfunction. In principle, transplanting or replacing a failing organ (e.g., the CP) or specific cell type is the most logical means of restoring lost function. Another possible strategy would be to harness the polypeptide synthesis of the CP for localized delivery after transplantation into an ectopic brain region. The endogenous role of the CP in growth factor and nutrient production creates the potential for these cells to serve as a viable source of stable and dose-controlled protein delivery.

Delivery of neurotrophic factors for CNS diseases has considerable therapeutic promise but is limited because of difficulties devising strategies to circumvent the BBB *(56)*. Clinically, controlled administration of trophic factors is restricted to intraventricular infusions using pumps or cannulae. These routes require repeated injections or refilling to maintain specific drug levels and to avoid protein degradation. Additionally, chronic low-dosage infusion is difficult to sustain using current pump technology and is not appropriate for intraparenchymal delivery. An alternative is the implantation of cells that have been genetically modified to produce a specific protein. Although the use of immortalized cell lines avoids many constraints associated with mechanical delivery forms, the incorporation of a transforming element, poor stability of protein expression, and safety concerns designate the use of genetically modified cells as currently untenable *(57)*. CP cells, alternatively, are primary cells with a minimal tumorigenicity risk. The endogenous influence of the CP in growth factor production makes these cells a source of stable, dose-controlled protein delivery simply by modifying the number of cells implanted. The notion of using CP grafts to deliver neurotrophic factors to damaged brain regions is in its infancy; yet, the studies conducted to date clearly support the concept and warrant further investigation.

Potent Neuroprotective Effects In Vitro From CP-Secreted Factors

CP isolated and maintained in vitro exerts potent neuronal effects. Conditioned media obtained from cultured CP was placed onto primary day-15 embryonic cortical neurons that were deprived of serum. The quantitative analysis of neuronal viability confirmed that molecules secreted from the CP exerted potent neurotrophic effects. Virtually all the neurons died when deprived of serum, whereas conditioned media collected from CP significantly protected against this same serum deprivation-induced cell death. This effect was dose-dependent, with nearly complete protection evident when neurons were cultured with 10–30% conditioned media. Moreover, conditioned media from porcine CP exerted significant, dose-dependent trophic effects on cultured neuroblastoma cells (Sks cell line) and neurons derived from embryonic mesencephalic tissue, where conditioned media enhanced neurite outgrowth and dopamine uptake, respectively.

These findings are complemented by a study in which mouse CP epithelial cells were cultured with dorsal root ganglion (DRG) neurons. After 4–5 h, the DRG neurons developed elongated neuronal processes, with elaborate branching patterns over the surface of the epithelial cells. The observation that the CP cells appeared to provide a scaffold for the extension of neurites is consistent with the well-known ability of CP to produce extracellular matrices, including laminin and fibronectin. Thus, the data demonstrate that CP secretes molecules capable of protecting neurons that are otherwise destined to die. These molecules appear biologically active across a number of different species, experimental paradigms, and cell types, raising the possibility that similar effects could be achieved in animals with neurological disorders. The trophic and tropic effects of CP provide potentially excellent circumstances for protecting and repairing damaged CNS architecture.

CP Transplantation for Stroke

Stroke is the third leading cause of death in the majority of developed countries and is a prominent health care burden in all developed countries *(59)*. There are still no effective treatments for mitigating the loss of neurons in stroke patients. Central delivery of neurotrophic factors may represent an means to enable neuroprotection following stroke. Indeed, several neurotrophic molecules have been reported to provide varying degrees of neuroprotection in animal models of stroke *(60–65)*.

Given these considerations, isolated CP obtained from rodents was tested for its neuroprotective effects in a conventional rodent model of stroke *(58)*. Rats received a 1-h MCA occlusion, immediately followed by transplantation of CP cells that had been placed into alginate microcapsules to facilitate handling, to minimize migration, and to isolate it from the recipient immune system. Instead of stereotaxically injecting the CP into the area of ischemic damage, the CP cells were placed on the cortex overlying the striatal region that would be normally infarcted following MCA occlusion. Using the elevated body swing test and Bederson neurological examination, behavioral testing 1–3-d posttransplantation revealed profound impairments in control animals that were significantly improved by CP transplants. Histological analysis at 3-d posttransplantation indicated that the behavioral improvements were accompanied by a significant decrease (approx 35–40%) in the volume of striatal infarction. This paradigm provided a fairly stringent test of the ability of the molecules secreted from the CP to exert a neuroprotective effect, because the molecules were required to diffuse out of the capsules and through several millimeters of cortical tissue. Accordingly, the concentrations of therapeutic molecules reaching the infarcted region were modest, compared to those that might be achieved locally. Nonetheless, even under these less-than-ideal conditions, a significant structural and functional benefit was produced by the CP transplants *(58)*. However, future studies should carefully consider alternative transplant sites.

CP Transplantation for Traumatic Brain and Spinal Cord Injury

No class of CNS injuries is more embedded in historic frustration than traumatic brain injuries. Like stroke, the delivery of neurotrophic factors to the site of injury offers theoretical promise for treating trauma. Recently, Ide and colleagues *(66)* demonstrated that CP grafted into a damaged spinal cord promoted regeneration of spinal cord axons. The CP was excised from the fourth ventricle of adult rats, minced into small fragments, and grafted into the dorsal funiculus at the C2 level in adult rat spinal cord from the same strain (i.e., a syngeneic transplant). At various times posttransplant, subsets of animals were evaluated histologically to confirm cell survival and determine any regenerative effect on the damaged spinal cord. Electron microscopy and fluorescence histochemistry showed that ependymal cells of the grafted CP survived well and induced a robust regeneration of the damaged axons of the spinal cord. Injections of horseradish peroxidase into the sciatic nerve labeled numerous regenerating fibers that extended from the fasciculus gracilis into the graft within 7-d posttransplantation. This effect was evident for at least 10 mo, with some axons elongating rostrally into the dorsal funiculus. Evoked potentials of long duration were recorded 5-mm rostral to the lesion in the rats 8–10 mo after grafting. These findings indicate that CP ependymal cells can facilitate axonal growth in vivo, suggesting that they are capable of exerting both trophic and tropic effects in vivo.

Kitada et al. *(67)* also demonstrated that grafted CP could survive in the damaged spinal cord but further showed that the surviving cells differentiated into astrocytes. The CP was excised from the fourth ventricle of GFP-transgenic mice, and the cells were dissociated and cultured for 4–6 wk. CP cells were injected into the prelesioned spinal cords of wild-type mice of the same strain using a Hamilton syringe. One week after injection, some GFP-positive transplanted cells became immunohistochemically positive for GFAP but negative for neurofilament and myelin basic protein. All the GFAP-positive transplanted cells were negative for vimentin. Two weeks after grafting, immunoelectron microscopy showed that the GFP-positive transplanted cells that had gained GFAP immunoreactivity contained numerous bundles of intermediate filaments, a morphological characteristic similar to that of astrocytes, and were in close contact with adjacent host tissue *(67)*. When grafted into the spinal cord, at least some cultured CP ependymal cells can differentiate into astrocytes.

The observation that CP ependymal cells differentiate into glia within the damaged spinal cord suggests that CP might be a unique source of progenitor cells. CP epithelial cells have the same origin as ventricular ependymal cells and are regarded as modified ependymal cells. Li and colleagues *(68)* demonstrated that CP cells proliferate and differentiate after stroke in adult rats. Following MCA occlusion in rats, increased bromodeoxyuridine immunoreactivity colocalized with neuronal nuclear antigen and GFAP immunoreactivity in cells in the CP of the ischemia-affected hemisphere. Along with the evidence that transplanted circulating bone marrow–derived endothelial progenitor cells incorporate into the CP *(69,70)*, it seems that at least some portion of cells within the CP may differentiate into neurons and glia. Future studies should focus on isolating these cells, evaluating their multilineage potential, and exploring pharmacological techniques relating to the native neurogenesis response that occurs following stroke to further optimize the endogenous repair mechanisms of the CP.

CONCLUSION

In general, CSF production and distribution by the CP provide buoyancy for the brain and spinal cord, serving as a nutritive milieu for neurons and glial cells by secreting numerous polypeptides and growth factors. In addition, CP acts as a vehicle for removing waste products of cellular metabolism. As critical as these functions are, the CP goes largely unnoticed in the neuroscience research community. The vast functions and potential of choroidal cells has only just begun to become apparent and appreciated. Several convergent data sets are documenting an increasingly detailed role of CP secretory products in normal brain development and functioning. Within the CNS, the CP is perhaps the most prolific producer of compounds that are vital for growth and function. The impact of the CP in maintaining extracellular concentrations of an array of proteins puts this tiny collection of cells at the center of understanding the complex intertwining of nutritive and protective factors in developmental biology. Also, profound changes in CP secretion and function occur with many pathological conditions, including aging and degeneration. Because the polypeptides carried by the CSF are distributed throughout the neuro-axis, changes in the basic secretory and transport functions of CP lead to consequences that are both local and comprehensive. A greater understanding of the CP in CNS diseases will bring forth a concomitantly greater elucidation of the biological underpinnings of disease and endogenous repair processes. Finally, refining the knowledge of CP's effect in CNS diseases will suggest new strategies to prevent or minimize disease processes. Based on the multitude of products secreted from CP, strategies may emerge that stray from current approaches that deliver a single neuroprotective protein to orchestrated polypharmacy techniques that take advantage of the endogenous growth factor symphony already perfected by the CP. Indeed, the biological effects of neurotrophic factors are so intertwined across and within cellular systems that it is probably simplistic to suggest manipulating cellular machinery via treatment with a single factor. Although from a pharmacological point of view, this "multimolecule therapeutic" approach is the most favorable method, it appears nonphysiological from a cell biological perspective. Using transplantable CP cells may be one way of harnessing the secretory potential of CP for local protein delivery.

REFERENCES

1. Chodobski, A. and Szmydynger-Chodobska. (2001) Choroid plexus, Target for polypeptides and site of their synthesis. *Microsc. Res. Techniq.* **52,** 65–82.
2. Stopa, E. G., Berzin, T. M., Kim, S., et al. (2001) Human choroid plexus growth factors, what are the implications for CSF dynamics in Alzheimer's disease. *Exp. Neurol.* **167,** 40–47.
3. Serot, J. M., Bene, M. C., and Faure, G. C. (2003) Choroid plexus, ageing of the brain, and Alzheimer's disease. *Front. Biosci.* **8,** 515–521.
4. Preston, J. E. (2001) Ageing choroid plexus-cerebrospinal fluid system. *Microsc. Res. Tech.* **52,** 31–37.
5. Ferrand-Drake, M. (2001) Cell death in the choroid plexus following transient forebrain global ischemia in the rat. *Microsc. Res. Techniq.* **52,** 130–136.
6. Engelhardt, B., Wolburg-Buchholz, K., and Wolburg, H. (2001) Involvement of the choroid plexus in central nervous system inflammation. *Microsc. Res. Techniq.* **52,** 112–129.
7. Korzhevskii, D. E. (2002) Structural organization of choroid plexus primordium in human telencephalon. *Morfologiia* **121,** 63–67.
8. Dziegielewska, K. M., Ek, J., Habgood, M. D., and Saunders, N. R. (2001) Development of the choroid plexus. *Microsc. Res. Techniq.* **52,** 5–20.
9. Dohrmann, G. J. (1970) The choroid plexus, a historical review. *Brain Res.* **18,** 197–218.
10. Cornford, E. M., Varesi, J. B., Hyman, S., et al. (1997) Mitochondrial content of choroid plexus epithelium. *Exp. Brain Res.* **116,** 399–405.
11. Faraci, F. M., Mayhan, W. G., and Heistad, D. D. (1989) Effect of serotonin on blood flow to the choroid plexus. *Brain Res.* **478,** 121–126.
12. Nilsson, C., Ekman, R., Lindvall-Axelsson, M., and Owman, C. (1990) Distribution of peptidergic nerves in the choroid plexus, focusing on coexistence of neuropeptide Y, vasoactive intestinal polypeptide and peptide histidine isoleucine. *Regul. Pept.* **27,** 11–26.
13. Speake, T., Whitwell, C., Kajita, H., et al. (2001) Mechanisms of CSF-secretion by the choroid plexus. *Microsc. Res. Tech.* **52,** 49–59.
14. Nilsson, C., Stahlberg, F., Gideon, P., et al. (1994) The nocturnal increase in human cerebrospinal fluid production is inhibited by a beta 1-receptor antagonist. *Am. J. Physiol.* **267,** 1445–1448.
15. Rall, D. P. (1964) The structure and function of the cerebrospinal fluid. In *Cellular Functions of Membrane Transport* (Hoffman, J., ed.), Prentice-Hall, NJ, pp. 269–282.
16. Segal, M. B. (2000) The choroid plexuses and the barriers between the blood and the cerebrospinal fluid. *Cell Mol. Neurobiol.* **20,** 183–196.
17. Felgenhauer, K. (1986) The blood-brain barrier redefined. *J. Neurol.* **233,** 193–194.
18. Spector, R. (1977) Vitamin homeostasis in the central nervous system. *N. Engl. J. Med.* **296,** 1393–1398.
19. Weisner, B. and Roethig, H. J. (1983) The concentration of prealbumin in cerebrospinal fluid (CSF), indicator of CSF circulation disorders. *Eur. Neurol.* **22,** 96–105.

20. Bruni, J. E. (1996) Cerebral ventricular system and cerebrospinal fluid. In *The Encyclopedia of Human Biology, 2nd ed.* (Dulbecco, R., ed.), Academic Press, San Diego, CA.

21. Black, P. L. and Ojemann, R. G. (1990) Hydrocephalus in adults. In *Neurological Surgery, 3rd edition, vol. 2* (Youmans, J. R., ed.), WB Saunders Co., Philadelphia, PA.

22. Vorbrodt, A. W. and Dobrogowska, D. H. (2003) Molecular anatomy of intercellular junctions in brain endothelia and epithelial barriers, electron microscopist's view. *Brain Res. Rev.* **42,** 221–242.

23. Graff, C. L. and Pollack, G. M. (2004) Drug transport at the blood-brain barrier and the choroid plexus. *Curr. Drug Metab.* **5,** 95–108.

24. Serot, J. M., Béné, M. C., Foliguet, B., and Faure, G. C. (2001) Choroid plexus and ageing, a morphometric and ultrastructural study. *Eur. J. Neurosci.* **14,** 794–798.

25. Sturrock, R. R. (1988) An ultrastructural study of the choroid plexus of aged mice. *Anat. Anz.* **165,** 379–385.

26. Serot, J. M., Béné, M. C., Foliguet, B., and Faure, G. C. (2000) Morphological alterations of the choroid plexus in late-onset alzheimer's disease. *Acta Neuropathol.* **99,** 105–108.

27. Wen, G. Y., Wisniewski, H. M., and Kascsak, R. J. (1999) Biondi ring tangles in the choroid plexus of Alzheimer's disease and normal aging brains, a quantitative study. *Brain Res.* **832,** 40–46.

28. Shuangshoti, S. and Netsky, M. G. (1970) Human choroid plexus, morphologic and histochemical alterations with age. *Am. J. Anat.* **128,** 73–96.

29. Ferrante, F. and Amenta, F. (1987) Enzyme histochemistry of the choroid plexus in old rats. *Mech. Ageing Dev.* **41,** 65–72.

30. Cottrell, D. A., Blakely, E. L., Johnson, M. A., et al. (2001) Cytochrome oxidase deficient cells accumulate in the hippocampus and choroid plexus with age. *Neurobiol. Aging* **22,** 265–272.

31. Kvitnitskaia-Ryzhova, T. I. and Shkapenko, A. L. (1992) A comparative ultracytochemical and biochemical study of the ATPases of the choroid plexus in aging. *Tsitologiia* **34,** 81–87.

32. Preston, J. E. (1991) Age-related reduction in rat choroid plexus chloride efflux and CSF secretion rate. *Soc. Neurosci.* **25,** P697–698.

33. May, C., Kaye, J. A., Atack, J. R., et al. (1990) Cerebrospinal fluid production is reduced in healthy aging. *Neurology* **40,** 500–503.

34. Miklossy, J., Kraftsik, R., Pillevuit, O., et al. (1998) Curly fiber and tangle-like inclusions in the ependyma and choroid plexus—a pathogenetic relationship with the cortical Alzheimer-type changes? *J. Neuropathol. Exp. Neurol.* **57,** 1202–1212.

35. Jellinger, K. (1976) Neuropathological aspects of dementias resulting from abnormal blood and cerebrospinal fluid dynamics. *Acta Neurol. Belg.* **76,** 83–102.

36. Serot, J. M., Béné, M. C., and Faure, G. C. (1994) Comparative immunohistochemical characteristics of human choroid plexus in vascular and Alzheimer's dementia. *Hum. Pathol.* **25,** 1185–1190.

37. Serot, J. M., Christmann, D., Dubost, T., and Couturier, M. (1997) Cerebrospinal fluid transthyretin, aging and late onset Alzheimer's disease. *J. Neurol. Neurosurg. Psychiatry* **63**, 506–508.
38. Tohgi, H., Abe, T., Nakanishi, M., et al. (1994) Concentrations of α-tocopherol and its quinone derivative in cerebrospinal fluid from patients with vascular dementia of the Binswanger type and Alzheimer type dementia. *Neurosci. Lett.* **174**, 73–76.
39. Schippling, S., Kontush, A., Arlt, S., et al. (2000) Increased lipoprotein oxidation in Alzheimer's disease. *Free Rad. Biol. Med.* **28**, 351–360.
40. Ikeda, T., Furukawa, Y., Mashimoto, S., et al. (1990) Vitamin B12 levels in serum and cerebrospinal fluid of people with Alzheimer's disease. *Acta Psychiatr. Scand.* **82**, 327–329.
41. Serot, J. M., Christmann, D., Dubost, T., et al. (2001) CSF-folate levels are decreased in late-onset AD patients. *J. Neural Transm.* **108**, 93–99.
42. Selley, M. L., Close, D. R., and Stern, S. E. (2002) The effect of increased concentrations of homocysteine on the concentration of (E)-4-hydroxy-2-nonenal in the plasma and cerebrospinal fluid of patients with Alzheimer's disease. *Neurobiol. Aging* **232**, 383–388.
43. Palm, D., Knuckey, N., Guglielmo, M., et al. (1995) Choroid plexus electrolytes and ultrastructure following transient forebrain ischemia. *Am. J. Physiol.* **269**, 73–79.
44. Gillardon, F., Lenz, C., Kuschinsky, W., and Zimmermann, M. (1996) Evidence for apoptotic cell death in the choroid plexus following focal cerebral ischemia. *Neurosci. Lett.* **29**, 113–116.
45. Ferrand-Drake, M. and Wieloch, T. (1999) The time-course of DNA fragmentation in the choroid plexus and the CA1 region following transient global ischemia in the rat brain. The effect of intra-ischemic hypothermia. *Neuroscience* **93**, 537–549.
46. Palm, D. E., Knuckey, N. W., Primiano, M. J., et al. (1995) Cystatin C, a protease inhibitor, in degenerating rat hippocampal neurons following transient forebrain ischemia. *Brain Res.* **691**, 1–8.
47. Scheepens, A., Sirimanne, E. S., Breier, B. H., et al. (2001) Growth hormone as a neuronal rescue factor during recovery from CNS injury. *Neuroscience* **104**, 677–687.
48. Knuckey, N. W., Finch, P., Palm, D. E., et al. (1996) Differential neuronal and astrocytic expression of transforming growth factor beta isoforms in rat hippocampus following transient forebrain ischemia. *Mol. Brain Res.* **40**, 1–14.
49. Klempt, N. D., Sirimanne, E., Gunn, A. J., et al. (1992) Hypoxia-ischemia induces transforming growth factor β1 mRNA in the infant rat brain. *Mol. Brain Res.* **13**, 93–101.
50. Prehn, J. H. M., Bindokas, V. P., Marcuccilli, C. J., et al. (1994) Regulation of neuronal Bcl2 protein expression and calcium homeostasis by transforming growth factor type β confers wide-ranging protection on rat hippocampal neurons. *Proc. Natl. Acad. Sci. USA* **91**, 12599–12603.
51. Henrich-Noack, P., Prehn, J. H. M., and Krieglstein, J. (1994) Neuroprotective effects of TGF-β1. *J. Neural Transm.* Suppl. **43**, 33–45.

52. Henrich-Noack, P., Prehn, J. H. M., and Krieglstein, J. (1996) TGF-β1 protects hippocampal neurons against degeneration caused by transient global ischemia. Dose-response relationship and potential neuroprotective mechanisms. *Stroke* **27,** 1609–1614.

53. Agnati, L. F., Zoli, M., Stromberg, I., and Fuxe, K. (1995) Intercellular communication in the brain, wiring versus volume transmission. *Neuroscience* **69,** 711–726.

54. Ferguson, I. A., Schweitzer, J. B., Bartlett, P. F., and Johnson, E. M., Jr. (1991) Receptor-mediated retrograde transport in CNS neurons after intraventricular administration of NGF and growth factors. *J. Comp. Neurol.* **313,** 680–692.

55. Mufson, E. J., Kroin, J. S., Sendera, T. J., and Sobreviela, T. (1999) Distribution and retrograde transport of trophic factors in the central nervous system, functional implications for the treatment of neurodegenerative diseases. *Prog. Neurobiol.* **57,** 451–484.

56. Flanagan, T. R., Emerich, D. F., and Winn, S. R., eds. (1994) *Providing Therapeutic Access to the Brain, New Approaches, Methods in Neuroscience, vol. 21,* Academic Press, San Diego, CA.

57. Lasic, D. and Templeton, N. S., eds. (2000) *Gene Therapy, Therapeutic Mechanisms and Strategies.* Marcel Dekker, New York.

58. Borlongan, C. V., Elliott, R. B., Skinner, S. J. M., et al. Intraparenchymal grafts of rat choroid plexus protect against cerebral ischemia in adult rats. *NeuroReport,* in press.

59. American Heart Association. (2002) Stroke Statistics.

60. Hoffer, B. and Olson, L. (1997) Treatment strategies for neurodegenerative diseases based on trophic factors and cell transplantation techniques. *J. Neural Transm.* Suppl. **49,** 1–10.

61. Wang, Y., Chang, C. F., Morales, M., et al. (2002) Protective effects of glial cell line-derived neurotrophic factor in ischemic brain injury. *Ann. NY Acad. Sci.* **962,** 423–437.

62. Lin, S. Z., Hoffer, B. J., Kaplan, P., and Wang, Y. (1999) Osteogenic protein-1 protects against cerebral infarction induced by MCA ligation in adult rats. *Stroke* **1,** 126–133.

63. Johnston, R. E., Dillon-Carter, O., Freed, W. J., and Borlongan, C. V. (2001) Trophic factor secreting kidney cell lines, in vitro characterization and functional effects following transplantation in ischemic rats. *Brain Res.* **900,** 268–276.

64. Borlongan, C. V., Yamamoto, M., Takei, N., et al. (2000) Glial cell survival is enhanced during melatonin-induced neuroprotection against cerebral ischemia. *FASEB* **14,** 1307–1317.

65. Cairnes, K. and Finklestein, S. P. (2003) Growth factors and stem cells as treatments for stroke recovery. *Phys. Med. Rehabil. Clin. N. Am.* **Suppl,** 135–142.

66. Ide, C., Kitada, M., Chakrabortty, S., et al. (2001) Grafting of choroid plexus ependymal cells promotes the growth of regenerating axons in the dorsal funiculus of rat spinal cord, a preliminary report. *Exp. Neurol.* **167,** 242–251.

67. Kitada, M., Chakrabortty, S., Matsumoto, N., et al. (2001) Differentiation of choroid plexus ependymal cells into astrocytes after grafting into the pre-lesioned spinal cord in mice. *Glia* **36,** 364–374.

68. Li, Y., Chen, J., and Chopp, M. (2002) Cell proliferation and differentiation from ependymal, subependymal and choroid plexus cells in response to stroke in rats. *J. Neurol Sci.* **193,** 137–46.

69. Eglitis, M. A. and Mezey, E. (1997) Hematopoietic cells differentiate into both microglia and macroglia in the brains of adult mice. *Proc. Natl. Acad. Sci. USA* **94,** 4080–4085.

70. Zhang, Z. G., Zhang, L., Jiang, Q., and Chopp, M. (2002) Bone marrow-derived endothelial progenitor cells participate in cerebral neovascularization after focal cerebral ischemia in the adult mouse. *Circ. Res.* **90,** 284–288.

11

Progress and Challenges in Immunoisolation for CNS Cell Therapy

Christopher G. Thanos and Dwaine F. Emerich

ABSTRACT

The blood–brain barrier (BBB) hinders the delivery of potentially therapeutic drugs to the brain by restricting the diffusion of drugs from the vasculature to the brain parenchyma. One method to overcome the BBB is with
cellular implants that produce and deliver therapeutic molecules directly to
the brain region of interest. Immunoisolation is based on the observation that
xenogeneic cells can be protected from host rejection by encapsulating, or
surrounding, them within an immunoisolatory, semipermeable membrane.
Cells can be enclosed within a selective, semipermeable membrane barrier
that admits oxygen and required nutrients and releases bioactive cell secretions but restricts passage of larger cytotoxic agents from the host immune
defense system. The selective membrane eliminates the need for chronic
immunosuppression of the host and allows the implanted cells to be obtained
from nonhuman sources. This chapter discusses cell immunoisolation for
treating central nervous system (CNS) diseases, from concept to preclinical
evaluation, in a wide range of animal models and relating to the clinical trials
conducted to date.

Key Words: Polymer encapsulation; xenotransplantation; neurodegeneration.

INTRODUCTION

Numerous central nervous system (CNS) diseases are characterized by
the degeneration of specific populations of cells, resulting in continuous
deterioration of vital function. Advances in molecular biology, genetic
engineering, proteomics, and genomics are providing an increasing number
of potentially efficacious proteins, peptides, and other compounds. Unfortunately, most of these compounds are not active following systemic administration, largely because the blood–brain barrier (BBB) modulates the local

From: *Contemporary Neuroscience: Cell Therapy, Stem Cells, and Brain Repair*
Edited by: C. D. Sanberg and P. R. Sanberg © Humana Press Inc., Totowa, NJ

and global exchange between the vasculature and brain parenchyma *(1,2)*. A number of strategies have been described to circumvent the BBB, including: (1) carrier- or receptor-mediated transcytosis *(3,4)*; (2) osmotic opening *(5,6)*; (3) direct infusion with stereotactic guidance *(7–9)*; (4) osmotic pumps *(10,11)*; (5) sustained-release polymer systems *(12–14)*; (6) cell replacement/cell therapy *(15–20)*; (7) direct gene therapy *(21–25)*.

One iteration of cell-based therapy proposes to encase xenogeneic cells within a selectively permeable polymeric membrane. The polymer membrane containing the cells can then be implanted into the brain, allowing cell-based delivery directly into the target site. This process, known as *immunoisolation*, is enabled because xenogeneic cells are protected from host rejection by the immunoisolatory, semipermeable membrane. Single cells, or small clusters of cells, can be enclosed within a selective, semipermeable membrane barrier, which admits oxygen and required nutrients and releases bioactive cell secretions but restricts passage of larger cytotoxic agents from the host immune defense system. The selective membrane eliminates the need for chronic immunosuppression of the host and allows the implanted cells to be obtained from nonhuman sources, thus avoiding the constraints associated with cell sourcing that have limited the application of unencapsulated cell transplantation. This chapter discusses the preclinical and clinical evaluation of encapsulated cells across a wide range of CNS diseases, highlighting the therapeutic potential of genetically modified, encapsulated cells for Alzheimer' disease (AD), Parkinson's disease (PD), and Huntingtons's disease (HD).

CELL IMMUNOISOLATION

There are generally two categories for cell immunoisolation by encapsulation, micro- and macro-, each with some benefits and limitations, as shown in Table 1 *(26–29)*. This chapter focuses primarily on macroencapsulation. Macroencapsulation involves filling a hollow, usually cylindrical, selectively permeable membrane with cells, generally suspended in a matrix, and then sealing the ends to form a capsule. Polymers used for macroencapsulation are biodurable, with a thicker wall than that found in microencapsulation. Although thicker wall and larger implant diameters can enhance long-term implant stability, these features may also impair diffusion, compromise the viability of the tissue, and slow the release kinetics of desired factors. In theory, macrocapsules can be retrieved from the recipient and replaced if necessary.

Macroencapsulation is generally achieved by filling preformed thermoplastic hollow fibers with a cell suspension. The hollow fiber is formed by

Table 1
Advantages and Disadvantages of Unencapsulated Micro- and Macroencapsulated Implants

	Advantages	Disadvantages
Unencapsulated cells or tissue	Permits anatomical integration between the host and transplanted tissue Good cell viability and neurochemical diffusion	Likely requires immunosuppression Limited tissue availability Difficult retrieval Societal and ethical issues
Microencapsulation	Permits use of allo- and xenografts without immunosuppression Thin wall and spherical shape are optimal for cell viability and neurochemical diffusion	Mechanically and chemically fragile Multiple implant sites Limited retrievability
Macroencapsulation	Permits us of allo- and xenografts without immunosuppression Reasonable mechanical stability Adequate cell viability and neurochemical diffusion Retrievable	Dimensions may limit neurochemical diffusion and cell viability Multiple implant sites May produce significant tissue damage/displacement during implantation

pumping a solution of polymer in a water-miscible solvent through an outer annular region of a nozzle, while an aqueous solution is pumped concurrently through a central bore. The polymer precipitates upon contact with the water and forms a cylindrical hollow fiber with a permselective inner membrane, or "skin." Further precipitation of the polymer occurs as the water moves through the polymer wall, forcing the organic solvent out and forming a trabecular wall structure. The hollow fiber is collected in a large aqueous water bath, where complete precipitation of the polymer and dissolution of the organic solvent occurs. The ends of the hollow fiber are then sealed to form macrocapsules. This final step is not trivial, as sealing the ends of capsules reliably can be extremely difficult and represents the barrier paramount for successful immunoisolation.

A second method of macroencapsulation, known as *coextrusion*, avoids the sealing problem by entrapping cells within the lumen of a hollow fiber during the fabrication process *(30)*. Pinching the fiber before complete precipitation of the polymer causes fusion of the walls, providing closure of the extremities while the cells are inside. The advantages of coextrusion, compared to loading-preformed capsules, include cell distribution more uniformly along the entire fiber length, reduction of shear stresses on the cells during the loading process, and the potential for mass production.

Although polymer capsules are relatively sturdy, it is critical to ensure that the encapsulating membrane is compliant enough to meet the dynamics of the surrounding tissue and remain mechanically resilient to resist failure during device implantation/retrieval. To accomplish this, the macrocapsule can be designed to offer added strength and device integrity. One approach includes the addition of a crosslinked hydrogel (e.g., a 2% alginate solution) within the device. This modification was observed to enhance structural support during the implantation procedure *(31)*. However, with time, the hydrogel loses structural integrity and does not provide added strength, especially regarding tensile strength—an important consideration for device retrieval. Mechanical supports can also be used for additional strength. Such supports are better served from within the device to prevent impeding diffusion between the encapsulating membrane and surrounding tissue. Titanium wires and braided materials are a few approaches that may be used to provide more tensile strength.

Cells and Extracellular Matrices Used in Encapsulation

Cells placed within encapsulation devices generally fall into one of three categories. The first category is represented by primary postmitotic cells, such as islets of Langerhans (for diabetes), adrenal chromaffin cells (for chronic pain), or hepatocytes (for liver devices). Second, immortalized (or dividing) cells, e.g., PC12 cells, have been utilized to deliver dopamine for PD. Typically, the third category is a cell line that has been genetically engineered to secrete a bioactive substance, like baby hamster kidney (BHK) cells to secrete such factors as nerve growth factor (NGF) for a potential therapy in AD. Dividing tissue has advantages over postmitotic tissue; it can be expanded, banked, and is thus more easily tested for sterility and contaminants. However, dividing tissue is also constrained by potential overgrowth within the capsule environment, resulting in an accumulation of necrotic tissue that could diminish the membrane's permeability characteristics, further reducing cell viability and neurochemical output.

In vivo, extracellular matrices (ECMs) provide control of cell function through the regulation of morphology, proliferation, differentiation, migration, and metastasis *(32,33)*. Within a capsule, ECMs were originally employed simply to prevent aggregation of cells (immobilization) and resultant central necrosis, but they have since been found beneficial to the viability and function of cells that require immobilization and serve as a scaffolding for anchorage-dependent cell lines. For example, adrenal chromaffin cells have been immobilized in alginate to prevent aggregation that, in turn, reduces central necrotic cores from forming *(34)*. The Chromaffin cells appear to thrive in alginate, whereas mitotically active fibroblasts do not. In this case, the use of alginate is essential to the optimal functioning of this device, because some anchorage-dependent cells (e.g., fibroblasts or endothelial cells) are present with the adrenal chromaffin cells. In the absence of alginate or similar immobilizing matrices, the fibroblasts can expand and outgrow the encapsulated milieu, resulting in a device deficient in bioactive factors produced from chromaffin cells *(35)*. In contrast, BHK cells are a fibroblastic cell line and prefer collagen, whereas PC12 cells exhibit a preference for distribution within precipitated chitosan that provides a scaffolding structure on which the cells anchor *(36)*.

To provide a substratum designed for more specific functions, the matrix material can be manipulated chemically or mechanically, which then may influence cell attachment, differentiation, and/or proliferation. For example, such peptides as arginine-glycine-aspartic acid (RGD) have been immobilized on a variety of surfaces to promote cell adhesion *(37)*. Integrin receptors on the cell surface membrane interact with the RGD sequence—a known ligand for fibronectin receptors. Glass microbeads have been modified by attaching RGD or tyrosine-Isoleucine-Glycine-Serine-Arginine (YIGSR) motifs to provide sites for cell adhesion *(37)*. Spherical ferromagnetic beads have been coated with specific receptor ligands to mediate cell attachment *(38)*. With competitive binding assays and a mechanical stress-testing apparatus, the endothelial cell's interaction with the ECM receptor integrin β-1 supported a force-dependent stiffening response, whereas nonadhesion receptors did not. CDPG-YIGSR, a laminin-derived oligopeptide sequence, has also been derivitized within an agarose hydrogel and permitted a dose-dependent increase in neurite outgrowth of neuronal cell bioassays *(39)*. Similarly, YIGSR and isoleucine-lysine-valine-alanine-valine, both of which are found in laminin, have been immobilized on surfaces to promote neuronal cell adhesion and neurite outgrowth *(40)*. Polyethylene oxide (PEO)–star copolymers have been fabricated as a potential synthetic ECM *(41)*. The star copolymers provide many hydroxyl groups, where various synthetic oligopeptides can be attached to desired specifications.

The survival and differentiation of encapsulated cells can be influenced by the matrix interactions. Various matrices for use in immunoisolatory devices (e.g., alginate, rat tail collagen extracts, gelatin shards, porous gelatin or collagen microcarriers, carrageenan, chondroitin sulfate, fibrin, hyaluronic acid, the positively charged substrate chitosan, and an acrylamide-based thermoresponsive gel) are available and were recently reviewed *(42)*. Overall, successful cell encapsulation involves the choice of cells to be encapsulated, the type of intracapsular matrix used, and the ability to control membrane properties, such as geometry, morphology, and transport. The interactions between the encapsulating membrane characteristics and the capsule core, or matrix within, should be rigorously characterized to determine the optimal configuration for each cell type.

Validation of the Concept of Immunoisolation

In Vitro Studies

The maintenance of immunoisolation (i.e., capsule integrity) can be easily confirmed in vitro. In one study, a polydisperse (10^3–10^6 g/mol) dextran solution was encapsulated into hollow fibers, and the flux of the molecules across the semipermeable membrane into a surrounding reservoir was monitored over time. A dextran rejection curve was produced from the filtrate, and reservoir concentrations were measured using gel permeation chromatography *(43)*. With control devices that had been damaged, the large-molecular-weight dextran species escaped into the surrounding reservoir. In contrast, intact capsules retained the encapsulated dextran. Capsule integrity can also be confirmed using standard immunological assays, such as measuring the protection of encapsulated cells against the cytotoxic killing of antibody immunoglobulin G (IgG)–mediated complement lysis *(29)*. With integral PC12 cell–loaded capsules, in the presence of antibody and complement, the capsular membrane prevented antibody-mediated complement lysis (<10% cell death), whereas complete killing (100%) was observed in cases of damaged capsules or with PC12 cells not encapsulated.

In Vivo Studies

The importance of polymer capsule integrity for xenografted cell survival is illustrated in studies where unencapsulated PC12 cells, or cells encapsulated in intentionally damaged membranes, have been implanted in the brains of guinea pigs *(30)*. Intact PC12 cell–loaded capsules implanted into the guinea pig striatum showed no lymphocytic infiltration and a minimal astrocytic reaction by glial fibrillary acid protein (GFAP) staining. In contrast, cell survival was poor in capsules that were intentionally damaged, with marked inflammation and heavy lymphocytic invasion into the capsule. Parallel studies confirmed that unencapsulated PC12 cells do not survive following implantation into either the guinea pig or the monkey striatum. However, encapsulated PC12 cells have demonstrated viability for 6 mo in monkeys *(44)*. Similar results have been obtained in rats that received intraventricular implants of bovine adrenal chromaffin cells when implanted in the ventricular space *(45)*. There was no evidence of elevated levels of rat antibovine adrenal chromaffin cell IgG or IgM levels in serum from rats implanted with encapsulated xenogeneic adrenal chromaffin cells for nearly 1.5 yr in vivo. In contrast, a robust host immune response was induced in animals after implantation of unencapsulated bovine adrenal chromaffin cells.

Biocompatibility

Transplant survival, with and without an encapsulating membrane, is mediated by many factors. The cellular/tissue reaction generated by a host in response to a foreign body, typically referred to as *biocompatibility*, impacts the success of the transplant. Factors affecting biocompatibility include the implantation method, site, and properties, such as composition of the polymer, potential residual processing agents, surface integrity and microgeometry, and the size and shape of the implant. Constituents of the implants should be assessed rigorously, both in vitro and in vivo, to determine the safety of the materials. The CNS tissue is not only privileged from an immunologic perspective, but it also lacks the primary reactive cells, fibroblasts, and macrophages found in peripheral locations. Therefore, the brain offers a unique environment in terms of the inflammatory response, as well as the cellular constituents that comprise the reactive cells. Immuno-specific antibodies are available to delineate the roles of the brain reactive cells, astroctyes, and microglia, with respect to their reactivity. Nevertheless, few studies have systematically examined the reaction of host brain tissue to the presence of polymeric devices.

Early investigations utilized electron microscopic techniques to characterize the brain tissue reaction to plastic-embedded metal electrodes and polymer implants *(46–48)*. Necrosis of the tissue surrounding the polymer capsules implanted into the striatum of rodents was minimal, with small Nissl-positive cells and capillaries invading the open trabeculae in the wall of the macrocapsules *(48)*. GFAP immunocytochemistry revealed local reactive astrocytes 1–2 wk after implantation. The intensity of the GFAP reaction diminished so rapidly that by 4-wk postimplantation, the gliotic reaction surrounding the polymer implant was minimal. No significant changes in myelin basic protein–reactive oligodendroglia were observed, and neuron-specific enolase-reactive neurons were readily identifiable adjacent to the implant. Subsequent studies with an immunospecific antisera against rat microglia OX-42 revealed a reaction in magnitude and with a time course similar to that seen for the astrocytes.

The lack of a significant host tissue reaction to the implant is crucial for the initial viability of the encapsulated cells, as well as for diffusion from the capsule. Although capillary invasion into the capsule walls helps provide nutrients and oxygen in proximity to the encapsulated cells, the process of angiogenesis for neovascularization typically evolves into a 4–7-day period *(49)*. Therefore, the encapsulated cells must endure an initial period of nutrient and oxygen deprivation when obtaining these essential factors only by diffusion. Moreover, because the only delivery mechanism of the desired cellular products from an encapsulated cell implant is via diffusion, any reaction around the capsule might diminish the diffusion of therapeutic products from the encapsulated cells. These studies clearly demonstrated biocompatibility within the host nervous system, suggesting that the bidirectional transport of low-molecular-weight solutes across the permselective membrane can be maintained in vivo. Many biocompatibility studies have been conducted with empty polymer macrocapsules, but the inclusion of such cells as bovine adrenal chromaffin cells or PC12 cells does appear to significantly impact the host reaction to the polymer device.

Notably, in an effort to maintain, or even further enhance, the biocompatibility for cell line–containing implants, or reduce protein adsorption that may negatively impact the ability to maintain adequate long-term diffusive characteristics, several postsynthesis modifications have been attempted. Poly(acrylonitrile-covinyl chloride) (PAN/VC) hollow fiber membranes, which were surface-modified by grafting PEO groups, exhibited improved biocompatibility in brain tissue over the unmodified PAN/VC controls *(50)*.

Similar observations were made with PEO-modified polyhydroxyethyl methacrylate-comethyl methacrylate membranes that were utilized extensively in cellular microencapsulation *(26)*.

Long-Term Product Secretion and Delivery

Before patients suffering from chronic CNS diseases can be routinely implanted with encapsulated cells, long-term survival of the encapsulated cells and continued release of the therapeutic molecule must be demonstrated. Although effective immunoisolation should result in long-term survival of encapsulated cells, surprisingly few studies have examined implant viability for more than a few months. Notable exceptions exist that provide compelling evidence about the potential long-term survival and release of molecules from the cells. Following encapsulation, PC12 cells have been maintained in vitro and in vivo for at least 6 mo, while maintaining a typical morphology and clustered arrangement along the lumen of the device *(30,44,51)*. The cells remain tyrosine hydroxylase (TH)–immunoreactive and mitotically active, with necrosis primarily in regions of high cell density. Electron microscopy confirms the presence of numerous mitochondria, polysomes, and electron-dense secretory vesicles distributed within the cytoplasm. Spontaneous and evoked release of dopamine can be detected from capsules maintained both in vitro and following explantation from the CNS. Both rodent microdialysis and positron emission tomography (PET) studies in primates have confirmed that encapsulated PC12 cells continue to produce levodopa (L-dopa)/dopamine *in situ (51,52)*.

Other cell types, including encapsulated bovine adrenal chromaffin cells, also survive for prolonged periods of time *(45,53)*. Intraventricular implants of encapsulated bovine chromaffin cell implants survived for nearly 1.5 yr and continued to produce catecholamines and met-enkephalin *(45)*. Polymer-encapsulated, genetically modified cells also survived and continued to secrete trophic factors, such as NGF, for 12–13.5 mo in rats *(54,55)*. The cells remained viable, and the NGF secreted from the encapsulated cells was 64% higher following removal from the rat lateral ventricles *(55)*. The NGF transgene copy number was equivalent to preimplant levels, indicating NGF gene stability. No deleterious effects from long-term NGF were detectable on body weight, mortality rate, motor/ambulatory function, or cognitive function, as assessed via the Morris water maze and delayed matching to position *(55)*. There was no evidence of NGF-induced hyperalgesia; however, tests of somatosensory thresholds did reveal effects related to NGF delivery. NGF from the encapsulated cells produced a marked hypertrophy of cholinergic neurons within the striatum and nucleus basalis, as well as a robust sprouting of cholinergic fibers within the frontal cortex and lateral septum proximal to the implant site. Although no deleterious behavioral effects were observed, the profound anatomical changes, and their relationship to functional alterations in normal and diseased brain, warrant additional study.

Host-Specific Effects on Output of Encapsulated Cells

Available data suggests substantial variability in the in vivo performance of encapsulated cells, even during the first few months. For example, there was a large range in dopamine and L-dopa output from explanted rodent-sized devices, from 0 to more than 50 pmol per device *(56)*. In fact, 15% of the devices from this study had no detectable output after only 4 mo in vivo. Similar variability in device performance has been observed in the majority of in vivo studies. In a primate study that produced therapeutic effects in two thirds of the 1-methyl-4-phenyl 1,2,3,6-tetrahydropyridine (MPTP) monkeys implanted with PC12 cells, all five devices implanted in one monkey had virtually 0 output after explantation, whereas all five devices implanted in another monkey had relatively high output, and all five devices in the third monkey had catecholamine output in the midrange *(44)*. These results indicate that some variability in device performance may be attributable to individual differences between hosts—a result consistent with that reported for NGF output from encapsulated BHK cells that were implanted into the lateral ventricle of rodents *(57)*. The exact mechanism for these individual differences remains undetermined but deserves significant attention.

Diffusion of Molecules From Polymer Devices

Tresco et al. *(51)* conducted a series of in vivo experiments to elucidate the relationship between diffusion of dopamine from encapsulated PC12 cells and behavioral recovery in dopamine-depleted rodents. As determined by microdialysis, dopamine was detectable by up to 200 μm from PC12 cell–loaded macrocapsules, in concentrations similar to those seen in unlesioned control striatum. In contrast, dopamine levels in the perfusate of animals that did not exhibit behavioral recovery were undetectable. Immunocytochemistry was used to estimate the diffusion of NGF from encapsulated cells implanted into the striatum of rats *(58)*. One month after implantation, the diffusion of NGF was estimated to be approx 1 mm. Ciliary neurotrophic factor (CNTF) has also been reported to be detectable in the cerebrospinal fluid (CSF) of amyotrophic lateral sclerosis patients who have received intrathecal implants of encapsulated CNTF-producing BHK cells *(59)*. However, CSF levels of NGF were not detectable in nonhuman primates that received intraventricular grafts of NGF-producing cells *(60)*. At best, these data suggest that diffusion of molecules from encapsulated cells is limited. Moreover, the majority of degenerative CNS diseases will likely not be treatable by delivering drugs from the ventricular space, given that diffusion of compounds in CNS tissue is generally severely limited when only governed by passive diffusion. Future work, particularly clinical studies, must consider this issue when determining the optimal numbers and spacing of polymer devices.

CURRENT THERAPY USING ENCAPSULATED CELLS

Table 2 summarizes the diseases currently being targeted by encapsulated cell therapy in different stages of investigation. This list is quite extensive; thus, the most relevant therapies, with the highest level of characterization, will be discussed.

PD

Effects of Catecholamine-Producing Cells

PD is an age-related neurodegenerative disorder characterized by hypokinesia, rigidity, and tremor secondary to the loss of dopaminergic neurons in the *pars compacta* of the substantia nigra. Replacing or increasing striatal dopamine levels with oral L-dopa significantly improves the motor deficits in the early stages of PD. Unfortunately, this pharmacological approach has limitations. Systemic administration results in drug distribution to extrastriatal dopamine receptors that may produce psychoses and vomiting. The therapeutic window of L-dopa's beneficial effects becomes progressively limited as the disease continues its degenerative course. Compared to continuous delivery, pulsatile delivery seems to be associated with more adverse effects *(61)*. Studies that use pumps to deliver dopamine continuously and directly to the striatum in both rodent and primate models of PD have reported improved motor function with few adverse effects *(62–64)*.

RODENT STUDIES

Given the promise of continuous local L-dopa delivery to the striatum, several studies have detailed the effects of implanting encapsulated dopamine-producing cells into the striatum in both rodent and primate models of PD. Encapsulated PC12 cells secrete high levels of catecholamines under both basal and evoked conditions. Rats with PC12 cell–loaded capsules implanted in a 6-hydroxydopamine (6-OHDA)–lesioned striatum exhibit fewer rotations after apomorphine administration than nonimplanted rats, which suggests that the devices are releasing catecholamines at levels sufficient to reduce the degree of synaptic supersensitivity that develops after

Table 2
Application of Cell Encapsulation

Disease/Model	Encapsulated cell/experimental paradigm	Results
	Hormonal and Whole-Organ Diseases	
Diabetes	Islets in rodents and dogs	Normoglycemia for 2 yr
Hypoparathyroidism	Parathyroid tissue in rats	Normocalcemia for 30 wk
Kidney failure	Orally delivered *Escherichia coli* bacteria to rats	Normalized urea metabolism
Growth hormone deficiency	Growth hormone–producing cells in dogs	Hormone secretion for 1 yr
	Single-Gene Diseases	
Hemophilia	Factor 9 cells in rats	Cell survival and secretion
Lysosomal storage disease	β-glucuronidase cells in mice	Behavioral normalization
	Age-Related/Neurodegenerative Diseases	
Age-related motor decline	Catecholamine and GDNF cells in rats	Improvement in motor function
ALS	CNTF-producing cells in mice	Protection of motor neurons
AD	NGF cells in rat and primates	Protection of cholinergic neurons, improved memory
HD	NGF and CNTF cells in rat and primates	Protection of neurons, improved behavior
PD	Catecholamine and GDNF cells in rat and primate brain	Improved behavior, protection of dopaminergic neurons
RP	Human cells secreting human CNTF into the vitreous	Rod preservation
Spinal cord damage	BDNF cells in rats	Outgrowth of neurites

Oncology		
Colon cancer	iNOS cells (tet-regulated system) in mice	Enhanced survival
Glioblastoma	Endostatin cells in mice and rats	Reduced tumor growth, enhanced survival
HER-2/neu-positive tumors	Interleukin-2 fused with anti HER-2/neu antibody in mice	Modest survival benefit
Leukemia	Hybridoma-producing antibodies to p15E in mice	Enhanced survival
Ovarian cancer	iNOS cells (tet-regulated system) in mice	Enhanced survival, cures
Other		
Acute and chronic pain	Chromaffin cells in rats	Reduced pain
Clinical trials		
ALS	CNTF cells intrathecally	Sustained delivery, no toxicity
Chronic pain	Chromaffin cells in subarachnoid space	Prolonged cell survival, no pain reduction in phase II trials
Diabetes	Human islets intraperitoneally	Insulin independence for 9 mo
HD	CNTF cells into ventricles	Delivery for 6 mo
Hypoparathyroidism	Parathyroid tissue	Successful in two patients
Pancreatic cancer	CYP2B1 cells in tumor vessel	Local tumor growth controlled, well tolerated
RP	CNTF cells in eye	Ongoing

dopamine-depleting lesions. Within 2 wk following implantation, the number of apomorphine-induced rotations was decreased by approx 40–50% and remained at that level for up to 6 mo *(51)*. Reductions in rotational behavior do not occur following implants of empty polymer devices and are only evident in rats implanted in the denervated striatum, not in rats with devices implanted into the lateral ventricles *(56,65)*. Behavioral effects persisted only as long as the devices remained in the striatum.

Measures of drug-induced rotations provide a convenient method for assessing potential efficacy, and significant information has been acquired using this initial preclinical screen. However, relying exclusively on changes in drug-induced rotations has limited clinical relevance and specificity. Accordingly, the effects of PC12 cells on a battery of nondrug-induced behaviors have been examined in 6-OHDA-lesioned rats. Neurological testing revealed behavioral deficits in the affected forelimb that were significantly attenuated by oral Sinemet *(56,65)*. Considering any transplantation procedure will be utilized as an adjunct to L-dopa administration, these data provided the opportunity to investigate the effects of PC12 cells on both relevant behavioral measures and the therapeutic window of oral Sinemet. Rats with severe unilateral dopamine depletions received striatal implants of encapsulated PC12 cells and were evaluated on a series of behavioral tests, with a range of oral doses of Sinemet. Delivery of L-dopa and dopamine from the encapsulated PC12 cells to the denervated striatum attenuated parkinsonian symptoms. The magnitude of the therapeutic effect produced by continuous site-specific delivery of catecholamines was greater than the effect produced by acute and systemic oral Sinemet. The beneficial effects of oral Sinemet and striatal implants of PC12 cells were additive, but there were no adverse effects related to the implantation of the PC12 cells, and these devices did not increase the adverse effects related to oral Sinemet *(67)*. Therefore, striatal implants of catecholamine-producing devices have direct therapeutic effects and, perhaps more importantly, may widen the therapeutic window of oral Sinemet.

In addition to motor deficits produced experimentally by depleting nigrostriatal dopamine systems in young rodents, motor deficits were also observed with increased age. Age-related deficits in motor functions include deficits in balance, coordinated movement, and generalized locomotion. Enhancement of striatal dopamine function in aged animals by induction of dopamine receptor upregulation, or administration of dopaminergic agonists, can reduce age-related motor deficits. As the motor deficits observed in aged rodents appeared to be mediated partially by alterations in striatal dopamine systems, studies were conducted to evaluate the potential efficacy of striatal implants of polymer-encapsulated PC12 cells on age-related motor dysfunction in rats *(68)*. In these studies, aged rats were significantly hypoactive relative to young animals. Moreover, compared to young rats, the aged rats: (1) remained suspended from a horizontal wire for less time; (2) were unable to descend a wooden pole covered with wire mesh in a coordinated manner; (3) fell more rapidly from a rotating rod; and (4) were unable to maintain their balance on a series of wooden beams, with either a square or rounded top of varying widths. Following baseline testing, aged rats received bilateral striatal implants of empty capsules or PC12 cell-loaded capsules. After 3 wk, the aged rats that received PC12 cells showed a robust improvement in performance on the rotarod task and balance on the wooden beams. Recovery was not observed regarding any other motor task in these animals based on any behavioral measure in those animals that received empty capsules.

PRIMATE STUDIES

The studies conducted in the 6-OHDA rat model of PD generally support the clinical utility of encapsulated catecholamine-producing cells. Nonhuman primates are more relevant models owing to their size and complexity of their nervous system, which more closely resembles humans. In terms of tissue volume, diffusion through a rat brain is easier to accomplish than adequate diffusion through the much larger human brain. The nonhuman primate model allows the assessment of therapeutic potential at this level in a way that cannot be approximated in rodents. The potential efficacy of encapsulated PC12 cells has been evaluated in a unilateral MPTP-lesioned primate model. Cynomolgus monkeys trained to perform a task that involved picking food from small food wells were unilaterally lesioned with an injection of MPTP in the right carotid artery. The resulting MPTP-induced lesion produced a significant and stable impairment in the animals' ability to use the contralateral limb to retrieve food rewards from the wells. The times required to empty the wells were measured for 3-mo postlesion, and the monkeys were then implanted with a U-shaped device that was immediately filled with a suspension of PC12 cells *(69)*. Following implantation of the cells, manual dexterity improved, and the time required for the monkeys to empty the tray using the impaired hand gradually decreased. Although the PC12 cells attenuated the parkinsonian deficit in this task, some tremor remained, and the animals did not recover to prelesion levels. Prior to cell implantation, the monkeys' left arms were essentially immobile. After implantation, the monkeys could consistently move both their arms and use their fingers. When the cells were flushed out of the device, performance declined to preimplant levels. Together, these data indicated that encapsulated cells survived, were functional, and promoted behavioral recovery, even in primate models of PD.

Similar results were obtained in a study by Kordower et al. *(44)*, where four cynomolgus primates were trained on a skilled reaching task similar to that described above, then rendered hemiparkinsonian with an intracarotid injection of MPTP. Three animals received implants of encapsulated PC12 cells into both the caudate and putamen, and one animal, which received implants of empty capsules, served as a control. After a transient improvement, limb use in the control monkey dissipated and returned to post-MPTP levels of disability. Two of the three PC12 cell implanted monkeys recovered on the task to near-normal levels for up to 6.5-mo posttransplantation. Capsules retrieved from the monkeys that recovered limb function contained abundant viable PC12 cells that continued to release L-dopa and dopamine. In contrast, capsules retrieved from the monkey that did not recover contained few viable PC12 cells. Neuroanatomical and neurochemical evaluation of the implanted striatum failed to reveal any host-derived sprouting of catecholaminergic or indolaminergic fibers, which further suggested that the observed behavioral recovery was because of secretion of catecholamines from the encapsulated PC12 cells.

Effects of Neurotrophic Factor–Producing Cells

Several recent studies have investigated the ability of encapsulated neurotrophic factor–secreting cells to exert neurotrophic effects in rodent models of PD *(70–72)*. In an initial study *(71)*, encapsulated cells releasing approx 5 ng of glial-derived neurotrophic factor (GDNF) per day were implanted immediately rostral to the substantia nigra. The medial forebrain bundle was transected 1 wk later, and the ability of encapsulated GDNF-producing cells to minimize the behavioral effects of the lesion and prevent the degeneration of dopaminergic neurons was determined. GDNF treatment significantly reduced the number of amphetamine-induced rotations in lesioned animals. GDNF treatment also attenuated the loss of neurons in the substantia nigra but had no effect on dopamine within the denervated striatum. Using the same model system, neurturin-producing cells—a homolog of GDNF—was investigated for its neurotrophic activity *(72)*. Neurturin-treated animals had substantially more TH-positive neurons in the substantia nigra (51% vs. 16% in controls) but failed to show any behavioral improvement, as measured by rotational behavior. Together, these data suggest that encapsulated cells may have a role in neurotrophic therapy for PD. However, additional studies in animal models are required to determine the relationship between the anatomical and behavioral consequences of cell-based delivery.

ENCAPSULATED NEUROTROPHIC FACTOR–PRODUCING CELLS
AND SURVIVAL OF UNENCAPSULATED COGRAFTS

The transplantation of encapsulated genetically modified cells also represents a potential method to deliver trophic factors to the brain to support the survival of cografted cells. In a series of studies, BHK cells were genetically modified to secrete NGF. Following polymer encapsulation, these cells were implanted into the left lateral ventricle or the left striatum approx 1.5 mm away from cografted unencapsulated adrenal medullary chromaffin cells in hemiparkinsonian rats *(73)*. Although the animals receiving adrenal medulla alone or adrenal medulla with intraventricular NGF-secreting cell grafting did not show recovery of apomorphine-induced rotational behavior, the animals receiving adrenal medulla with intrastriatal NGF-secreting cell implants showed a significant recovery of rotational behavior 2 and 4 wk after transplantation. Histological analysis revealed that intraventricular NGF increased the number of surviving chromaffin cells five to six times above that seen in animals receiving adrenal medulla alone. Even more impressively, intrastriatal NGF-secreting cells increased the quantity of surviving chromaffin cells by more than 20 times that in animals receiving adrenal medullary cells alone. Additional studies determined that the beneficial effects of NGF-producing cells were evident for as long as 12-mo postgrafting (i.e., the longest time point examined) and were independent of age of the chromaffin cell donor *(54,74)*. These results indicate the potential use of intrastriatal implantation of encapsulated NGF-secreting cells for augmenting the survival of cografted chromaffin cells, as well as promoting the functional recovery of hemiparkinsonian rats.

Encapsulated NGF–Producing Cells in Animal Models of AD

AD affects approx 5% of people over age 65 and is the most prevalent form of adult-onset dementia. The most prominent feature of AD is a progressive deterioration of cognitive and mnemonic ability, which is at least partially related to the degeneration of basal forebrain cholinergic neurons. At present, treatments do not slow or prevent cholinergic neuron loss or the associated memory deficits. Several converging lines of evidence indicate that NGF has potent target-derived trophic and tropic effects upon cholinergic basal forebrain neurons *(75,76)*.

Rodent Studies

Although no model faithfully recapitulates the complex etiology and time-dependent loss of cholinergic neurons seen in AD patients, model systems have been developed to determine if NGF prevents the death of damaged cholinergic neurons following trauma. Initial studies determined whether encapsulated BHK cells modified to produce NGF could prevent cholinergic neuron loss following aspiration of the fimbria/fornix *(77)*. Rats received lesions of the fimbria/fornix, followed by intraventricular implants of either NGF-producing or control (nontransfected) cells. Control-implanted animals had an extensive loss (88%) of ChAT-positive cholinergic neurons ipsilateral to the lesion that was prevented by NGF cell implants (14% loss).

One cardinal behavioral symptom of AD is a progressive loss of cognitive ability. Just as no animal model reliably mimics the complex etiology and pathophysiology of AD, comparable behavioral abnormalities are difficult to reproduce in animal models. However, the aged rodent shows a progressive degeneration of basal forebrain cholinergic neurons, together with marked cognitive impairments that are partly reversible by administering NGF. Lindner et al. *(57)* trained 3-, 18-, and 24-mo-old rats on a spatial learning task in a Morris water maze. Cognitive function, as measured in this task, declined with age. Following training, animals received bilateral intraventricular implants of encapsulated NGF or control cells. The 18- and 24-month old animals receiving NGF cells showed a significant improvement in cognitive function. No improvements or deleterious effects were observed in the young nonimpaired animals. There was no evidence that the NGF cells produced changes in mortality, body weights, somatosensory thresholds, potential hyperalgesia, or activity levels, suggesting that the NGF levels produced were neither toxic nor harmful to the aged rats. Evidence of age-related atrophy of cholinergic neurons was observed in the striatum, medial septum, nucleus basalis, and vertical limb of the diagonal band. These anatomical changes were most severe in animals with the greatest cognitive impairments, suggesting a link between the two pathological processes. Anatomically, the NGF released from the encapsulated cells increased the size of the atrophied basal forebrain and striatal cholinergic neurons to the size of the neurons in the young healthy rats.

Primate Studies

Results similar to those obtained in rodents were obtained in nonhuman primates *(60)*, an essential prerequisite to human clinical trials. In these studies, cynomolgus primates received transections of the fornix, followed by placement of encapsulated NGF or control cells into the lateral ventricle. In the control animals, the number of cholinergic neurons was reduced significantly in the medial septum and vertical limb of the diagonal band of Broca. Again, loss of cholinergic neurons was prevented by implants of NGF-secreting cells. Also, cholinergic neurons within the medial septum of NGF-treated animals appeared larger, more intensely labeled, and elaborated more extensive proximal dendrites than those displayed by BHK control animals.

In addition to the effects on cell viability, NGF implants induced a robust sprouting of cholinergic fibers proximal to the implant site *(60)*. All monkeys receiving NGF implants displayed dense collections of NGF receptor–immunoreactive fibers throughout the dorsoventral extent of the lateral septum. This effect was unilateral, as the contralateral side displayed only a few cholinergic fibers in a manner similar to that seen in control-implanted monkeys. The cholinergic nature of this sprouting was confirmed by an identical pattern of fibers that were ChAT-immunoreactive and AChE-positive. These fibers ramified against the ependymal lining of the lateral ventricle, adjacent to the transplant site, and were particularly prominent within the dorsolateral quadrant of the septum, corresponding to the normal course of the fornix. The cell sparing and sprouting results have been replicated in a group of aged nonhuman primates *(78)*.

Polymer-Encapsulated Cells to Deliver Neurotrophic Factors in Animal Models of HD

HD is an inherited, progressive neurological disorder characterized by a severe degeneration of basal ganglia neurons, particularly the intrinsic neurons of the striatum. A progressive dementia accompanies the pathological changes, coupled with uncontrollable movements and abnormal postures. From the time of onset, an intractable course of mental deterioration and progressive motor abnormalities begins, with death usually occurring within 15–17 yr. Overall, the prevalence rate of HD in the United States is approx 50 per 1,000,000. At present, no treatment effectively addresses the behavioral symptoms or slows the inexorable neural degeneration in HD.

Intrastriatal injections of excitotoxins, such as quinolinic acid (QA), have become a useful model of HD and can evaluate novel therapeutic strategies aimed at preventing, attenuating, or reversing neuroanatomical and behavioral changes associated with HD *(79–82)*. The use of trophic factors in a neural protection strategy may be particularly relevant for the treatment of HD. Unlike other neurodegenerative diseases, genetic screening can identify individuals at risk, providing a unique opportunity to design treatment strategies to intervene prior to the onset of striatal degeneration.

Rodent Studies

Infusions of such trophic factors as NGF or implants of cells genetically modified to secrete NGF have been proven effective in preventing the neuropathological sequelae that results from intrastriatal injections of excitotoxins, including QA *(83–85)*. We *(87,88)* examined the ability of encap-

sulated trophic factor–secreting cells to affect central striatal neurons in a
series of defined animal models of HD. In these experiments, rats received
implants of NGF- or CNTF-producing cells into the lateral ventricles.
One week later, the same animals received unilateral injections of QA
(225 nmol) or the saline vehicle into the ipsilateral striatum. An analysis of
Nissl-stained sections demonstrated that the size of the lesion was significantly reduced in those animals receiving NGF and CNTF cells, compared
to those animals receiving control implants. Moreover, CNTF cells attenuated the extent of host neural damage produced by QA, as assessed by
sparing specific populations of striatal cells, including cholinergic, diapho-
rase-positive, and GABAergic neurons. Neurochemical analyses have confirmed the protection of multiple striatal cell populations using this strategy.
Importantly, behavioral studies offer additional and compelling evidence of
neuronal protection that can be produced in animal models of HD. Trophic
factor–secreting cells have provided extensive behavioral protection, as
measured by tests that examine both gross and subtle movement abnormalities. Moreover, these same animals show improved performance on learning and memory tasks, indicating the anatomical protection afforded by
trophic factors in this model is paralleled by a robust and relevant behavioral protection *(89)*.

Primate Studies

The ability of cellularly delivered trophic factors to preserve neurons
within the striatum in a rodent model of HD led to similar studies in nonhuman primates *(90)*. Polymer capsules containing CNTF-producing cells
were grafted into the striatum of Rhesus monkeys. After 1 wk, a QA injection was placed into the putamen and caudate proximal to the capsule
implants. As seen in the rodent studies, the volume of striatal damage was
decreased, and both GABAergic and cholinergic neurons destined to degen-
erate were spared in CNTF-grafted animals. Although all animals had significant lesions, there was a three- and sevenfold increase in GABAergic
neurons in the caudate and putamen, respectively, in CNTF-grafted animals,
compared to controls. Similarly, there was a 2.5- and fourfold increase in
cholinergic neurons in the caudate and putamen, respectively, in CNTF-
grafted animals.

The ability to preserve GABAergic neurons in animal models of HD is an
important, although not entirely sufficient, step to develop a useful therapeutic. If the perikarya are preserved without sustaining their innervation,
then the experimental therapeutic strategy under investigation is not likely
to yield notable value. The striatum is a central station in a series of loop
circuits that receive inputs from all the neocortex, project to a number of

subcortical sites, and then return information flow to the cerebral cortex. One critical part of this circuitry are the GABAergic projections to the globus pallidus and substantia nigra pars reticulata—components of the direct and indirect basal ganglia loop circuits. The integrity of this circuit can be evaluated immunocytochemically using an antibody that recognizes GABAergic terminals (DARPP-32) to determine if the preservation of GABAergic somata within the striatum also results in preservation of the axons of these neurons to critical extrastriatal sites. Using quantitative morphological assessment of DARPP-32 optical density, monkeys receiving QA lesions had major reductions in DARPP-32 immunoreactivity within the globus pallidus and substantia nigra. The lesion-induced decrease in GABAergic innervation for both these regions was prevented in CNTF-grafted monkeys, demonstrating that this treatment strategy protected GABAergic neurons and sustained their normal projection systems *(90)*.

The intrinsic striatal cytoarchitecture can be preserved in monkeys by CNTF grafts; once exposed to these grafts, the cells apparently maintain their projections. But are the afferents to the striatum, specifically from the cerebral cortex, also influenced by these grafts? This may be particularly critical if some of the more devastating nonmotor symptoms seen in HD result from cortical changes secondary to striatal degeneration. Because layer V neurons from the motor cortex send a dense projection to the postcommissural putamen, a region that was severely impacted by the QA lesion, the effects of QA lesions and CNTF implants on the number and size of cortical neurons in this region were examined. Although the QA lesion did not impact the amount of neurons in this cortical area, layer V neurons were significantly reduced in cross-sectional area on the side ipsilateral to the lesion in control-grafted monkeys. This atrophy of cortical neurons was virtually completely reversed by CNTF grafts *(90)*.

A recent set of studies using CNTF-producing cells in 3NP-treated monkeys have replicated and extended these results *(91)*. Following 10 wk of 3NP treatment, monkeys displayed pronounced chorea and severe deficits in frontal lobe cognitive performance, as determined by the object retrieval detour test. Following implantation of CNTF-producing cells, a progressive and significant recovery of motor and cognitive recovery occurred. Histological analysis demonstrated that CNTF was neuroprotective and spared NeuN and calbindin-positive cells in the caudate and putamen.

The sparing of striatal neurons and maintenance of intrinsic circuitry is impressive; however, the magnitude of the effect is less than that seen in rodents. In primates, robust protection is limited to the area of the capsules, and the total area of the lesion remains extensive, likely because the diffusion of CNTF from the capsule is not sufficient to protect more distant striatal regions undergoing degeneration. This concept is supported by a recent experiment examining the effects of intraventricular grafts of encapsulated CNTF grafts in the nonhuman primate model of HD *(92)*. Contrasting when the capsules were placed directly within brain parenchyma, intraventricular placements failed to engender neuroprotection for any striatal cell types. The complete lack of neuroprotection provided by intraventricular implants in primates should be considered more carefully in the current clinical trials conducted in which encapsulated cells are placed into the lateral ventricles of HD patients *(93,94)*. If human trials yield clinically relevant positive effects, the vehicle of CNTF delivery utilized in these studies needs to be improved. Whether this entails grafting more capsules, enhancing the CNTF delivery from the cells by changing the vector system or cell type employed, or changing the characteristics of the polymer membrane, remains to be determined.

Encapsulated Cells for Treating Brain Tumors

Malignant brain tumors represent 2–3% of all adult neoplasms, occurring in up to nine in 100,000 individuals. Gliomas account for approximately half of all primary brain tumors. Despite the continued refinement of neuroimaging techniques, progress in microsurgery, and evolving chemotherapeutic drugs, the prognosis for glioma patients is poor, with a median survival less than 1 yr. Surgical debulking of the growing tumor mass is the primary treatment modality. Although surgery provides short-term improvement for subsequent treatments, it is ultimately ineffective because of the infiltrative nature of glioma, resulting in tumor recurrence within 2 cm of the resection margin.

If a vehicle delivering therapeutic molecules directly to the region of tumor recurrence could be developed, then the prognosis for glioma patients might be significantly improved. Cell encapsulation has recently been suggested as one method to achieve this goal *(95–102)*. The majority of the data obtained thus far utilizes microencapsulation techniques, but the data sufficiently intriguing are presented here. Particularly relevant to glioma, recent studies demonstrate that experimental gliomas could be successfully treated with endostatin-producing alginate encapsulated cells. In one study, BT4C glioma cells and endostatin-producing capsules were implanted into the brains of syngeneic rats *(99)*. The encapsulated cells survived well and continued to secrete endostatin for up to 4 mo in vivo. Survival was increased by 84%, compared to control animals. Histological assessment of the transplant revealed that the endostatin-producing cells produced cellular apoptosis, as well as the formation of large necrotic areas within the treated tumors. The treatment was otherwise reportedly well tolerated, with no serious side effects. These data were supported by a similar study in nude mice. Endostatin-producing BHK cells were encapsulated into alginate microspheres and injected 10 d prior to a subcutaneous injection of U87 glioma cells. At 21-d posttreatment, growth of the gliobastoma xenografts was notably reduced, with markedly less neovascularization surrounding the tumor *(96)*. Similar results have been achieved using encapsulated kidney 293 cells that are modified to produce endostatin *(103)*. Intravital video microscopy revealed reductions in the density, diameter, and function of tumor-associated microvessels.

Another approach to treat glioma uses encapsulated cells to enable gene transfer in vivo. Martinet and colleagues *(98)* encapsulated mouse psi20VIK-packaging cells into alginate microspheres. The capsules were then stereotaxically injected into established C6 glioblastomas, and the animals were treated with gancyclovir. After 14 d, the tumors were harvested, and 3–5% of the tumor cells had been transduced, resulting in a 45% reduction in tumor volume.

Together, these data signify a young, but encouraging, application of encapsulated cell therapy. The ability of cell-based therapies to provide continuous local delivery of a wide range of therapeutic molecules will likely lead to further refinements in this area. Still, as recently pointed out *(101)*, optimization of capsule biostability, tissue biocompatibility, cell choice and viability, dose-release, and biodistribution of therapeutic molecules represent unresolved areas of preclinical research.

Encapsulated Cells for Retinitis Pigmentosa

Retitinis pigmentosa (RP)—an inherited degenerative disease of the retina—affects roughly one out of 4000 individuals worldwide and is the sixth leading cause of blindness *(104)*. The disease typically progresses slowly, from mild degeneration of the rod photoreceptors in the periphery to a more pronounced loss of vision involving the cones in the central retina. The end result is complete blindness that occurs within months or years, depending on the etiology of the disease.

RP is a disease with an extremely complex genetic profile, and over 100 rp-inducing mutations have been identified to date *(105)*. Although there has been progress in this regard, the influence of genetic medicine on the development of therapeutics for RP continues to be minimal. Current clinical strategies focus on optimization of the light path (lens replacements) and treatment of secondary effects of the disease. Work in the field of protein therapeutics, specifically neurotrophic factors, has led to the use of CNTF as a potential neuroprotective mediator of RP. For example, in the *rd* rodent model of retinal degeneration, it was shown that gene transfer of CNTF retarded the degeneration of photoreceptors *(106,107)*. Purified CNTF also slows photoreceptor degeneration in the *rd/rd* and *nr/nr* murine models of the disease *(108)*. Because of the relatively short half-life of the recombinant protein, along with the immunoprivileged environment of the eye, encapsulated cell technology is an extremely appealing mode of delivery for CNTF and other potential protein therapeutics to the eye.

More recently, retinal epithelial cells have been transfected with a plasmid containing the CNTF gene. CNTF is released in vivo at a rate of approx 1.5 ng/d. Initial efficacy studies with the CNTF-producing cell line were carried out in s334ter-3 transgenic rats *(109)*. Naked cells were injected into the vitreous of rat eyes, and after 20 d, eyes were enucleated and processed for histology. Untreated eyes or eyes injected with the nontransfected parental cell line showed severe progression of retinal degeneration (about one row of the outer nuclear layer [ONL] remaining), whereas eyes treated with the CNTF-secreting cell line showed five to six times thicker ONL throughout the retina.

The canine *rcd1* model for retinal degeneration *(110,111)* was used to evaluate the ECT device loaded with the CNTF-secreting cell line. In this experiment, various amounts and combinations of parental and secreting cells were used to achieve five different doses between less than 100 pg/day to 15 ng/day. Again, as with the work in rodents, neuroprotection was shown to occur; in this case, it was dose-dependent. Untreated eyes showed an average of three layers of the ONL, whereas treated eyes showed between three and six layers. In all cases, the distribution of retinal preservation was homogenous throughout the retina.

INITIAL CLINICAL TRIALS

Amyotropic Lateral Sclerosis (ALS)

Neuromuscular disorders, such as ALS, are marked by a progressive degeneration of spinal motor neurons. Different families of neurotrophic factors demonstrate therapeutic potential in vitro and in animal models of motor neuron disease *(112–119)*. The cytokine CNTF has neuroprotective effects for motor neurons in *wobbler* mice *(116)* and homozygote pmn (progressive motorneuropathy) mice *(117,118)*. The delivery of CNTF to motor neurons by peripheral administration proved difficult owing to severe systemic side effects, short half-life of CNTF, and the inability of CNTF to cross the BBB *(120–122)*.

Continuous intrathecal delivery of CNTF proximal to the nerve roots in the spinal cord is a practical alternative that could result in fewer side effects and better efficacy of CNTF in ALS patients. After safety, toxicology, and preclinical evaluation *(123)*, a clinical trial to establish safety has been performed in ALS patients using polymer-encapsulated cells genetically modified to secrete CNTF *(124)*. A total of six ALS patients with early stage disease, indicated by a forced vital capacity greater than 75% with no other major illness or previous treatment with any investigational drugs for ALS, were included. These patients were baseline-tested for Tufts Quantitative Neuromuscular Evaluation, the Norris scale, blood levels of acute reactive protein, and CNTF levels in the serum and CSF. BHK cells were encapsulated into 5-cm long by 0.6-mm diameter hollow membranes and implanted into the lumbar intrathecal space. The device included a silicone tether that was sutured to the lumbodorsal fascia, and the skin was closed over the device. CNTF concentrations in the CSF were not detectable prior to implantation but were found in all six patients at 3–4-mo postimplantation. All six explanted devices had viable cells and CNTF secretion of approx 0.2–0.9 μg/day. No CNTF was detected in the serum. More recently, a phase I/II clinical trial was initiated in 12 ALS patients using the same approach. Again, CNTF was detectable for several weeks in the CSF of nine out of 12 patients. Concurrent in vitro studies using CSF samples from these same patients revealed only a very weak antigenic immune response, with bovine fetuin as the main antigenic component *(125)*.

Chronic Pain

Numerous studies with rodent models of acute and chronic pain have suggested that adrenal chromaffin cells implanted into the intrathecal space, and in the periaqueductal gray, reliably produce significant analgesic effects *(126,127)*. Although the majority of these studies have used unencapsulated cells, recent studies suggested that encapsulated cell implants also produce analgesia in rats. The analgesic effects of adrenal chromaffin cells in the rodent model have provided the rationale to pursue clinical trials in patients with chronic pain. Small open-label trials demonstrated that the implantation procedure was minimally invasive and well tolerated *(128,129)*. Neurochemical and histological studies determined that the encapsulated cells survived and were biochemically functional for up to 1 yr. Because reductions in morphine intake were noted following implantation (suggesting efficacy), larger scale, randomized studies were initiated in a collaborative study between CytoTherapeutics and Astra Pharmaceuticals. Unfortunately, the trials were recently halted because the efficacy achieved did not reach a level high enough to warrant further study *(130)*.

Interestingly, several recent reports indicate encapsulated adrenal chro-maffin cell implants may not produce efficacy, as originally suggested *(131–133)*. Extensive studies in acute and chronic rodent pain models have failed to find any evidence of analgesia. This lack of effect occurred under condi-tions that were apparently designed to reproduce previous testing proce-dures exactly that did demonstrate efficacy. Among the variables examined were the location of implant (intrathecal vs intraventricular), a wide range in cell preparation techniques, and an exhaustive battery of acute and chronic pain tests, with and without nicotine stimulation. Importantly, the authors reported that systemic administration of morphine produced significant analgesia when tested in parallel in the same models. Subtle testing differ-ences cannot be eliminated as contributing factors in the differences between these recent and previous studies, but together with the only well-controlled clinical trial conducted to date, at the least, it appears that adrenal chromaf-fin cells do not produce analgesic effects as consistently as previous reported.

HD

Recently, clinical trials were initiated to determine the safety and toler-ability of CNTF-producing cells implanted into the lateral ventricle of HD patients *(93)*. Although the case for clinical evaluation is compelling, sev-eral issues are apparent with the design of ongoing clinical trials. The spar-ing of striatal neurons and maintenance of intrinsic circuitry in monkeys is impressive; yet, the magnitude of the effect is less than that seen in rodents. Robust protection is largely limited to the area of the capsules. However, the area of the lesion remains extensive, and it is likely that diffusion of CNTF from the capsule may not be sufficient to protect more distant striatal regions undergoing degeneration. This concept is supported by a recent experiment that examined the effects of intraventricular grafts of encapsulated CNTF grafts in the QA monkey model of HD. In contrast to when the capsules were placed directly within brain parenchyma, intraventricular placements failed to engender neuroprotection for any striatal cell types, again suggest-ing that diffusion is a key factor in the efficacy of this experimental thera-peutic strategy *(92)*. The complete lack of neuroprotection provided by intraventricular implants in monkeys should be considered more carefully in ongoing clinical trials. If human trials are to yield clinically relevant posi-tive effects, the means of CNTF delivery utilized in these studies needs to be improved. Whether this entails changing the site of implantation from the ventricle to the parenchyma, grafting more capsules, enhancing the CNTF delivery from the cells by changing the vector system or cell type employed, or changing the characteristics of the polymer membrane, remains to be determined.

CONCLUSIONS AND FUTURE DIRECTIONS

There is considerable promise of encapsulated cell therapy to treat a wide range of CNS disorders. Still, a number of research avenues exist that are incompletely explored and deserve attention prior to wide-scale clinical use of this technology. In some preclinical studies, the extent of diffusion from the implants appears to limit the therapeutic effectiveness of the encapsulated cells *(51,58)*. Given the size of the human brain relative to the rodent and nonhuman primate brain, the potential problems related to limited tissue diffusion should be examined empirically. Studies in larger animals should be conducted to assess the optimal spacing and distribution of multiple implants and information regarding the relative risks of repeated tissue penetrations, tissue damage, and potential infection.

Encapsulation allows the use of cells from human and animal sources with and without genetic modification. In theory, the capsule should isolate the cells from the surrounding tissue. Still, if a capsule ruptures during implantation or retrieval, a deleterious host immunological response could be induced. Although the host immune system should reject any released cells following capsule damage, the potential for tumorous growth remains. Alterations in the ability of the host immune system to reject cells following damage to implants could also change upon long-term residence of the cells within the host. To date, no studies have systematically evaluated these risks, particularly regarding the long-term effects of encapsulated cell implants. Again, large animal studies using intact and intentionally damaged devices would provide a useful starting point for evaluating these issues. These studies could use normal and immunosuppressed animals to evaluate potential tumorigenicity and changes in the host immune system over short and long periods of time.

Regulation of dosage is another area that deserves attention. In its most basic iteration, varying the numbers of cells within an implant, implant size, or the use of multiple implants may permit a range of doses to be delivered. Some long-term cell survival studies have been conducted, but they tend to utilize only CSF-filled spaces and have not systematically examined cell survival and output of the desired molecule over the long term. Instead, studies have provided a "snapshot" of survival and output at a single time point. Large, long-term, and well-controlled studies need to be conducted to examine the relationship between such variables as time, cell survival, gene expression (when modified cells are used), neurochemical output, initial numbers of cell encapsulated, and type of semipermeable membrane and ECM used for encapsulation. Obviously, these studies are time-consuming and expensive. But without them, the conditions optimal for successful cell encapsulation will remain speculative. It should be pointed out that some efforts are continuing in this area, and a recent study raised the interesting possibility that dose control for dividing cells could be accomplished with the use of cell-containing microcarriers in nonmitotic hydrogels *(80)*.

Another area that has attracted little attention concerns the variability in the in vivo performance of encapsulated cells and the possible role that the host tissue environment has in this variability. As discussed earlier, it appears that at least some variability in device performance is attributable to differences between hosts. The mechanism(s) underlying these individual differences remain undetermined; however, several potential candidates exist, including variations in the general health of animals between animal differences in immune function and undetected microbreaches in the polymer membrane prior to or during implantation. The notion that the viability of grafted cells may depend partly on host-related variability in the CNS environment has only been implicated for encapsulated cells to date *(56)*. However, this emerging concept might also prove to be relevant for all CNS transplantation approaches that are cellular-based. Indeed, the entire field of neural transplantation might benefit from this new perspective uncovered using encapsulated cells.

Finally, few clinical studies have been conducted thus far. Several small safety studies have been completed, but only one large controlled clinical study has been performed using encapsulation technology. This study evaluated the use of encapsulated adrenal chromaffin cells for the treatment of pain but failed to reveal analgesia sufficient enough to continue the trials. As discussed in a previous section, the selection of pain as an initial indication for detailed study might have been an unfortunate choice, given that recent preclinical data using encapsulated chromaffin cells is mixed at best. The only other clinical targets under investigation are ALS and HD, which are apparently modest efforts. Until larger controlled clinical trials are performed, the potential of this technology will not be fully realized.

In conclusion, it appears that the implantation of encapsulated cells may provide an effective means of alleviating the symptoms of numerous human conditions/diseases. One particularly appealing avenue of research continues to be the application of trophic factors to minimize or halt the progression of neural degeneration or to promote regeneration of damaged central nerves. However, caution must be taken when considering any novel therapy for treating brain disorders, and the wide-scale use of polymer neural implants should be an option only after rigorous scientific experimentation in animal models and their demonstrated efficacy and safety in human clinical trials.

REFERENCES

1. Cervos-Navarro, J., Kannuki, S., and Nakagawa, Y. (1998) Blood-brain barrier (BBB). Review from morphological aspect. *Histol. Histopathol.* **3,** 203–213.
2. Rubin, L. L. and Staddon, J. M. (1999) The cell biology of the blood-brain barrier. *Ann. Rev. Neurosci.* **22,** 11–28.
3. Friden, P. M., Walus, L. R., Watson, P., et al. (1993) Blood-brain barrier penetration and in vivo activity of an NGF conjugate. *Science* **259,** 373–377.
4. Friden, P. M. (1994) Receptor-mediated transport of therapeutics across the blood-brain barrier. *Neurosurgery* **35,** 294–298.
5. Jiao, S., Miller, P. J., and Lapchak, P. A. (1996) Enhanced delivery of [^{125}I] glial cell line-derived neurotrophic factor to the rat CNS following osmotic blood-brain barrier modification. *Neurosci. Lett.* **220,** 187–190.
6. Kroll, R. A. and Neuwelt, E. A. (1998) Outwitting the blood-brain barrier for therapeutic purposes: osmotic opening and other means. *Neurosurgery* **42,** 1083–1099.
7. Kordower, J. H., Mufson, E. J., Granholm, A. C., et al. (1993) Delivery of trophic factors to the primate brain. *Exp. Neurol.* **124,** 21–30.
8. Mufson, E. J., Kroin, J. S., Sendera, T. J., and Sobreviela, T. (1999) Distribution and retrograde transport of trophic factors in the central nervous system: functional implications for the treatment of neurodegenerative diseases. *Prog. Neurobiol.* **57,** 451–484.
9. Riddle, D. R., Katz, L. C., and Lo, D. C. (1997) Focal delivery of neurotrophins into the central nervous system using fluorescent latex microspheres. *Biotechniques* **23,** 928–937.
10. Olson, L., Backlund, E. O., Ebendal, T., et al. (1991) Intraputaminal infusion of nerve growth factor to support adrenal medullary autografts in Parkinson's disease. One-year follow-up of first clinical trial. *Arch. Neurol.* **48,** 373–381.
11. Vahlsing, H. L., Varon, S., Hagg, T., et al. (1989) An improved device for continuous intraventricular infusions prevents the introduction of pump-derived toxins and increases the effectiveness of NGF treatments. *Exp. Neurol.* **105,** 233–243.
12. Hoffman, D., Wahlberg, L., and Aebischer, P. (1990) NGF released from a polymer matrix prevents loss of ChAT expression in basal forebrain neurons following a fimbria-fornix lesion. *Exp. Neurol.* **110,** 39–44.
13. Langer, R. and Moses, M. (1991) Biocompatible controlled release polymers for delivery of polypeptides and growth factors. *J. Cell Biochem.* **45,** 340–345.
14. Winn, S. R., Wahlberg, L., Tresco, P. A., and Aebischer, P. (1989) An encapsulated dopamine-releasing polymer alleviates experimental Parkinsonism in rats. *Exp. Neurol.* **105,** 244–250.
15. Dunnett, S. B. and Bjorklund, A., eds. (1994) *Functional Neural Transplantation,* Raven, New York.
16. Freed, W. J., Poltorak, M., and Becker, J. B. (1990) Intracerebral adrenal medulla grafts: a review. *Exp. Neurol.* **110,** 139–166.

17. Olanow, C. W., Goetz, C. G., Kordower, J. H., et al. (2003) A double-blind controlled trial of bilateral fetal nigral transplantation in Parkinson's disease. *Ann. Neurol.* **5,** 403–414.

18. Shoichet, M. S. and Winn, S. R. (2000) Cell delivery to the central nervous system. *Adv. Drug Deliv. Rev.* **20,** 81–102.

19. Winn, S. R., Zielinski, B., Tresco, P. A., et al. (1991) Behavioral recovery following intrastriatal implantation of microencapsulated PC12 cells. *Exp. Neurol.* **113,** 322–329

20. Yurek, D. M. and Sladek, J. R. (1990) Dopamine cell replacement: Parkinson's disease. *Ann. Rev. Neurosci.* **13,** 415–440.

21. Barkats, M., Bilang-Bleuel, A., Buc-Caron, M. H., et al. (1998) Adenovirus in the brain: recent advances of gene therapy for neurodegenerative diseases. *Prog. Neurobiol.* **55,** 333–341.

22. Bowers, W., Howard, D., and Federoff, H. (1997) Gene therapeutic strategies for neuroprotection: implications for Parkinson's disease. *Exp. Neurol.* **144,** 58–68.

23. Kaplitt, M., Darakchiev, B., and During, M. (1998) Prospects for gene therapy in pediatric neurosurgery. *Ped. Neurosurg.* **28,** 3–14.

24. Kordower, J. K. (2003) In vivo gene delivery of glial cell line-derived neurotrophic factor for Parkinson's disease. *Ann. Neurol.* **53,** 120–132.

25. Zlokovic, B. V. and Apuzzo, M. L. J. (1997) Cellular and molecular neurosurgery: Part II: vector systems and delivery methodologies for gene therapy of the central nervous system. *Neurosurgery* **40,** 805–812.

26. Crooks, C. A., Douglas, J. A., Broughton, R. L., and Sefton, M. V. (1990) Microencapsulation of mammalian cells in a HEMA-MMA copolymer: effects on capsule morphology and permeability. *J. Biomed. Mater. Res.* **24,** 1241–1262.

27. Gentile, F. T., Doherty, E. J., Rein, D. H., et al. (1995) Polymer science for macroencapsulation of cells for central nervous system transplantation. *J. Reactive Polymers* **25,** 207–227.

28. Shoichet, M. S., Gentile, F. T., and Winn, S. R. (1995) The use of polymers in the treatment of neurological disorders. *Trends Poly. Sci.* **3,** 374–380.

29. Winn, S. R. and Tresco, P. A. (1994) Hydrogel applications for encapsulated cellular transplants. In *Providing Pharmacological Access to the Brain*, Methods in Neuroscience, vol. 21 (Flanagan, T. F., Emerich, D. F., and Winn, S. R., eds.), Academic Press, Orlando, FL, pp. 387–402.

30. Aebischer, P., Tresco, P., Winn, S. R., et al. (1991) Long-term cross-species brain transplantation of a polymer encapsulated dopamine-secreting cell line. *Exp. Neurol.* **111,** 269–275.

31. Shoichet, M. S. and Rein, D. H. (1996) In vivo biostability of a polymeric hollow fiber membrane for cell encapsulation. *Biomaterials* **17,** 285–290.

32. Dunn, J. C. Y., Tompkins, R. G., and Yarmush, M. L. (1991) Long-term in vitro function of adult hepatocytes in a collagen sandwich configuration. *Biotechnol. Prog.* **7,** 237–245.

33. Emerich, D. F., Frydel, B., McDermott, P., et al. (1993) Polymer-encapsulated PC12 cells promote recovery of motor function in aged rats. *Exp. Neurol.* **122,** 37–47.

34. Aebischer, P., Tresco, P. A., Sagen, J., and Winn, S. R. (1991) Transplantation of microencapsulated bovine chromaffin cells reduces lesion-induced rotational asymmetry in rats. *Brain Res.* **560,** 43–49.
35. Tresco, P. A., Winn, S. R., and Aebischer, P. (1992) Polymer encapsulated neurotransmitter secreting cells: potential treatment for Parkinson's disease. *ASAIO* **38,** 17–23.
36. Emerich, D. F., Frydel, B., Flanagan, T. R., et al. (1993) Transplantation of polymer encapsulated PC12 cells: Use of chitosan as an immobilization matrix. *Cell Transplant.* **2,** 241–249.
37. Massia, S. P. and Hubbell, J. A. (1990) Covalent surface immobilization of Arg-Gly-Asp- and Tyr-Ile-Gly-Ser-Arg-containing peptides to obtain well-defined cell adhesive substrates. *Anal. Biochem.* **187,** 292–299.
38. Wang, N., Butler, J. P., and Ingber, D. E. (1993) Mechanotransduction across the cell surface and through the cytoskeleton. *Science* **260,** 1124–1127.
39. Bellamkonda, R., Ranieri, J. P., and Aebischer, P. (1995) Laminin oligopeptide derivatized agarose gels allow three-dimensional neurite extension in vitro. *J. Neurosci. Res.* **41,** 501–509.
40. Tong, Y. W. and Shoichet, M. S. (1998) Peptide surface modification of poly(tetrafluoroethylene-co-hexafluoroethylene) enhances its interaction with central nervous system neurons. *J. Biomed. Mater. Res.* **42,** 85–95.
41. Cima, L. G., Lopina, S. T., Kaufamn, M., and Merrill, E. W. (1994) Polyethylene oxide hydrogels modified with cell attachment ligands. ASAIO National Meeting, San Francisco, CA.
42. Emerich, D. F. and Winn, S. R. (1999) Application of polymer-encapsulated cell therapy for CNS diseases. In *Neuromethods: Neural Transplantation Methods* (Dunnett, S. B., Boulton, A. A., and Baker, G. B., eds.), Humana Press, NJ, pp. 233–277.
43. Dionne, K. E., Cain, B. M., Li, R. H., et al. (1996) Transport characterization of membranes for immunoisolation. *Biomaterials* **17,** 257–266.
44. Kordower, J. H., Liu, Y.-T., Winn, S. R., and Emerich, D. F. (1995) Encapsulated PC12 cell transplants into hemiparkinsonian monkeys: a behavioral, neuroanatomical and neurochemical analysis. *Cell Transplant.* **4,** 155–171.
45. Lindner, M. D., Plone, M. A., Frydel, B. R., et al. (1997) Intraventricular encapsulated bovine adrenal chromaffin cells: viable for at least 500 days in vivo without detectable host immune sensitization or adverse effects on behavioral/cognitive function. *Restor. Neurol. Neurosci.* **11,** 21–35.
46. Rauch, H. C., Ekstrom, M. E., Montgomery, I. N., et al. (1986) Histopathologic evaluation following chronic implantation of chromium and steel based metal alloys in the rabbit central nervous system. *J. Biomed. Mater. Res.* **20,** 1277–1293.
47. Stensass, S. S. and Stensass, L. J. (1978) Histopathological evaluation of materials implanted into the cerebral cortex. *Acta Neuropathol.* **41,** 145–155.
48. Winn, S. R., Aebischer, P., and Galletti, P. M. (1989) Brain tissue reaction to permselective polymer capsules. *J. Biomed. Mater. Res.* **23,** 31–44.
49. Clark, R. A. F., ed. (1996) *The Molecular and Cellular Biology of Wound Repair.* Plenum Press, New York.

50. Shoichet, M. S., Winn, S. R., Athavale, S., et al. (1994) Poly(ethylene oxide)-grafted thermoplastic membranes for use as cellular hybrid bioartificial organs in the central nervous system. *Biotechnol. Bioeng.* **43,** 563–572.

51. Tresco, P. A., Winn, S. R., Jaeger, C. B., et al. (1992) Polymer-encapsulated PC12 cells: Long-term survival and associated reduction in lesioned-induced rotational behavior. *Cell Transplant.* **1,** 255–264.

52. Subramanian, T., Emerich, D. F., Bakay, R. A. E., et al. (1997) Polymer-encapsulated PC12 cells demonstrate high affinity uptake of dopamine in vitro and 18F-dopa uptake and metabolism after intracerebral implantation in non-human primate. *Cell Transplant.* **6,** 469–477.

53. Sagen, J., Wang, H., Tresco, P. A., and Aebischer, P. (1993) Transplants of immunologically isolated xenogeneic chromaffin cells provide a long-term source of pain-reducing neuroactive substances. *J. Neurosci.* **13,** 2415–2423.

54. Date, I., Shingo, T., Ohmoto, T., and Emerich, D. F. (1997) Long-term enhanced chromaffin cell survival and behavioral recovery in hemiparkinsonian rats with co-grafted polymer-encapsulated human NGF-secreting cells. *Exp. Neurol.* **147,** 10–17.

55. Winn, S. R., Lindner, M. D., Haggett, G., et al. (1996) Polymer-encapsulated genetically-modified cells continue to secrete human nerve growth factor for over one year in rat ventricles: behavioral and anatomical consequences. *Exp. Neurol.* **140,** 126–138.

56. Lindner, M. D. and Emerich, D. F. (1998) Therapeutic potential of a polymer-encapsulated L-DOPA and dopamine-producing cell line in rodent and primate models of Parkinsons disease. *Cell Transplant.* **7,** 65–174.

57. Lindner, M. D., Kearns, C. E., Winn, S. R., et al. (1996) Effects of intraventricular encapsulated hNGF-secreting fibroblasts in aged rats. *Cell Transplant.* **5,** 205–223.

58. Kordower, J. H., Chen, E.-Y., Mufson, E. J., et al. (1996) Intrastriatal implants of polymer-encapsulated cells genetically modified to secrete human NGF: trophic effects upon cholinergic and noncholinergic neurons. *Neuroscience* **72,** 63–77.

59. Aebischer, P., Schleup, M., Deglon, N., et al. (1996) Intrathecal delivery of CNTF using encapsulated genetically modified xenogeneic cells in amyotrophic lateral sclerosis patients. *Nature Med.* **2,** 696–699.

60. Emerich, D. F., Winn, S. R., Harper, J., et al. (1994) Implants of polymer-encapsulated human NGF-secreting cells in the nonhuman primate: rescue and sprouting of degenerating cholinergic basal forebrain neurons. *J. Comp. Neurol.* **349,** 148–164.

61. Fahn, S. (1982) Fluctuations of disability in Parkinson's disease: pathophysiological aspects. In *Movement Disorders* (Marsden, C. D. and Fahn, S., eds.), Butterworth Scientific, London, pp. 123–145.

62. Becker, J., Robinson, T. E., Barton, P., et al. (1990) Sustained behavioral recovery from unilateral nigrostriatal damage produced by the controlled release of dopamine from a silicone polymer pellet placed into the denervated striatum. *Brain Res.* **508,** 60–64.

63. DeYebens, J. G., Fahn, S., Mena, M. A., et al. (1998) Intracerebroventricular infusion of dopamine and its agonists in rodents and primates: an experimental approach to the treatment of Parkinson's disease. *Trans. Am. Soc. Artif. Intern. Organs* **34,** 951–957.

64. Hargraves, R. and Freed, W. J. (1987) Chronic intrastriatal dopamine infusions in rats with unilateral lesions of the substantia nigra. *Life Sci.* **40,** 959–966.

65. Emerich, D. F., Winn, S. R., and Lindner, M. D. (1997) Continued presence of intrastriatal but not intraventricular polymer-encapsulated PC12 cells is required for alleviation of behavioral deficits in Parkinsonian rats. *Cell Transplant.* **5,** 589–596.

66. Lindner, M. D., Plone, M. A., Francis, J. M., and Eamerich, D. F. (1996) Validation of a rodent model of Parkinson's disease: Evidence of a therapeutic window for oral Sinemet. *Brain Res. Bull.* **39,** 367–372.

67. Lindner, M. D., Plone, M. A., Mullins, T. D., et al. (1997) Somatic delivery of catecholamines in the striatum attenuate parkinsonian symptoms and widen the therapeutic window or oral Sinemet in rats. *Exp. Neurol.* **14,** 130–140.

68. Emerich, D. F., Frydel, B., McDermott, P., et al. (1993) Polymer-encapsulated PC12 cells promote recovery of motor function in aged rats. *Exp. Neurol.* **122,** 37–47.

69. Aebischer, P., Goddard, M. B., Signore, P., and Timpson, R. (1994) Functional recovery in hemiparkinsonian primates transplanted with polymer encapsulated PC12 cells. *Exp. Neurol.* **126,** 1–12.

70. Hoane, M. R., Puri, K. D., Xu, L., et al. (2000) Mammalian-cell-produced Neurturin (NTN) is more potent that purified *Escherichia coli*-produced NTN. *Exp. Neurol.* **162,** 189–193.

71. Tseng, J. L., Baetge, E. E., Zurn, A. D., and Aebischer, P. (1997) GDNF reduces drug-induced rotational behavior after medial forebrain bundle transection by a mechanism not involving striatal dopamine. *J. Neurosci.* **1,** 325–333.

72. Tseng, J. L., Bruhn, S. L., Zurn, A. D., and Aebischer, P. (1998) Neurturin protects dopaminergic neurons following medial forebrain bundle axotomy. *NeuroReport* **9,** 1817–1822.

73. Date, I., Ohmoto, T., Ono, T., et al. (1996) Cografting with polymer-encapsulated human nerve growth factor-secreting cells and chromaffin cell survival and behavioral recovery in hemiparkinsonian rats. *J. Neurosurg.* **84,** 1006–1012.

74. Date, I., Ohmoto, T., Imaoka, T., et al. (1996) Chromaffin cell survival from both young and old donors is enhanced by co-grafts of polymer-encapsulated human NGF-secreting cells. *Neuroreport* **7,** 1813–1818.

75. Hefti, F. (1994) Neurotrophic factor therapy for nervous system degenerative diseases. *J. Neurobiol.* **25,** 1418–1435.

76. Hefti, F. (1997) Pharmacology of neurotrophic factors. *Ann. Rev. Pharmacol. Toxicol.* **37,** 239–267.

77. Winn, S. R., Hammang, J. P., Emerich, D. F., et al. (1994) Polymer-encapsulated cells genetically modified to secrete human nerve growth factor pro-

mote the survival of axotomized septal cholinergic neurons. *Proc. Natl. Acad. Sci. USA* **91,** 2324–2328.

78. Kordower, J. H., Winn, S. R., Liu, Y.-T., et al. (1994) The aged monkey basal forebrain: rescue and sprouting of axotomized basal forebrain neurons after grafts of encapsulated cells secreting human nerve growth factor. *Proc. Natl. Acad. Sci.* **91,** 10898–10902.

79. Beal, M. F., Kowall, B. W., Ellison, D. W., et al. (1986) Replication of the neurochemical characteristics Huntington's disease by quinolinic acid. *Nature* **321,** 168–171.

80. Beal, M. F., Mazurek, M. F., Ellison, D. W., et al. (1988) Somatostatin and neuropeptide Y concentrations in pathologically graded cases of Huntington's disease. *Ann. Neurol.* **23,** 562–569.

81. Beal, M. F., Kowall, N. W., Swartz, K. J., et al. (1989) Differential sparing of somatostatin-neuropeptide Y and cholinergic neurons following striatal excitotoxin lesions. *Synapse* **3,** 38–47.

82. Sanberg, P. R., Calderon, S. F., Giordano, M., et al. (1989) The quinolinic acid model of Huntington's disease: locomotor abnormalities. *Exp. Neurol.* **105,** 45–53.

83. Frim, D. M., Schumacher, J. M., Short, M. P., et al. (1992) Local response to intracerebral grafts of NGF-secreting fibroblasts: Induction of a peroxidative enzyme. *Soc. Neurosci. Abstr.* **18,** 1100.

84. Frim, D. M., Simpson, J., Uhler, T. A., et al. (1993) Striatal degeneration induced by mitochondrial blockade is prevented by biologically delivered NGF. *J. Neurosci. Res.* **35,** 452–458.

85. Frim, D. M., Uhler, T. A., Short, M. P., et al. (1993) Effects of biologically delivered NGF, BDNF, and bFGF on striatal excitotoxic lesions. *Neuroreport* **4,** 367–370.

86. Schumacher, J. M., Short, M. P., Hyman, B. T., et al. (1991) Intracerebral implantation of nerve growth factor-producing fibroblasts protects striatum against neurotoxic levels of excitatory amino acids. *Neuroscience* **45,** 561–570.

87. Emerich, D. F., Hammang, J. P., Baetge, E. E., and Winn, S. R. (1996) Implantation of polymer-encapsulated human nerve growth factor-secreting fibroblasts attenuates the behavioral and neuropathological consequences of quinolinic acid injections into rodent striatum. *Exp. Neurol.* **130,** 141–150.

88. Emerich, D. F., Lindner, M. D., Winn, S. R., et al. (1996) Implants of encapsulated human CNTF-producing fibroblasts prevent behavioral deficits and striatal degeneration in a rodent model of Huntington's disease. *J. Neurosci.* **1,** 5168–5181.

89. Emerich, D. F., Cain, C. K., Greco, C., et al. (1997) Cellular delivery of human CNTF prevents motor and cognitive dysfunction in a rodent model of Huntington's disease. *Cell Transplant.* **6,** 249–266.

90. Emerich, D. F., Winn, S. R., Hantraye, P. M., et al. (1997) Protective effects of encapsulated cells producing neurotrophic factor CNTF in a monkey model of Huntington's disease. *Nature* **386,** 395–399.

91. Mittoux, V., Joseph, J. M., Monville, C., et al. (2000) Restoration of cognitive and motor function with ciliary neurotrophic factor in a primate model of Huntington's disease. *Hum. Gene Ther.* **11,** 1177–1187.

92. Kordower, J. H., Isacson, O., and Emerich, D. F. (1999) Cellular delivery of trophic factors for the treatment of Huntington's disease: Is neuroprotection possible? *Exp. Neurol.* **59,** 4–20.

93. Bachoud-Levi, A. C., Deglon, N., Nguyen, J.-P., et al. (2000) Neuroprotective gene therapy for Huntington's disease using a polymer encapsulated BHK cell line engineered to secrete human CNTF. *Hum. Gene Ther.* **11,** 1723–1729.

94. Emerich, D. F. (2000) Encapsulated CNTF-producing cells for Huntington's disease. *Cell Transplant.* **8,** 581–582.

95. Cirone, P., Bourgeois, J. M., Austin, R. C., and Chang, P. L. (2000) A novel approach to tumor suppression with microencapsulated recombinant cells. *Hum. Gene Ther.* **13,** 1157–1166.

96. Joki, T., Machluf, M., Atala, A., et al. (2001) Continuous release of endostatin from microencapsulated engineered cells for tumor therapy. *Nat. Biotech.* **19,** 35–39.

97. Lang, M. S., Hovenkanmp, E., Savelkoul, H. F., et al. (1995) Immunotherapy with monoclonal antibodies directed against the immunosuppressive domain of p15E inhibits tumour growth. *Clin. Exp. Immunol.* **102,** 468–475.

98. Martnet, O., Schreyer, N., Reis, E. D., and Joseph, J. M. (2003) Encapsulation of packaging cell results in successful retroviral-mediated transfer of a suicide gene in vivo in an experimental model of glioblastoma. *Eur. J. Surg. Oncol.* **29,** 351–357.

99. Read, T. A., Sorensen, D. R., Mahesparen, R., et al. (2001) Local endostatin treatment of gliomas administered by microencapsulated producer cells. *Nat. Biotech.* **19,** 29–34.

100. Thorsen, F., Read, T. A., Lund-Johansen, M., et al. (2000) Alginate-encapsulated producer cells: a potential new approach for the treatment of malignant brain tumors. *Cell Transplant.* **9,** 773–783.

101. Visted, T. and Lund-Johansen, M. (2003) Progress and challenges for cell encapsulation in brain tumor therapy. *Expert Opin. Biol. Ther.* **3,** 551–561.

102. Xu, W., Liu, L., and Charles, I. G. (2002) Microencapsulated iNOS-expressing cells cause tumor suppression in mice. *FASEB J.* **16,** 213–215.

103. Bjerkvig, R., Read, T. A., Vajkoczy, P., et al. (2003) Cell therapy using encapsulated cells producing endostatin. *Acta Neurochir. Suppl.* **88,** 137–141.

104. Boughman, J. A., Conneally, P. M., and Nance, W. E. (1980) Population genetic studies of retinitis pigmentosa. *Am. J. Hum. Genet.* **32,** 223–235.

105. Farrar, G., Kenna, P., and Humphries, P. (2002) On the genetics of retinitis pigmentosa and on mutation-independent approaches to therapeutic intervention. *EMBO* **21,** 857–864.

106. Cayouette, M. and Gravel, C. (1997) Adenovirus-mediated gene transfer of ciliary neurotrophic factor can prevent photoreceptor degeneration in the retinal degeneration (rd) mouse. *Hum. Gene Ther.* **8,** 423–430.

107. Cayouette, M., Behn, D., Sendtner, M., et al. (1998) Intraocular gene transfer of ciliary neurotrophic factor prevents death and increases responsiveness of rod photoreceptors in the retina degeneration slow mouse. *J. Neurosci.* **18,** 9282–9293.

108. LaVail, M., Asumura, D., and Matthes, M. (1998) Protection of mouse photoreceptors by survival factors in retinal degenerations. *Invest. Ophthalmol. Vis. Sci.* **39,** 592–602.

109. Tao, W., Wen, R., Goddard, M., et al. (2002) Encapsulated cell-based delivery of CNTF reduces photoreceptor degeneration in animal models of retinitis pigmentosa. *Invest. Ophthalmol. Vis. Sci.* **43,** 3292–3298.

110. Aguirre, G., Farber, D., and Lolley, R. (1982) Retinal degenerations in the dog. III: abnormal cyclic nucleotide metabolism in rod-cone dysplasia. *Exp. Eye Res.* **35,** 625–642.

111. Schmidt, S. and Aguirre, G. (1985) Reduction in taurine secondary to photoreceptor loss in Irish setters with rod-cone dysplasia. *Invest. Ophthalmol. Vis. Sci.* **26,** 679–683.

112. Henderson, C. E. (1994) GDNF: A potent survival factor of motoneurons present in peripheral nerve and muscle. *Science* **266,** 1062–1064.

113. Hughes, R. A., Sendtner, M., and Thoenen, H. (1993) Members of several gene families influence survival of rat motoneurons in vitro and in vivo. *J. Neurosci. Res.* **36,** 663–671.

114. Kato, A. C. and Lindsay, R. M. (1994) Overlapping and additive effects of neurotrophins and CNTF on cultured human spinal cord neurons. *Exp. Neurol.* **130,** 196–201.

115. Lewis, M. E., Neff, N. T., Contreras, P. C., et al. (1993) Insulin-like growth factor-I: Potential for treatment of motor neuronal disorders. *Exp. Neurol.* **124,** 73–88.

116. Mitsumoto, H., Ikeda, K., Klinkosz, B., et al. (1994) Arrest of motor neuron disease in wobbler mice cotreated with CNTF and BDNF. *Science* **265,** 1107–1110.

117. Sagot, Y., Tan, S. A., Baetge, E. E., et al. (1995) Polymer encapsulated cell lines genetically engineered to release ciliary neurotrophic factor can slow down progressive motor neuronopathy in the mouse. *Eur. J. Neurosci.* **7,** 1313–1320.

118. Sendtner, M., Holtmann, B., Kolbeck, R., et al. (1992) Brain-derived neurotrophic factor prevents the death of motor neurons in newborn rats after nerve section. *Nature* **360,** 757–759.

119. Sendtner, M., Schmalbruch, H., Stöckli, K. A., et al. (1992) Ciliary neurotrophic factor prevents degeneration of motor neurons in mouse mutant progressive motor neuronopathy. *Nature* **358,** 502–504.

120. Dittrich, F., Thoenen, H., and Sendtner, M. (1994) Ciliary neurotrophic factor: Pharmacokinetics and acute-phase response in rat. *Ann. Neurol.* **35,** 151–163.

121. The ALS CNTF Treatment Study (ACTS) Phase I-II Study Group (1995). The pharmacokinetics of subcutaneously administered recombinant human ciliary neurotrophic factor (rhCNTF) in patients with amytrophic lateral

sclerosis: Relationship to parameters of the acute phase response. *Clin. Neuropharmacol.* **18,** 500–514.

122. The ALS CNTF Treatment Study (ACTS) Phase I-II Study Group (1995). A phase I study of recombinant human ciliary neurotrophic factor (rHCNTF) in patients with amyotrophic lateral sclerosis. *Clin. Neuropharmacol.* **18,** 515–532.

123. Tan, S. A., Deglon, N., Zurn, A. D., et al. (1996) Rescue of motoneurons from axotomy-induced cell death by polymer encapsulated cells genetically engineered to release CNTF. *Cell Transplant.* **5,** 577–587.

124. Aebischer, P., Schleup, M., Deglon, N., et al. (1996) Intrathecal delivery of CNTF using encapsulated genetically modified xenogeneic cells in amyotrophic lateral sclerosis patients. *Nature Med.* 696–699.

125. Zurn, A. D., Henry, H., Schluep, M., et al. (2000) Evaluation of an intrathecal immune response in amyotrophic lateral sclerosis patients implanted with encapsulated genetically-engineered xenogeneic cells. *Cell Transplant.* **9,** 471–484.

126. Czech, K. A. and Sagen, J. (1995) Update on cellular transplantation into the rat CNS as a novel therapy for chronic pain. *Prog. Neurobiol.* **46,** 507–529.

127. Sagen, J. (1998) Cellular Transplantation for intractable pain. *Adv. Pharmacol.* **42,** 579–582.

128. Aebischer, P., Buschser, E., Joseph, J. M., et al. (1994) Transplantation in humans of encapsulated xenogeneic cells without immunosuppression— a preliminary report. *Transplantation* **58,** 1275–1277.

129. Buschser, W., Goddard, M., Heyd, B., et al. (1996) Immunoisolated xenogenic chromaffin cell therapy for chronic pain. Initial clinical experience. *Anesthesiology* **85,** 1005–1012.

130. CytoTherapeutics Press Release, Providence, RI, June 24, 1999.

131. Lindner, M. D., Francis, J. M., McDermott, P. E., et al. (2000) Numerous adrenal chromaffin cell preparations fail to produce analgesic effects in the formalin test or in tests of acute pain even with nicotine stimulation. *Pain* **88,** 177–188.

132. Lindner, M. D., Francis, J. M., Plone, M. A., et al. (2000) The analgesic potential of intraventricular polymer-encapsulated adrenal chromaffin cells in a rodent model of chronic neuropathic pain. *Exp. Clin. Psychopharmacol.* **8,** 524–538.

133. Lindner, M. D., Francis, J. M., and Saydoff, J. A. (2000) Intrathecal polymer-encapsulated bovine adrenal chromaffin cells fail to produce analgesic effects in the hotplate and formalin test. *Exp. Neurol.* **165,** 370–383.

12

Evidence-Based Methodology for Advancing Neural Reconstruction

Stephen Polgar

ABSTRACT

The purpose of this chapter is to identify recent developments in health research methodology that might be useful for ensuring progress in cellular therapy for brain repair. Recently, commentators suggested that the need for rapid scientific and commercial successes has privileged approaches, which are inadequate for solving the problems inherent in the project of repairing the human brain. Critical analysis of recent clinical trials for the treatment of Parkinson's disease (PD) revealed a number of unresolved conceptual and methodological issues. These issues included: a lack of consensus concerning treatment goals, the absence of clearly defined effect sizes for clinical significance, and the weak communication of research findings for the accurate synthesis of evidence. A particular problem has been the lack of interest in collecting qualitative data regarding the experiences and values of patients participating in the research. It is recommended that the explicit adoption of so-called "evidence-based" approaches to research design, data collection, and analysis will ensure optimal outcomes for using stem cells for cellar therapies.

Key Words: Cellular therapy; brain repair; progress; methodology; evidence-based medicine; qualitative research.

INTRODUCTION

Recent placebo-controlled, randomized clinical trials (RCTs) failed to demonstrate clinically meaningful benefits for grafting embryonic cells in patients with Parkinson's disease (PD; *1–3*). Moreover, the emergence of late-onset, off-medication dyskinesias in two trials *(1,3)* resulted in the temporary cessation of clinical trials. Given that the research for the treatment of PD was the most advanced arm of the overall project of brain repair, current progress has not met original expectations *(4)*.

From: *Contemporary Neuroscience: Cell Therapy, Stem Cells, and Brain Repair*
Edited by: C. D. Sanberg and P. R. Sanberg © Humana Press Inc., Totowa, NJ

Current research is focused on investigating a number of cell lines, particularly stem cells as alternatives for the embryonic cells previously used for neural transplantation *(5,6)*. For example, preclinical studies indicated that stem cells from umbilical cord blood appears to have behavioral benefits in animal models and might be suitable for clinical transplantation *(7)*. However, using alternatives to embryonic cells might not be sufficient for ensuring better clinical outcomes, as demonstrated by the weaker behavioral recovery using stem cells, compared to embryonic cells *(8)*. There are both conceptual and methodological problems entrenched in the research program for neural reconstruction *(4,9,10)*. It is essential that these problems are clearly identified, discussed, and resolved to ensure the future optimal progress of the research program.

NEURAL RECONSTRUCTION IN CONTEXT

The methodology (i.e., design, data collection, and analysis) of health research is primarily determined by the logic and principles of science applied to solving medical problems. Health research is conducted out in social settings; the values and resources of a community also shape the methodology of a research project *(11)*. In western societies, the health care industry is a significant source of employment and wealth. Commercial considerations are therefore an integral part of the social context for the development of new medical technologies. The advantage of commercial enterprises' involvement in cellular therapies has been via the funding of preclinical and clinical research.

The availability of resources is the central issue. Freeman and Vawter *(12)* pointed out the "unprecedented costs" associated with implementing RCTs to evaluate the clinical efficacy of novel biotechnologies, such as brain repair. An RCT involving neurosurgery, with modest sample sizes, can cost tens of millions of dollars to complete in the United States *(12)*. These costs have dramatic implications for deciding which of the many innovative preclinical projects can be selected for clinical development. Even with the relatively generous public support for stem cell research, the "bottleneck" for transforming laboratory evidence into clinical trials has decisive consequences for both the laboratory and clinical phases of the research program for neurological advancements.

Paul and Brundin *(13)* identified problems in balancing commercial confidentiality against the open flow of information necessary for scientific advancement. An issue that must be addressed is the need for profits (within a reasonable period of time), resulting in methodological decisions that are inconsistent with traditional scientific values and practices.

Freed *(4)* concluded that progress of cells for central nervous system (CNS) disorders had been slower and more difficult than anticipated 25 yr ago. Crucial questions regarding the implementation of the procedure are still unanswered. For example, which cells are suitable for cellular therapies? How accurate are animal models to predict the harm and benefits of cellular therapy in humans? Which circuits are repairable in the CNS, and which repairs are beyond biological limits? Although there have been remarkable scientific discoveries and technological advancements, the reconstruction of the human brain has revealed itself as a yet unmet challenge to neuroscientists *(9)*. Regardless of the pressure on researchers to achieve clinically and commercially meaningful outcomes, Freed *(4)* advocated a more gradual "speculative" approach for effectively solving the multifaceted problems inherent to cellular therapies for CNS disorder.

"These more speculative approaches will require more patience and may be particularly unsuited to the kind of focused approaches, and requirement for rapid results, that are characteristic of the research programs of small private enterprises" *(4)*.

Recently published reports of clinical evaluations of neural transplantation for PD provided the basis for the present discussion of the methodological problems prevalent in the field. The aim of this chapter is to identify the current research methodologies that are most appropriate to resolve these problems and advance stem cell research.

THE GOALS FOR CELLULAR THERAPIES

The long-term goal of reconstructive neurosurgery is to provide a cure for patients with neurological disorders. Also, in degenerative conditions, such as PD or Alzheimer's disease, the expeditious use of cellular therapies might prevent the onset of disabling symptoms. The problem here is defining what constitutes a "cure" and "prevention." For example, cure and prevention for a genetic condition (e.g., Huntington's disease [HD]) refers to grafting cells that reverse or delay the signs and symptoms of the disease. Other interventions, such as genetic counseling or genetic engineering, might cure the disease in the sense of reducing population prevalence to zero or eliminating the underlying pathology. At present, more realistic goals, rather than a cure, need to be identified.

Expert Opinion

The traditional criterion to evaluate the efficacy of new surgical procedures is the extent to which it is adopted by individual practitioners. This is a "*laissez-faire*," free-market approach, leaving it up to individual service providers to make the decision of efficacy. Assessment of success relies simply on the uptake of the new procedure to patients. This is a reasonable policy, as it is the individual neurologist or neurosurgeon who ultimately decides whether to offer a procedure. However, this approach raises the question: What standards are being applied for decisions to be made by individual practitioners?

Several leading researchers and surgeons *(14–16)* voiced strong concerns that some surgical procedures may have been introduced on the basis of invalid evidence obtained from poorly controlled clinical trials. Lang *(16)* argued that:

> "In contrast to the intensive assessment required before a new drug is established as sufficiently safe and efficacious for widespread use, no such standards exist for surgery. As long as there are willing neurologists and surgeons desperate for patients, this problem will persist until the professional community decides to regulate the practice of its members or until external regulations are imposed."

Obviously, more explicit and rigorous standards are required to protect patients and the public.

Double-Blind RCTs

North American researchers have made vigorous representations for the use of sham neurosurgery to ensure the principled evaluation of the harm and benefits associated with cellular therapies for PD. The justification for using sham or placebo neurosurgery is that this procedure enables the implementation of prospective, double-blind RCTs to examine the efficacy of neural transplantation. Double-blind RCTs are seen as the most rigorous designs available for testing hypotheses on the safety and efficacy of neural transplantation—the gold-standard design for demonstrating causal effects *(15,17,18)*.

In the context of transplantation research, patients are randomly assigned to sham-operated or grafted groups. The type of procedure (i.e., placebo or actual) is carefully concealed from both the participants and assessors. The outcome variables that represent recovery are measured and compared between the control and treatment groups, thus identifying the difference attributable to the causal effects of treatment. The rules of evidence can be stated as: If there is a statistically significant difference in favor of the grafted group on key outcome measures, then treatment is considered effective. Although this is a principled decision and is commonly used in treatment evaluation, there are a number of problems with using statistically significant differences in a double-blind RCT as evidentiary basis for safety and efficacy *(19)*. The following points are most noteworthy:

1. The use of placebo neurosurgery has been strongly opposed on ethical grounds, as discussed later in this chapter.
2. Statistical significance indicates the probability that a difference or association reflects random error or a real population-based phenomenon. As it is influenced by factors other than effect size, such as sample size and variability, statistical significance is not a preferred indicator of efficacy. As Freed *(4)* stated: "We are not looking for small improvements but instead a clear and easily observed result that makes a difference in the life of the patient." Criteria of clinically significant effect sizes can be provided by experts, but they need to be cross-referenced to the goals and expectations of patients undergoing the procedure.
3. Patients and their treating clinicians need evidence for the anticipated benefits in comparison to standard treatments. Placebo control groups do not provide this information, as sham neurosurgery is not an alternative practice offered outside research settings.

Best Practice

The principles of evidence-based medicine provide solid methodological foundations for decisions concerning the efficacy of neural transplantation *(19)*. Briefly, evidence-based medicine *(20,21)* integrates public policy and health research methodology. The essential orientation of the policy is to identify and select the most beneficial, safe, and cost-effective treatment available for individuals or populations with a specific health problem. The methodology emphasizes the use of RCTs, the calculation of effect sizes and confidence intervals, specifying criteria for clinical significance, and the use of systematic reviews and meta-analyses to synthesize the overall evidence *(20,22)*.

Although placebo controls are frequently used in RCTs, a range of other control groups are also employed, including no-treatment and standard-treatment controls (Cochrane Collaboration, see cochrane.org). There is no implication that rigorous evaluations of treatment efficacy should (as a matter of necessity) include placebo controls *(20,22)*.

The key difference between "significantly better than placebo" and "evidence-based medicine" is that the latter approach requires a broader range of evidence. This evidence is analyzed for effect size and clinical significance to identify the best practice, even if evaluations of neural grafting produce consistent data of clinical benefits, it is unlikely to be introduced as a procedure if there are alternative developments. For example, in the case of PD, deep-brain stimulation *(23)* and neuroprotection *(24)* are strong competitors to neural reconstruction for the treatment of PD. The projected costs are a fundamental consideration; what is the point of developing neural transplantation as a procedure if relatively few patients or communities will have the resources to benefit?

In the context of evidence-based practice, the goal of cellular therapies is to develop best practice procedures for treatment and, where possible, prevention of a range of neurological disorders *(10)*. The identification of the best practice is a far more demanding policy to introduce a novel procedure for public use. "Significantly better than placebo" is advantageous commercially on the grounds that benefits in a single placebo-controlled RCT are sufficient for a new practice to be adopted. This policy does not provide protection for the public as is provided by the more detailed policy associated with the standards for evidence-based practice.

OUTCOME MEASURES

There are now a variety of standardized scales available for the measurement of functional changes in patients with neurological disorders *(25)*. Leading researchers recognized that the progress of cellular therapies for neurological disorders would be enhanced if research groups used similar assessment procedures. Langston and colleagues *(25)* constructed the Core Assessment Program for Intracerebral Transplantation (CAPIT) to guide research in the field. The CAPIT-PD contains a battery of functional assessments for PD, including the commonly used United Parkinson's Disease Rating Scale (UPDRS) *(25)*. The CAPIT-PD protocol specifies not only pre- and postgrafting of the tests but also the conditions ("on" or "off" medication) to conduct the tests. In addition, pre- and postoperative assessment of levels of dopaminergic activity are determined using ^{18}F-fluorodopa positron emission tomography (PET) scans. There are more recent developments of the CAPIT-PD protocol, as well as protocols for assessing recovery for HD and stroke *(26)*.

Following publication of the CAPIT-PD, most mainstream research groups *(27,28)* adopted the use of this protocol. However, the more recent double-blind RCTs using human embryonic cells did not explicitly follow core assessment protocols. Freed and colleagues *(1)* reported detailed results on the following outcome measures:

1. Primary outcome: a rating of the patient-perceived change in severity of the disease scored on an ordinal scale of −3.0 to 3.0 (global rating scale).
2. The total score on UPDRS measured under off medication conditions.
3. The Schwab-England Scale (off).
4. Striatal fluorodopa uptake PET scans were completed at 12 mo and two postmortem histological analyses were completed.
5. Detailed records of adverse events.

The authors gave no explicit justification why the subjective global rating scale was selected as the primary outcome measure. Also, no evidence was provided for the validity and reliability of this scale, which carried the primary evidentiary burden for this important RCT.

Olanow and colleagues *(3)* specified their primary outcome as the "motor" component of the UPDRS under practically defined off (no medication) state. Many outcome measures under the CAPIT protocol were monitored, and changes from baseline were reported both on and off at 2 yr after grafting. Also, flurodopa uptake PET results were reported from baseline.

In a recent systematic review *(26)* of PD outcomes, UPDRS motor (off) scores emerged as the preferred and most reported outcome measure to analyze clinical outcomes. Motor (off) is a conceptually valid measure to indicate benefits of neural grafts, given that PD is a movement disorder, and that the off condition provides a clearer indication of dopaminergic graft benefits than on medication measures. There are problems with motor (off) as a primary outcome measure. This is the measure on which statistically significant outcomes are most likely to be obtained. Although motor (off) is a sensitive behavioral indicator of recovery, there is a decision bias in selecting the measure that gives the best outcomes. Other measures might be just as valid and important for characterizing recovery following neural grafting. One cannot simply ignore unfavorable outcomes on variables that were consensually designated as relevant for quantifying changes in the signs and symptoms of a neurological disorder.

Clinical researchers should not slavishly comply with such protocols as CAPIT-PD. Rather, far more effort is needed to identify valid outcome measures and to define the effect sizes indicating clinical significance. Multivariate statistical techniques, e.g., factor analysis and path analysis, are more useful to examine the predictive value of individual outcome measures *(26)*, provided that a sufficient number of scores is available.

To ensure valid outcome measurement, it is also essential to collect evidence regarding the personal meaning of recovery on outcome measures from the patient's perspective. For example, PD patients continue taking medications following neural transplantation. Do patients value recovery on UPDRS on outcomes more than, or equally, to off conditions? In addition, how important are outcomes on other drug-related outcomes, such as "percentage of time with dyskinesia" or "percentage off time?" To answer these questions, evidence is needed from both quantitative and qualitative surveys. Clinically substantial improvements are required on all the key outcome measures identified through the combined use of statistical modeling and qualitative evidence on patient values. The problem is that the systematic evidence-based approach to assess recovery is a far more demanding process than simply nominating desirable outcome measures. Even with the risk of hindering rapid progress, the more complex consumer-oriented methodology will provide a more solid evidentiary basis for neural reconstruction.

HARMFUL AND ADVERSE SIDE EFFECTS

There are specified ways to determine the risk-to-benefit ratios associated with existing and novel treatments. Adverse events, including transient hallucination and confusion, as well as relatively minor, reversible postsurgical events were frequently reported following neural transplantation for PD. The most serious adverse side effects were intractable off-medication dyskinesias with late (approx 12 mo) onset. Freed and colleagues' *(1)* reported that five out of 33 grafted patients developed serious late-onset dyskinesias at 1 yr following transplantation.

Olanow et al. *(3)* found that 13 out of 23 patients developed off-medication dyskinesias. In three patients, the dyskinesias were so severe that the surgical intervention was indicated at the conclusion of the study. Olanow et al. *(3)* employed a video-based, standardized assessment of dyskinesia scored on a 0–28 scale to determine the severity of both on- and off-medication dyskinesias. At 2-yr postsurgery, the dyskinesia scores were 0 for placebo controls, 3.2 ± 12.3 for the donor-per-side group, and 2.7 ± 10.4 for the four-donors-per-side group results reported as means and estimated standard deviations). The highly skewed distributions for the grafted group reflect that 10 out of 23 patients with neural grafts had no dyskinesias, and that three patients who required surgery presumably had very high scores. It can be inferred that the majority of off-medication scores would be much lower than the averages for on-medication dyskinesias, calculated as 11.2 and 6.2.

In evaluating adverse effects, it is useful to differentiate between mild or nondisabling dyskinesias and those that distress or disable patients *(16)*. Despite that 56% of patients developed off-medication dyskinesias, this statistic might overestimate the severity of the problem. Pooled data for all acceptable studies for neural reconstruction would provide a more accurate estimate of harm. Also, it is important to investigate how serious the dyskinesias were from the patients' perspectives. Patients can be asked how they experience mild or severe dyskinesias. Without detailed patient reports, valid risk–benefit analyses are difficult, and principled decisions for implementing or discontinuing an experimental treatment becomes problematic *(20)*.

SYNTHESIS AND THE AVAILABILITY OF THE EVIDENCE

As researchers devising the CAPIT-PD protocol pointed out *(25)*, it is difficult for any one group to evaluate a sufficient number of patients to accurately identify the combination of variables that determine the risks and benefits of reconstructive neurosurgery. Even with the problems inherent in synthesizing data from diverse sources, meta-analyses improve the statistical power *(29)* for identifying emergent overall trends in a research program.

Cellular therapy for PD is the most advanced branch of the research program, with well over 300 patients receiving dopamine-rich embryonic grafts. However, using exclusion criteria, such as outmoded surgical techniques reduced the information that could be incorporated into a meta-analysis *(26)*. Disconcertingly, there are serious doubts surrounding the accuracy of reporting and analysis of some results, even those published in mainstream journals. The reviewers *(10)* suggested that one problem was the way in which conclusions were reported. This included the uninformative presentation of descriptive statistics and the failure to report outcomes that did not reach statistical significance. Also, the inferential analyses focused on statistical significance. There has been little interest in calculating effect sizes and related confidence intervals as indicators of clinical significance beyond reporting the percentage changes. Incomplete reporting of the data hinders effective synthesis of the evidence and prevents the testing of hypotheses on the conditions to minimize harm and optimize recovery. The difficulties involved with analyzing and reporting data are clearly illustrated in two key National Institutes of Health–funded RCTs *(1,3)*:

1. *Descriptive statistics* were not reported in sufficient detail. The reader is unable to construct a clear perspective of the research outcomes, and the reviewer is incapable of incorporating the results into a meta-analysis. For example, standard deviations were not reported for UPDRS and Schwab and England outcome measures *(1)*, rendering both the calculation of confidence intervals and valid synthesis of the results impossible. Olanow et al. *(3)* reported unadjusted changes on fluorodopa uptake at 24 mo but not at baseline. The authors also reported the statistical significance of changes in dopaminergic activity, but these probabilities were not directly helpful to estimate effect sizes.
2. *Comparison* of outcomes for the two RCTs is difficult. In some detail, Freed reported on three functional outcome measures: UPDRS total (off), global subjective measures, and Schwab and England measures (under off). Results were reported up to 12 mo. Olanow et al. *(3)* reported UPDRS motor off in detail up to 24 mo but UPDRS (total, off) scores did not include baseline scores. There is insufficient evidence to compare outcomes at crucial times, such as 6 mo or 12 mo following transplantation. The justified conclusion drawn was that neither study was successful in demonstrating the safety and efficacy of cellular

therapy for PD *(3)*. However, there appears to be no functional measures on which the two RCTs can be quantitatively compared or statistics that enable synthesis of the results with those from other related studies.

3. *Critical analysis* of a publication entails the reader's ability to follow the authors' reasoning when making serious decisions concerning the implications of the results. For example, in contrast to previous reports, Olanow et al. *(3)* "found no correlation between on and off medication dyskinesia scores." This conclusion is important, given the current questions relating to the mechanisms underlying delayed-onset, disabling dyskinesias. Unfortunately, the reader is unable to follow the logic of the analysis that leads to this conclusion. It is unclear how the correlations were calculated using the sample data. Also, given the markedly skewed distributions for both off- and on-medication dyskinesias, it is unknown which correlation coefficient was used and the power associated with the analysis.

One potentially valuable finding reported by Olanow and colleagues *(3)* was that the post hoc analysis of the data revealed significant transplant benefits in patients with lower UPDRS motor (off) scores. The effect was only evident for the four-grafts-per-side group, and the authors were careful to point out that this finding "reflects failure of transplanted patients to deteriorate, rather than improvement in parkinsonian features." The effect was only evident when compared to the control group ($n = 6$) with milder scores. In a later review, Olanow *(30)* suggested that, "Post hoc analysis demonstrated significant improvements with transplantation in patients with milder disease, but no age-related benefits were detected." The notion that patients with milder PD benefit more from transplantation than those with a more advanced form is an appealing hypothesis. Unfortunately, inspection of the evidence *(3)* does not appear to support this hypothesis. Both Graph A (mild) and Graph B (advanced) show very similar patterns of recovery following transplantation. The average scores of all four grafted groups show recovery up to 6–9 mo, then revert to baseline. Figure 2, Graph A indicates a remarkable, almost linear 50% deterioration in the performance of the "mild condition" placebo control groups ($n = 6$), supporting the idea that the apparent finding might be owing to sampling error.

Thus, it is difficult to conduct a systematic, statistically valid critique based on the results as presented to the reader. One must estimate true values from graphs, which can lead to both descriptive and inferential errors *(26)*. In the authors' defense, they cannot possibly predict exactly which detailed result will be required by their readers. There are space limitations when publishing research articles, even for recent electronic versions. In addition, it can be a nuisance for researchers when colleagues make uninvited email requests for large quantities of data. The most obvious approach to correct this problem is to construct a collaborative database for the various programs of brain repair. However, in the case of PD, there appears to be limited support for this initiative. In contrast, there is a strong initiative to ensure consistency of results with the creation of a database by European researchers working on cellular therapies for HD *(31,32)*.

Inadequate access to the results of PD trials constitutes both a methodological and ethical problem. First, the information obtained in clinical trials for PD remains uninterpreted in a critical and systematic fashion. There is a loss of accurate, pooled information, which is essential for the open, collegiate evaluation of progress and for identifying the factors that ensure the best outcomes for neural transplantation. Second, clinical evaluations are expensive and demanding enterprises. Public fund providers expect that these hard-gained results are broadly disseminated and effectively utilized. Even where researchers receive private funding, there is an obligation to ensure the accurate and timely dissemination of their results. Third, patients with PD volunteered for clinical research projects with the understanding that they were at risk, but this risk was balanced by their contribution to knowledge. If the participants consented for experimental neurosurgery on the grounds of advancing neurosciences, it is unethical not to make the best use of the information produced.

ETHICS

Controversial ethical, political, and medicolegal issues are associated with both preclinical and clinical research in brain repair. The most contested issue has been the use of cells from embryonic donors. This is a politically contentious and morally sensitive issue *(33)*, which will not be addressed in the present chapter. Another issue that has attracted attention has been the use of sham or placebo surgery for conducting double-blind RCTs to evaluate the efficacy of cellular therapy for PD *(17,34,35)*. Some authors characterized the use of neurosurgical procedures for the purpose of research, rather than healing, as contrary to professional and community values *(34,36,37)*. Other researchers *(17,18,38)* take an opposing view, contending that the use of placebo neurosurgery is an ethical requirement to conduct rigorous research and to protect the public from the introduction of useless procedures.

Miller *(39)* argued that there were no convincing ethical reasons for an absolute prohibition of sham surgery in clinical trials. A case-by-case risk–benefit assessment was suggested, where the burdens and potential harm associated with sham surgery can be balanced carefully against the methodological benefits of prospective double-blind placebo-controlled RCTs. A double-blind RCT enables researchers to estimate true efficacy by subtracting the component of recovery attributable to confounding extraneous variables, such as placebo reactions and assessor bias. The design reduces the probability of making false-positive decisions on the efficacy of the procedure and, as stated earlier, contributes to protecting the patients from useless and ineffectual surgical procedures *(15,17,38)*.

Polgar and Ng *(19)* pointed out that the methodological benefits of placebo surgery are theoretical, rather than factual, benefits in the context of reconstructive neurosurgery. Recent meta-analyses *(40)* demonstrated that placebo effects for nonsurgical treatments cannot be taken for granted. One key theoretical assumption is that sham neurosurgery will produce large and long-term functional improvements. Analysis of the evidence provided by recent RCTs *(1–3)* indicated that there were no long-term clinically substantial improvements in either the transplanted or sham-operated groups. It was evident that neural grafting was unsafe and ineffective, even without reference to the results of the sham-operated groups. Consistently with previous authors *(34,37)*, it was suggested that the best information for evaluating the risks and benefits of neural reconstruction was based on data from standard-treatment RCTs. It was contended that, as placebo controls were unnecessary, the use of sham surgery is unethical within a utilitarian framework *(19)*.

Notably, patients are required to consent to the risk of being assigned to a placebo group for time periods up to 2 yr. Macklin *(37)* argued that volunteers might not fully understand the medical and psychological consequences of being assigned to the placebo group. The problem is not the serious physical harm because of sham surgery. Data indicate none of the 39 sham-operated patients participating in the RCTs *(1–3)* suffered serious harm. Instead, the problem is that the patient is locked into a trial where the pharmacological treatments offered to placebo neurosurgery groups are limited by the research protocol.

The patients themselves and their advocates have the closest, most detailed knowledge of what it is like to participate in a neurosurgical trial. They have experienced what it is like to either to have their brain "repaired" or have experienced the uncertainty of being assigned to a sham surgery group. There appears to be no published evaluations of rigorous debriefings of previous research participants. Without such qualitative evidence, it is questionable if accurate risk–benefit analyses for neural reconstruction can be conducted, or even if truthful informed consent forms can be provided for future participants. Fletcher *(14)* stated that, "Society and researchers owe a great debt of gratitude" to the PD patients who participated in double-blind, randomized trials. "Some suffered more than the best-laid plans could anticipate." A genuine commitment to the use of qualitative methodology would ensure that the voices of the participants are heard and that the meaning of their experiences are formally incorporated into the literature.

CONCLUSIONS

The aim of the present chapter was to identify research methods that were consistent with what Freed *(4)* called a "speculative" approach to advance the program for brain repair. The following strategies were suggested to resolve the above methodological problems:

1. Define the goal of cellular therapies as providing best practice procedures for a subset of patients with targeted disorders.
2. Select key outcome measures and estimate clinically significant improvements required based on both qualitative and quantitative evidence.
3. Use effect sizes and associated confidence intervals to determine the harm and benefits.
4. Create databases to synthesize evidence for benchmarking progress and to test hypotheses concerning the optimal conditions for safety and efficacy.
5. Ensure that the experiences and values of the consumers are taken into account by collecting and systematically incorporating qualitative evidence into the literature.

There is nothing original or exotic about the above suggestions; they simply reflect the emerging evidence-based approach to health research evaluation and policy formulation *(20,22)*. The implementation of an evidence-based methodology would resolve the problems identified in the present chapter and would help ensure the optimal progress of the current research program using stem cells for brain repair.

REFERENCES

1. Freed, C., Greene, P., Breeze, R., et al. (2001) Transplantation of embryonic dopamine neurons for severe Parkinson's disease. *N. Engl. J. Med.* **344,** 710–719.
2. Freeman, T. B., Watts, R., Hauser, R., et al. (2002) A prospective, randomized, double-blind, surgical placebo-controlled trial of intrastriatal transplantation of fetal porcine ventral mesencephalic tissue (neurocell-PD) in subjects with Parkinson's disease. *Exp. Neurol.* **175,** 426.
3. Olanow, C. W., Goetz, C. G., Kordower, J. H., et al. (2003) A Double-Blind Controlled Trial of Bilateral Nigral Transplantation in Parkinson's disease. *Ann. Neurol.* **54,** 403–414.
4. Freed, W. A. (2004) Perspective on transplantation therapy and stem cells for Parkinson's disease. *Cell Transplant.* **13,** 319–327.
5. Yang, M., Donaldson, A. E., Marshall, C. E., et al. (2004) Studies on the differentiation of dopaminergic traits in human neural progenitor cells in vitro and in vivo. *Cell Transplant.* **13,** 535–547.

6. Zigova, T., Snyder, E., and Sanberg, P. R., eds. (2003) *Preface: Neural Stem Cells for Brain and Spinal Cord Repair.* Humana Press, Inc., Totowa, NJ, vii–ix.
7. Sapporta, S., Kim, J. J., Willing, A. E., et al. (2003) Human umbilical cord blood stem cells infusion in spinal cord injury: engraftment and beneficial influence on behaviour. *J. Hematother. Stem Cell Res.* **12,** 271–278.
8. Yurek, D. M. and Fletcher, A. (2004) Comparison of embryonic stem cell-derived dopamine neuron grafts and fetal ventral mesencephalic tissue grafts. *Cell Transplant.* **19,** 295–306.
9. Ourednik, J. and Ourednik, V. (2004) Review: Graft-induced plasticity in the mammalian host CNS. *Cell Transplant.* **13,** 307–318.
10. Polgar, S. (2004) Cells to best-practice: the structure of the research program for neural reconstruction. *Exp. Neurol.* **187,** 216–217.
11. Polgar, S. and Thomas, S., eds. (2000) *Introduction to Research in the Health Sciences, 4th ed.* Churchill Livingstone, Edinburgh, UK.
12. Freeman, T. B. and Vawter, D. E. (2004) From the laboratory to the clinic: unique issues in the clinical evaluation of neural reconstructive therapies. *Exp. Neurol.* **187,** 203.
13. Paul, G. and Brundin, P. (2002) Funding stem cell research: will commerce counteract collaboration. *Drug Discov. Today* **7,** 22–23.
14. Fletcher, J. C. (2003) Sham neurosurgery in Parkinson's disease: ethical at the time. *Am. J. Bioeth.* **3,** 54–56.
15. Freeman, T. B, Vawter, D. E., Leaverton, P. E., et al. (1999) Use of placebo surgery in controlled trials of a cellular-based therapy for Parkinson's disease. *N. Engl. J. Med.* **341,** 988–991.
16. Lang, A. (2000) Surgery for Parkinson's disease: a critical evaluation of the state of the art. *Arch. Neurol.* **57,** 1118–1125.
17. Vawter, D. E., Gervais, K. G., and Freeman, T. B. (2003) Does placebo surgery-controlled research call for new provisions to protect human research participants? *Am. J. Bioeth.* **3,** 50–53.
18. Vawter, D. E., Gervais, K. G., Prehn, A. W., et al. (2004) Placebo controlled surgical trials; perspectives of Parkinson's disease (PD) researchers. *Exp. Neurol.* **187,** 221–222.
19. Polgar, S. and Ng, J. (2005) Ethics, methodology and the use of placebo controls in surgical trials. *Brain Res. Bull.* **67,** 290–297.
20. Sackett, D., Straus, S., Richardson, W., et al., eds. (2000) *Evidence-Based Medicine: How to Practice and Teach EBM, 2nd ed.* Churchill Livingstone, Edinburgh, UK.
21. Schwartz, M. and Polgar, S., eds. (2003) *Statistics for Evidence-Based Health Care.* Tertiary Press, Australia.
22. Muir Gray, J. A., ed. (1997) *Evidence-based healthcare: How to Make Policy and Management Decisions.* Churchill Livingstone, Edinburgh, UK.
23. Obeso, J. A. (2001) Deep-brain stimulation of the subthalamic nucleus or the pars interna of the globus pallidus in Parkinson's disease. *N. Engl. J. Med.* **345,** 956–963.
24. Borlongan, C., Isacson, O., and Sanberg, P. R. (2003) *Immunosuppressant Analogs in Neuroprotection.* Humana Press Inc., Totowa, NJ.

25. Langston, J., Widner, H., Goetz, C., et al. (1992) Core Assessment Program for Intracerebral Transplantations (CAPIT). *Mov. Disord.* **7,** 2–13.
26. Polgar, S., Morris, M., Reilly, S., et al. (2003) Reconstructive neurosurgery for Parkinson's disease: a systematic review and preliminary meta-analysis. *Brain Res. Bull.* **60,** 1–24.
27. Hauser, R, Freeman, T., Snow, B., et al. (1999) Long-term evaluation of bilateral fetal nigral transplantation in Parkinson's disease. *Arch. Neurol.* **56,** 179–187.
28. Kopyov, O., Jacques, D., Lieberman, A., et al. (1996) Clinical study of fetal mesencephalic intracerebral transplants for the treatment of Parkinson's disease. *Cell Transplant.* **5,** 327–337.
29. Cohen, J. (1988) *Statistical Power Analysis for the Behavioural Sciences, 2nd ed.* Lawrence Erlbaum Associates, Inc., New York, NY.
30. Olanow, C. W. (2004) The scientific basis for the current treatment of Parkinson's disease. *Annu. Rev. Med.* **55,** 41–60.
31. Rosser, A. E., Barker, R. A., Armstrong, R. J. E., et al. (2003) Staging and preparation of human fetal striatal tissue for neural transplantation in Huntington's disease. *Cell Transplant.* **12,** 679–686.
32. Watts, C., Donovan., T., Gillard, J. H., et al. (2003) Evaluation of an MRI-based protocol for cell implantation in four patients with Huntington's disease. *Cell Transplant.* **12,** 697–704.
33. Savulescu, J. (2002) The embryonic stem cell lottery and the cannibalization of human beings. *Bioethics* **16,** 508–529.
34. Dekkers, W. and Boer, G. (2001) Sham surgery in patients with Parkinson's disease: is it morally acceptable? *J. Med. Ethics* **27,** 151–156.
35. London, A. J. and Kadane, J. B. (2003) Sham surgery and genuine standards of care: can the two be reconciled? *Am. J. Bioeth.* **3,** 61–64.
36. Leeds, H. S. (2003) Social aspects of sham surgeries. *Am. J. Bioeth.* **3,** 70–71.
37. Macklin, R. (1999) The ethical problems with sham surgery in clinical research. *N. Engl. J. Med.* **341,** 992–996.
38. Albin, R. L. (2002) Sham surgery controls: intracerebral grafting of fetal tissue for Parkinson's disease and proposed criteria for use of sham surgery controls. *J. Med. Ethics* **28,** 322–325.
39. Miller, F. G. (2003) Sham-surgery: an ethical analysis. *Am. J. Bioeth.* **3,** 41–48.
40. Hrobjartsson, A. and Gotzsche, P. C. (2001) Is the placebo powerless? An analysis of clinical trials comparing placebo with no treatment. *N. Engl. J. Med.* **344,** 1594–1602.

13

Hematopoietic Cell Therapy for Brain Repair

Martina Vendrame and Alison E. Willing

ABSTRACT

While the discovery of neural stem cells revolutionized the field of neural transplantation, ethical and funding limitations have made the search for alternative cell sources imperative for stem cell researchers. Bone marrow and umbilical cord blood both harbor a population of stem cells and transplantation studies in multiple models of brain injury and disease have demonstrated proof of principle for these cells as possible therapeutics. In this chapter, we discuss the characteristics of cells from bone marrow, umbilical cord blood and granulocyte colony stimulating factor exposed peripheral blood and summarize the recent literature on the use of these cells in experimental models of central nervous system injury and disease. We will discuss the neural potential of these hematopoietic cells and possible mechanisms underlying reported behavioral recovery.

Key Words: Stem cells; bone marrow; umbilical cord blood; GCSF stimulated peripheral blood; trophic factors; anti-inflammatory; angiogenesis; neural differentiation.

INTRODUCTION

The increasing evidence showing that most tissues possess an endogenous regenerative capacity mediated by mobilized and/or resident stem cells has expanded not only new perspectives on human biology but also novel approaches for cell therapy. Stem cells from several different tissues have been isolated and characterized phenotypically and functionally. Neural stem cells that are able to renew and commit to neuronal, astrocytic, or oligodendroglial lineages have been traditionally derived from embryos, fetuses, and adult brains (1,2). Although this revolutionary discovery introduced a new era of brain cell therapy, it has concurrently generated ethical concerns regarding the acquisition and use of these cells. Therefore, the

From: *Contemporary Neuroscience: Cell Therapy, Stem Cells, and Brain Repair*
Edited by: C. D. Sanberg and P. R. Sanberg © Humana Press Inc., Totowa, NJ

search for alternative cell sources has become a new imperative for stem cell researchers. The identification of stem cells isolated from hematopoietic tissues, such as adult bone marrow and cord blood, which also demonstrate the ability to attain neuronal and glial properties, provides a potential solution.

This chapter outlines the most relevant characteristics of cells harvested from bone marrow and cord blood that have been shown to develop neuronal or glial traits in vitro and to induce functional recovery in models of neurodegenerative diseases.

BONE MARROW AS A SOURCE OF CELLS FOR BRAIN REPAIR

Hematopoietic and Nonhematopoietic Stem Cells

Hematopoietic stem cells (HSCs) derived from bone marrow were first identified for their ability to reconstitute blood lineages; they were subsequently shown to differentiate into skeletal and cardiac myocytes, endothelial cells, and hepatocytes. Surprisingly, bone marrow–derived HSCs have been recently described to *trans*-differentiate into neuroectodermic cell types, e.g., neurons and glial cells *(3)*. Despite some skepticism regarding their capacity to generate cells of such dissimilar lineage, an ever-increasing body of literature suggests that development of neurons and astrocytes from bone marrow–derived HSCs is a bonafide phenomenon. These studies have introduced the option of using an ethical, and easily available, source of stem cells to treat neurological deficits.

In addition to HSCs, bone marrow contains nonhematopoietic precursors cells, referred to as *mesenchymal* or *bone marrow stromal stem cells* (BMSCs). This cell population is also known as colony-forming unit fibroblasts, because it is composed of fibroblast-like cells, which have been shown to contain progenitors capable of generating bone, cartilage, and adipocytes. Several reports demonstrated that BMSCs can differentiate in vitro into glia and neurons *(3–7)*.

BMSCs are generally separated by adherence to plastic after depletion of HSCs (CD34$^+$, CD45$^+$, and CD11bc$^+$) from the cultures *(3,4,7)*. Several factors are used to induce the neural differentiation: retinoic acid (RA; *3*); growth factors (e.g., brain-derived neurotrophic factor [BDNF], nerve growth factor [NGF], and fibroblast growth factor [FGF; *3,5,7*]); demethylating agents, such as 5-Aza-C *(5)*; cyclic adenosine monophosphate–inducing agents *(6)*; and inhibitors of specific phenotypes, such as noggin *(5)*. A number of different markers of neural phenotype have been used to demonstrate that BMSCs can attain a neuronal fate, at least in vitro. The most common markers that have been used are neuronal precursor, nestin *(3,4)*, nuclear factor neuronal N (NeuN; *3,5*), neuron-specific enolase (NSE; *4,6*), and class III β-tubulin *(3,7)*. Additionally, expression of the glial fibrillary astrocytic protein (GFAP) has been generally associated with a restricted glial fate *(3,7)*. Verfaillie and colleagues showed a very significant demonstration of the multipotent capability of a stem/progenitor cell generated from BMSCs, when cells derived from a single expanded clone developed into neurons and astrocytes throughout the brain after implantation into the blastocyst *(8)*.

Recently, two studies have been published that make these observations questionable. Lu et al. *(9)* indicated that in the presence of stressors, a number of different cell types, including BMSCs, adopted a neuronal morphology and expressed neuronal markers. However, this altered phenotype occurred as a function of cell shrinkage and in the absence of protein synthesis. Similarly, in the second study *(10)*, with the neural induction media used by Woodbury and associates *(4,11)*, there was a rapid disruption of the actin cytoskeleton and retraction of cellular cytoplasm. The processes did not have growth cones or express synaptophysin at any time. Furthermore, as soon as the induction media was removed, the cells reverted to their original morphology. Clearly, further study is needed to clarify these issues.

Bone Marrow Cell Transplantation for Brain Repair

Several independent reports have confirmed the observation that systemically infused BMSCs can repopulate nonhematopoietic tissues and acquire the cellular phenotypes of the tissues they repopulate *(12)*. To assess whether BMSCs could also reconstitute central nervous system (CNS) tissues, bone marrow cells genetically marked with a retroviral tag were injected intravenously in sublethally irradiated WBB6F1yJ-KitW/KitW-v mice (considered good recipients for bone marrow transplant because they possess genetically defective hematopoiesis; *13*). Weeks after the injection, tagged cells were seen in the brain, where they were widely distributed in the cortex, hippocampus, thalamus, cerebellum, and brain stem. These cells expressed either microglial/macrophage markers (F4/80) or GFAP, suggesting that precursors resident in the bone marrow were able to reconstitute CNS glial populations *(13)*. Others have confirmed this finding *(14,15)* and further demonstrated that infused bone marrow–derived cells, after grafting into the brain, can also develop neural traits, as seen with expression of the NeuN *(16,17)*, NSE *(17)*, class III β-tubulin, and the 200-kDa neurofilament *(16)*.

BMSCs have also been transplanted intracerebrally. When injected into the striatum of albino rat brain, BMSCs engrafted and migrated from the transplant site to different CNS areas along known pathways of neural stem cell migration *(18)*, indicating that these cells can respond to local cues. Moreover, after engraftment, these cells lost markers typical of cultured BMSCs and developed phenotypes similar to astrocytes.

Considering these results in nondiseased animals, bone marrow–derived cells have been subsequently studied in animal models of neurodegenerative disease, with the theory that these cells might also be able to reconstitute damaged tissues and subsequently restore brain function. The most relevant studies employing bone marrow–derived cells in models of neurodegenerative diseases are presented in Table 1.

In the middle cerebral artery occlusion (MCAO) rodent model of focal brain ischemia, transplantation of BMSCs either intrastriatally or intravascularly found that the BMSCs homed to the ischemic boundary zone and bore neuronal and glial antigens *(19)*. Another mechanism by which these cells may induce recovery is through the promotion of neovascularization processes in the infarct penumbra by inducing expression of vascular endothelial growth factor (VEGF) and its receptor *(20)*.

There has also been extensive study regarding the use of BMSCs in a rodent model of traumatic brain injury (TBI). Whole bone marrow, bone marrow cells cultured with BDNF, and NGF, bone marrow–derived stromal spheres cultured with embryonic neurospheres have been transplanted either intracerebrally or intravascularly in the controlled cortical impact injury mouse model *(21–26)*. The transplantation of these cells induced significant functional recovery, as seen with the rotarod test and the modified neurological severity score *(23,26)*. Additionally, BMSCs have been shown to integrate into the brain parenchyma and acquire a neuronal or glial phenotype, as shown by the expression of NeuN *(23–25)*, microtubule-associated protein (MAP-2) *(22,23)*, and GFAP *(22,23,26)*.

BMSCs have been recently tested in models of Parkinson's disease. Because of its defined pathogenesis, this disease has been considered a good target for cell replacement therapy. Transplantation of BMSCs in the corpus striatum of the 1-methyl-4-phenyl-1,2,3,6-tetrahydropyridine mouse model of Parkinson's disease has shown that BMSCs express the dopaminergic synthetic enzyme tyrosine hydroxylase and induce behavioral recovery in this model *(27)*.

Additionally, it is possible to engineer BMSCs to overexpress specific genes, thereby increasing their therapeutic efficiency. Given that the standard treatment for Parkinson's disease (3,4-dihydroxyphenylalanine [L-dopa]) has limited long-term benefit, therapy with transduced cells may provide a constant and well-tolerated source of the lacking compounds. Investigators have examined the possibility of using several different promoters of the two genes necessary for the cells to synthesize L-dopa, introducing them in a self-inactivating retrovirus (pSIR) or standard retroviruses. pSIR vectors are constructed using the mouse phosphoglycerate kinase-1 promoter or the cytomegalovirus promoter to drive expression of a GFP reporter gene or a bicistronic sequence containing the genes for human tyrosine hydroxylase type I and rat GTP cyclohydrolase I. Such transduced BMSCs express GFP and are able to synthesize and secrete L-dopa (89–283 pmol/10^6cells/h). Additionally, engineered BMSCs can be cultured and expanded more than 1000-fold in 4 wk while they continue to express GFP or produce L-dopa *(28)*. Transduced BMSCs have been transplanted into the corpus striatum of 6-hydroxydopamine-lesioned rats, where they engrafted, produced L-dopa and metabolites, and promoted functional recovery *(28)*.

Functional Recovery Induced by Bone Marrow–Derived Cells

The functional recovery induced by the transplantation of transfected cells can be easily attributable to their expression of introduced compounds. However, in the majority of experiments, naïve cells harvested from bone marrow have been used. The mechanism by which this bone marrow–based cell therapy appears to be beneficial is not fully understood. Not all the investigators believe that a real neuronal/glial differentiation of these cells is possible. Even accepting the belief that bone marrow cells truly develop neuronal or glial characteristics in vivo, the functionality of these cells as neurons or glia has not yet been proven. One emerging theory is that bone marrow–derived cells may induce the differentiation of resident totipotent stem cells into functional cells that reconstitute damaged neurons. An increase in endogenous neural stem cell proliferation within the subventricular zone and hippocampus has been observed after BMSC transplantation *(23)*. Alternatively, the observed functional recovery may be the result of the increased expression of neurotrophic/growth factors. This phenomenon could be owing to the endogenous production of growth factors induced by BMSCs or by the direct delivery of these factors from the transplanted cells *(29,30)*. In vitro studies employing TBI-conditioned BMSC cultures have shown a time-dependent increase in BDNF, NGF, VEGF, and hepatocyte growth factor, indicating a responsive production of these growth factors by the BMSCs *(30)*.

Recent reports suggest additional mechanisms that transplanted BMSCs may undergo. For instance, the expression of neural antigens by transplanted stem cells may be explained by their fusion with endogenous cells *(31)*. This hypothesis, known as the "fusion theory," is supported by in vitro coculture experiments that show embryonic stem cells are able to fuse with

Table 1
Summary of Bone Marrow Stromal Cell Transplantation Studies

Animal model	Type of cells	Route of delivery	Behavioral tests/benefit	Phenotype of cells in vivo	Reference
MCAO rats	Whole marrow from SHR rats	Intravenous (i.v.) tail vein (3.25×10^7)	None	GFAP	72
MCAO rats		Intracarotid		Localized in CNS by MAP-2, GFAP	73
MCAO rats	Human marrow stromal cells	i.v. femoral vein	None	HuNu (MAB1281), GFAP, VEGF	20
TBI rats	Whole bone marrow from adult Wistar rats labeled with BrdU	Intraparenchymal adjacent to contusion site (10^6)	Rotarod	Localized in CNS by NeuN, MAP-2, and GFAP	23
TBI rats	Mesenchymal stem cells (MSCs) from male Wistar rats selected by adherence to plastic	i.v. into tail vein (2×10^6)	Rotarod; modified (m) neurological severity score (NSS)	NeuN, GFAP	24
TBI rats	MSCs from Wistar rats selected by adherence to plastic	Into internal carotid artery		MAP-2, NeuN, and GFAP	21

TBI rats	MSCs from Wistar rats cultured with or without BDNF and NGF	i.v. into tail vein (10^6)	Rotarod; mNSS	NeuN, MAP-2, and GFAP	25
TBI rats	Embryonic neurospheres and bone marrow–derived stromal cell spheres	i.v. tail vein	Rotarod; NSS score		25
TBI rats	Human whole bone marrow	i.v. tail vein $(1–2 \times 10^6)$	Rotarod; mNSS	Expressing TuJ1, GFAP	26
TBI rats		i.v. tail vein	Rotarod; mNSS	Increased expression of NGF and BDNF	29
6-hydroxy-dopamine (Parkinson's)	L-dopa and GTP cyclohydroxylase I transfected	Intrastriatal	Reduction of apomorphine-induced rotations; L-dopa and metabolites detected with microdialysis		28
MPTP mice (Parkinson's)	MSCs from C57BL/6 mice labeled with 5-FU	Intrastriatal (3×10^5)	Rotarod	Tyrosine hydroxylase	19

bone marrow cells or neural stem cells *(31,32)*. However, this issue remains contentious. A recent study reported that as many as 6% of neural stem cells cocultured with endothelial cells transdifferentiated into endothelial cells, without any evidence of cell fusion *(33)*. If cell fusion does contribute to apparent expression of neural markers in the transplanted BMSCs, this should not discount any possible beneficial effect; e.g., transplanted cells may provide healthy genetic material to a cell otherwise undergoing necrosis. However, whether expression of aneuploid nuclear material may lead to tumorogenetic processes has yet to be resolved.

HUMAN CORD BLOOD CELLS

As cord blood has been used extensively as an alternative source of HSCs in allogeneic stem cell transplantation for the treatment of acquired and genetic diseases, using this new source of transplantable stem cell for CNS cell therapy has been very appealing. The first related human umbilical cord blood (HUCB) transplant was performed in the 1970s in a child with acute lymphoblastic leukemia *(34,35)*. Since then, thousands of HUCB transplants have been performed worldwide, with most performed as a therapy for hematological malignancies *(36)*. From these early experiences, some clear advantages using HUCB cells became evident. First, the HUCB is readily available. Second, there is a lower incidence of graft-versus-host disease (GVHD) with HUCB transplants, compared to bone marrow transplant. Third, the risk of viral transmission with HUCB transplant is minimal *(36)*.

Phenotypic and Functional Characteristics of Cord Blood Stem Cells

Stem cells isolated from cord blood have been identified as belonging to a subpopulation expressing CD34—a common marker for human HSCs. This population constitutes about 1% of the heterogeneous cell population within the mononuclear fraction of the HUCB. This population contains a set of subtypes that, when differentiated, bear different CD markers based on the type of cell they become (Table 2).

Accumulating evidence supports the notion that, when exposed to defined culture conditions, cord blood cells undergo a phenotypic conversion into neuronal and glial lineages *(37)*. Using a combination of reverse-transcriptase polymerase chain reaction (RT-PCR), western blots, and immunohistochemistry, these authors demonstrated that exposure of HUCB cells to RA and NGF increased the expression of many proteins associated with a neural phenotype, including musashi, class III β tubulin, glypican 4 and pleiotropin, and GFAP. Although this was the first demonstration that a population of HUCB could express neural proteins, other research groups

Table 2
Antigen Expression of HUCB Mononuclear Cells

Subtypes of mononuclear fraction	Properties	References
CD34$^+$ whole subtype	Heterogeneous population of progenitors cells with high proliferative capacity and long-term bone marrow reconstituting properties	74
CD34$^+$CD117$^+$	Population of progenitors cells with long-term engraftment potential. CD 117 (c-KIT) is a transmembrane, tyrosine kinase growth factor receptor present on several fetal and adult cells, like hematopoietic cells, mast cells, melanocytes, and germ cells. c-KIT and its ligand stem cell factor (SCF) are involved in the control of several tissues at different stages of life and in the adhesion of stem cells to the microenvironment.	75–77
CD34$^+$CD133$^+$	This subtype has been shown to express angiopoietin-1 (Ang-1), Ang-2, and VEGF and their receptors, indicating a role of these cells in control of angiopoiesis and hematopoiesis.	65,78
CD34$^+$CD38$^-$	The number of CD34$^+$CD38$^-$ cells is shown to correlate with the number of committed progenitors cells and the capability of producing CD34$^+$ cells.	79,80
CD31$^+$	CD31, also called PECAM-1, is the platelet-endothelial cell adhesion molecule expressed on several mature hemopoietic cell types and endothelia. It functions as an adhesion molecule in cell migration and inflammation. This fraction of MNC can express CD34, but it is generally believed to be a more mature subtype and have a role in the adhesion of CB cells to the microenvironment.	81,82
Monocytes	Express CD11, CD49, HLA-DR CD31, and CD62L, CD14, CD18	83
Lymphocytes	50% T lymphocytes (40% CD3$^+$/CD4$^+$ and 10% CD3$^+$/CD8$^+$). Expansion of specific T-cell subsets can be done by modulating cytokines SCF, IL-7, IL-2	82,84

have confirmed and expanded these results *(38–41)*. Jang et al. *(41)* isolated a CD133+ cell population that became bipolar and expressed nestin and musashi, as well as neural filament, class III β tubulin, NSE, MAP-2, and NeuN in response to RA exposure. The CD133− population never expressed these markers and only became burst-forming unit-erythrocyte colonies. Buzanska et al. *(38)* have used magnetic cell sorting and subfractionation based on the expression of cell surface antigens to isolate a HUCB clone that expresses nestin and can be differentiated to all three neural phenotypes. Interestingly, this study suggests that the stem/progenitor cell that gives rise to the neural phenotypes is derived from CD34−/nestin+ cells, not the CD34+ HSC. These latter results are consistent with the observation that umbilical cord blood harbors a population of mesenchymal stem cells similar to bone marrow stem cells that can be induced to express neural proteins *(42–45)*.

Cord Blood Cells for Therapy of Neurodegenerative Disorders

Cord blood mononuclear cells have been used for transplantation in models of stroke, brain injury, Parkinson's disease, and amyotrophic lateral sclerosis (ALS). According to these and other investigators, intracerebral transplant or intravenous injection of cord blood can significantly induce behavioral recovery in animal models (Table 3).

The most advanced work has been conducted in stroke animal models. The first report of benefit from HUCB transplant in an animal model of stroke occurred in 2001, when HUCB was infused in the MCAO model, and significant functional recovery was observed *(46)*. Later, intravenous and intracerebral delivery of HUCB cells were compared; a surprising revelation was that intravenous delivery was equal to, or even more effective at, enhancing stroke recovery than intracerebral delivery *(47)*.

The intravenous delivery route is a practical noninvasive route of administration that nonspecialized personnel can perform. Additionally, the efficacy of treatment is shown when cells are delivered 24–48 h after the cerebrovascular accident. This time window of effective treatment is longer than that for currently available acute stroke treatments, which require administration within minutes to hours after stroke. Although the full extent of the therapeutic window for HUCB administration after stroke has yet to be examined systematically, this longer treatment window means that time is available for careful screening and testing of potential subjects for stroke recovery trials, both in terms of size and location of cerebral infarcts and in type and severity of neurological deficits.

The first study to examine the use of HUCB cells in a mouse model of ALS occurred in 2000, when HUCB cells were delivered intravenously into the SOD1-G93A mice (overexpressing human SOD1 and carrying the Gly93 Ala mutation; *48*), and an increase in life span was observed. The same group of researchers had previously observed that HUCB could improve the course of autoimmune disease by giving HUCB to MRL Lpr/Lpr mice (that have an autoimmune disease similar to lupus in humans); the lifetime of the animals could be significantly increased, and the onset of pathological changes could be delayed *(49)*. Considering that ALS has also been considered by several authors to be an autoimmune disease *(50)*, the effect of HUCB could be exerted through an immune regulatory mechanism. Recently, other investigators have shown that HUCB can prolong the life span of ALS mice and further delay the onset of pathology *(51)*.

As with BMSCs, cord blood–derived cells have also been thought to be beneficial in models of TBI. Similar to the results of the previous study conducted by Chen, HUCB cell injection notably reduced motor and neurological deficits, as measured by the rotarod test and the neurological severity score *(52)*. Intravenously injected cells appeared to preferentially enter the brain, migrate into the parenchyma of the injured brain and expressed a neuronal trait, as seen by the expression of NeuN and MAP-2, as well as an astrocytic trait, as shown by positivity for GFAP. Some HUCB cells integrated into the vascular walls within the boundary zone of the injured area.

Mechanisms Underlying Behavioral Recovery Induced by Cord Blood Stem Cells

The mechanism(s) by which transplanted HUCB cells induce functional benefit in these animal models is not clear. The majority of the studies use intravenous injection as the route of cell transplantation. One concern about experiments involving systemic infusion of HUCBs is whether the cells can reach and engraft into the brain. This concern is particularly pertinent to reports in which HUCBs were detected in relatively small numbers within brain tissue.

Some studies have demonstrated the presence of HUCB cells in the brain after intravenous delivery *(46,51,52)*. The studies by Chen et al. and Lu et al. used models of stroke and TBI, respectively, and delivered approx 10^6 cells intravenously. Cells were localized in the CNS 2–4 wk after transplant through positive immunoreactivity for human nuclei (HuNu). Quantification of cells revealed the presence of about 30,000 cells in the hemisphere ipsilateral to the injury. When Garbuzova-Davis implanted less than half of the dose of the previous two studies in an ALS mouse model, cells were localized to a variety of brain regions through positive immunoreactivity for HuNu 10–12 wk after transplantation.

Table 3
Summary of HUCB Cell Transplantation Studies

Animal model	Number of cells delivered	Route of delivery	Behavioral tests/benefit	Localization of cells in vivo	Reference
SOD1 mice (ALS)	$3.4 \times 10^7 - 3.5 \times 10^7$	i.v. retro-ocular	Prolonged survival	By RT-PCR cells found in spleen, liver, and lung	48
SOD1 mice (ALS)	$7.0 \times 10^7 - 7.3 \times 10^7$	i.v. retro-ocular	Prolonged survival	By RT-PCR cells found in spleen, liver, and lung	85
G93A SOD1 mice (ALS)	1×10^6	i.v. jugular vein	Prolonged survival Reduced weight loss Delayed progression of disease	Localized in CNS by HuNu. Positive for nestin, CD45, Tuj1, and GFAP. Cells positive for HuNu, CD45, and CD43 found also in spleen, liver, lung, kidneys, and heart	51
MCAO rats	3×10^6	i.v. tail vein	Rotorod test; mNSS	Localized in CNS by HuNu. Positive for NeuN, MAP-2, and GFAP	46
MCAO rats	2.5×10^5 1×10^6	Intrastriatum; i.v. femoral vein	Reduction of spontaneous activity Passive avoidance test Step test	No cells localized in CNS in animals transplanted intrastriatally No cells localized in CNS in animals transplanted i.v.	47

			Rotorod test mNSS	Localized in CNS by HuNu. Positive for NeuN, MAP-2, and GFAP	
TBI rats	2×10^6	i.v. tail vein		Localized in CNS by HuNu. Positive for NeuN, MAP-2, and GFAP	52
APPsw2676 mice (Alzheimer's)	11×10^7	i.v. retro-ocular	Prolonged survival	Not reported	86
B6CBACa-AW-J/A-K cnj6 mice (Parkinson's)	$10–11 \times 10^7$	i.v. retro-ocular	Prolonged survival	Not reported	87
B6CBA-TgN (Hdexon1) 62Gpb mice (Huntington)	$7.1 \times 10^7 - 10 \times 10^7$	i.v. retro-ocular	Prolonged survival	Not reported	88

The results of these studies suggest that intravenously delivered HUCB cells home to the brain; however, several other studies employing intravenous injection of HUCB failed to find these cells within the CNS. In the two studies by Ende et al. in 2000 *(48,53)*, intravenous injections of approx 3×10^7 cells and a higher dose of approx 7×10^7 cells were administered in a model of ALS. The latter dose represents the highest dose of HUCB ever used in neurological studies. There was prolonged survival in the animals treated with the first dose and even more prolonged survival in animals treated with the second dose, showing a dose-dependent beneficial effect. RT-PCR was used for detecting human genes in the CNS and peripheral organs, and no human DNA was found in the brain, whereas human DNA was detected in the spleen and lymph nodes. The authors speculated that the increased survival of SOD1 mice receiving cord blood could be explained by the HUCB cells directly or indirectly providing adequate (nonmutant) superoxide dismutase, thereby delaying the onset of death. As discussed previously, ALS has also been considered an autoimmune disease *(50)*; thus, the finding of HUCB in secondary immune organs may suggest a regulatory role of HUCB in the immunopathogenesis of ALS.

The fact that behavioral benefits occur even though HUCB cells were not found in the CNS implies that mechanisms other than local effects of transplanted cells may be responsible for the observed behavioral phenomenon. For instance, HUCB cells may provide a source of growth factors or cytokines, and these released intercellular factors may be the cause of regulation of survival, proliferation, and likely differentiation of endogenous cells *(54)*. Moreover, cytokines from HUCB cells may regulate endogenous production of other cytokines, therefore contributing to complex intercellular communications regulating cell proliferation and the immune/inflammatory response. Such growth factors as epidermal growth factor and FGF have been shown to be involved in the proliferation and differentiation of brain-resident neuronal precursor cells *(55–60)*, and it has already been demonstrated that exogenously administered neurotrophic growth factors may limit the extent of acute ischemic neural injury *(59,61,62)*.

HUCB cells produce several growth factors and cytokines. Recently, CD34-positive cells from umbilical cord blood were shown to express NGF and its receptor, TrkA *(63)*. Umbilical cord blood cells have also been shown to selectively produce large amounts of interleukin (IL)-10 after stimulation with an anti-CD3 antibody and IL-2, therefore inducing a Th1/Th2 switch response *(64)*. This phenomenon has been thought to be responsible for the decreased severity and less frequent incidence of GVHD observed after cord blood transplants, compared to bone marrow transplants. Furthermore, other investigators have demonstrated that CD34$^+$/CD133$^+$ cord blood cells express such angiogenic factors as angiopoietin-1, angiopoietin-2, and VEGF, as well as their receptor mRNAs, suggesting these cells have a role in regulation of both angiopoiesis and hematopoiesis *(65)*. The importance of these findings is that it indicates a possible mechanism by which the behavioral benefits occur after transplantation of HUCB in MCAO models *(46,47,52)*.

G-CSF–STIMULATED PERIPHERAL BLOOD FOR BRAIN REPAIR

Similar to cord blood, peripheral blood progenitor cells (PBPCs) have been used during the last few decades as an alternative to bone marrow for hematogenic reconstitution. These cells are easily obtained by apheresis procedures after daily injections of the donor with granulocyte colony-stimulating factor (G-CSF; *66*), and their transplant leads to a faster engraftment and less therapeutic failure, compared to bone marrow transplants *(67)*.

Recently, researchers have reported that infusion of PBPCs can ameliorate functional deficits in the MCAO model of ischemic stroke, inducing benefits comparable to cord blood *(68)*. Although PBPCs can induce a significant recovery 1 mo after the transplant, it is not clear if the mechanism of action is similar to that of cord blood or bone marrow–derived cells.

Other investigators have attempted to mobilize endogenous hematopoietic progenitors from the bone marrow by stimulation of the recipient with G-CSF *(69,70)*. In the study by Six et al., mice receiving a subcutaneous injection of G-CSF (50 µg/kg) 24 h after MCAO had increased survival rates and decreased brain infarct volume 4 d after the injection *(70)*. However, although it has been speculated that these effects are because of the mobilization of endogenous progenitors from the bone marrow, others have demonstrated that G-CSF also displays a significant neuroprotective effect in vitro, and that this effect is mediated by the presence of G-CSF receptors on neurons *(69)*.

CONCLUSIONS

Clearly, hematotherapy holds promise as a therapeutic intervention strategy for neurological patients, as the cells from hematogenic sources are easily obtainable, can be expanded in culture, and can be delivered without the necessity of invasive surgical procedures. However, a limited understanding still exists regarding the biology of these cells in vivo in response to brain injury and disease; much more work lies ahead to further characterize their plasticity and behavior. Whether cells harvested from bone marrow, cord blood, or stimulated peripheral blood can improve neurological outcome by replacing dead neurons, providing trophic support, and/or by enhancing endogenous mechanisms of recovery, remains unknown. Given the heterogeneity of the pathogenesis of neurodegenerative disorders and injuries, hematotherapy's efficacy may have to be assessed by examining each disorder separately.

Other issues need to be further addressed, such as the safety of these transplants. One major risk of all allotransplants is the risk of host immune reactions or GVHD. However, the use of immunosuppressive therapies to counteract these risks puts the patients at risk for serious side effects and opportunistic infections. Their use in clinical trials will complicate the interpretation of the observed therapeutic effect from the transplants. Fortunately, the possibility of harvesting these cells from the patients themselves (or having it banked from birth, as for the cord blood) will circumvent all issues related to allogenic transplantations. Alternatively, it is not yet clear that immunosuppression will be necessary. In those studies that have been completed in animals, some have used immunosuppression and others have not, but all have shown improved functional outcome. There are also promising results from ongoing clinical trials employing BMSCs in ALS patients that show a low incidence of complications *(71)*.

It is important to realize that although experimental results are very promising, the effectiveness of these transplants in humans remains unknown. The benefit of the transplant will be obviously affected by such factors as the severity, stage, and site of the pathology of the neurodegenerative disorder affecting the recipient. For instance, in diseases like stroke, the location and extent of the neuronal damage represent factors that need to be assessed before the transplant is performed. It will be essential to establish criteria based on clinical and pathological parameters to screen which patients will be suitable candidates for transplantation.

REFERENCES

1. Reynolds, B. A., Tetzlaff, W., and Weiss, S. (1992) A multipotent EGF-responsive striatal embryonic progenitor cell produces neurons and astrocytes. *J. Neurosci.* **12,** 4565–4574.
2. Weiss, S., Reynolds, B. A., Vescovi, A. L., et al. (1996) Is there a neural stem cell in the mammalian forebrain? *Trends Neurosci.* **19,** 387–393.
3. Sanchez-Ramos, J., Song, S., Cardozo-Pelaez, F., et al. (2000) Adult bone marrow stromal cells differentiate into neural cells in vitro. *Exp. Neurol.* **164,** 247–256.
4. Woodbury, D., Schwarz, E. J., Prockop, D. J., and Black, I. B. (2000) Adult rat and human bone marrow stromal cells differentiate into neurons. *J. Neurosci. Res.* **61,** 364–370.

5. Kohyama, J., Abe, H., Shimazaki, T., et al. (2001) Brain from bone: efficient "meta-differentiation" of marrow stroma-derived mature osteoblasts to neurons with Noggin or a demethylating agent. *Differentiation* **68,** 235–244.

6. Deng, W., Obrocka, M., Fischer, I., and Prockop, D. J. (2001) In vitro differentiation of human marrow stromal cells into early progenitors of neural cells by conditions that increase intracellular cyclic AMP. *Biochem. Biophys. Res. Commun.* **282,** 148–152.

7. Reyes, M. and Verfaillie, C. M. (2001) Characterization of multipotent adult progenitor cells, a subpopulation of mesenchymal stem cells. *Ann. NY Acad. Sci.* **938,** 231–233; discussion 233–235.

8. Keene, C. D., Ortiz-Gonzalez, X. R., Jiang, Y., et al. (2003) Neural differentiation and incorporation of bone marrow-derived multipotent adult progenitor cells after single cell transplantation into blastocyst stage mouse embryos. *Cell Transplant.* **12,** 201–213.

9. Lu, P., Blesch, A., and Tuszynski, M. H. (2004) Induction of bone marrow stromal cells to neurons: differentiation, transdifferentiation, or artifact? *J. Neurosci. Res.* **77,** 174–191.

10. Neuhuber, B., Gallo, G., Howard, L., et al. (2004) Reevaluation of in vitro differentiation protocols for bone marrow stromal cells: disruption of actin cytoskeleton induces rapid morphological changes and mimics neuronal phenotype. *J. Neurosci. Res.* **77,** 192–204.

11. Woodbury, D., Reynolds, K., and Black, I. B. (2002) Adult bone marrow stromal stem cells express germline, ectodermal, endodermal, and mesodermal genes prior to neurogenesis. *J. Neurosci. Res.* **69,** 908–917.

12. Pereira, R. F., O'Hara, M. D., Laptev, A. V., et al. (1998) Marrow stromal cells as a source of progenitor cells for nonhematopoietic tissues in transgenic mice with a phenotype of osteogenesis imperfecta. *Proc. Natl. Acad. Sci. USA* **95,** 1142–1147.

13. Eglitis, M. A. and Mezey, E. (1997) Hematopoietic cells differentiate into both microglia and macroglia in the brains of adult mice. *Proc. Natl. Acad. Sci. USA* **94,** 4080–4085.

14. Kopen, G. C., Prockop, D. J., and Phinney, D. G. (1999) Marrow stromal cells migrate throughout forebrain and cerebellum, and they differentiate into astrocytes after injection into neonatal mouse brains. *Proc. Natl. Acad. Sci. USA* **96,** 10711–10716.

15. Nakano, K., Migita, M., Mochizuki, H., and Shimada, T. (2001) Differentiation of transplanted bone marrow cells in the adult mouse brain. *Transplantation* **71,** 1735–1740.

16. Brazelton, T. R., Rossi, F. M., Keshet, G. I., and Blau, H. M. (2000) From marrow to brain: expression of neuronal phenotypes in adult mice. *Science* **290,** 1775–1779.

17. Mezey, E., Chandross, K. J., Harta, G., et al. (2000) Turning blood into brain: cells bearing neuronal antigens generated in vivo from bone marrow. *Science* **290,** 1779–1782.

18. Azizi, S. A., Stokes, D., Augelli, B. J., et al. (1998) Engraftment and migration of human bone marrow stromal cells implanted in the brains of albino rats—similarities to astrocyte grafts. *Proc. Natl. Acad. Sci. USA* **95,** 3908–3913.
19. Li, Y. and Chen, J., and Chopp, M. C. (2001) Adult bone marrow transplantation after stroke in adult rats. *Cell Transplant.* **10,** 31–40.
20. Chen, J., Zhang, Z. G., Li, Y., et al. (2003) Intravenous administration of human bone marrow stromal cells induces angiogenesis in the ischemic boundary zone after stroke in rats. *Circ. Res.* **92,** 692–699.
21. Lu, D., Mahmood, A., Wang, L., et al. (2001) Adult bone marrow stromal cells administered intravenously to rats after traumatic brain injury migrate into brain and improve neurological outcome. *Neuroreport* **12,** 559–563.
22. Lu, D., Li, Y., Wang, L., et al. (2001) Intraarterial administration of marrow stromal cells in a rat model of traumatic brain injury. *J. Neurotrauma* **18,** 813–819.
23. Mahmood, A., Lu, D., Yi, L., et al. (2001) Intracranial bone marrow transplantation after traumatic brain injury improving functional outcome in adult rats. *J. Neurosurg.* **94,** 589–595.
24. Mahmood, A., Lu, D., Wang, L., et al. (2001) Treatment of traumatic brain injury in female rats with intravenous administration of bone marrow stromal cells. *Neurosurgery* **49,** 1196–1203; discussion 1203–1204.
25. Mahmood, A., Lu, D., Wang, L., and Chopp, M. (2002) Intracerebral transplantation of marrow stromal cells cultured with neurotrophic factors promotes functional recovery in adult rats subjected to traumatic brain injury. *J. Neurotrauma* **19,** 1609–1617.
26. Mahmood, A., Lu, D., Lu, M., and Chopp, M. (2003) Treatment of traumatic brain injury in adult rats with intravenous administration of human bone marrow stromal cells. *Neurosurgery* **53,** 697–702; discussion 702–703.
27. Li, Y., Chen, J., Wang, L., et al. (2001) Intracerebral transplantation of bone marrow stromal cells in a 1-methyl-4-phenyl-1,2,3,6-tetrahydropyridine mouse model of Parkinson's Disease. *Neurosci. Lett.* **316,** 67–70.
28. Schwarz, E. J., Alexander, G. M., Prockop, D. J., and Azizi, S. A. (1999) Multipotential marrow stromal cells transduced to produce L-DOPA: engraftment in a rat model of Parkinson disease. *Hum. Gene Ther.* **10,** 2539–2549.
29. Mahmood, A., Lu, D., and Chopp, M. (2004) Intravenous administration of marrow stromal cells (MSCs) increases the expression of growth factors in rat brain after traumatic brain injury. *J. Neurotrauma* **21,** 33–39.
30. Chen, X., Katakowski, M., Li, Y., et al. (2002) Human bone marrow stromal cell cultures conditioned by traumatic brain tissue extracts: growth factor production. *J. Neurosci. Res.* **69,** 687–691.
31. Terada, N., Hamazaki, T., Oka, M., et al. (2002) Bone marrow cells adopt the phenotype of other cells by spontaneous cell fusion. *Nature* **416,** 542–545.
32. Ying, Q. L., Nichols, J., Evans, E. P., and Smith, A. G. (2002) Changing potency by spontaneous fusion. *Nature* **416,** 545–548.
33. Wurmser, A. E., Nakashima, K., Summers, R. G., et al. (2004) Cell fusion-independent differentiation of neural stem cells to the endothelial lineage. *Nature* **430,** 350–356.

34. Ende, M. and Ende, N. (1972) Hematopoietic transplantation by means of fetal (cord) blood. A new method. *Virginia Med. Monthly* **99,** 276–280.

35. Bandini, G., Bonifazi, F., and Baccarani, M. (2003) 15 or 33 years of cord-blood transplantation? *Lancet* **361,** 1566–1567.

36. Lewis, I. D. (2002) Clinical and experimental uses of umbilical cord blood. *Int. Med. J.* **32,** 601–609.

37. Sanchez-Ramos, J. R., Song, S., Kamath, S. G., et al. (2001) Expression of neural markers in human umbilical cord blood. *Exp. Neurol.* **171,** 109–115.

38. Buzanska, L., Machaj, E. K., Zablocka, B., et al. (2002) Human cord blood-derived cells attain neuronal and glial features in vitro. *J. Cell Sci.* **115,** 2131–2138.

39. Ha, Y., Choi, D. H., Yeon, D. S., et al. (2001) Neural phenotype expression cultured human cord blood cells in vitro. *Neuroreport* **12,** 3523–3527.

40. Ha, Y., Lee, J. E., Kim, K. N., et al. (2003) Intermediate filament nestin expressions in human cord blood monocytes (HCMNCs). *Acta Neuro-chirurgica.* **145,** 483–487.

41. Jang, Y. K., Park, J. J., Lee, M. C., et al. (2004) Retinoic acid-mediated induction of neurons and glial cells from human umbilical cord-derived hematopoietic stem cells. *J. Neurosci. Res.* **75,** 573–584.

42. Lee, O. K., Kuo, T. K., Chen, W. M., et al. (2004) Isolation of multipotent mesenchymal stem cells from umbilical cord blood. *Blood* **103,** 1669–1675.

43. Lee, M. W., Choi, J., Yang, M. S., et al. (2004) Mesenchymal stem cells from cryopreserved human umbilical cord blood. *Biochem. Biophys. Res. Commun.* **320,** 273–278.

44. McGuckin, C. P., Forraz, N., Allouard, Q., and Pettengell, R. (2004) Umbilical cord blood stem cells can expand hematopoietic and neuroglial progenitors in vitro. *Exp. Cell Res.* **295,** 350–359.

45. Minjun, Y., Zhifeng, X., Shen, L., and Li, L. (2004) Mid-trimester fetal blood-derived adherent cells share characteristics similar to mesenchymal stem cells but full-term umbilical cord blood does not. *Br. J. Haematol.* **124,** 666–675.

46. Chen, J., Sanberg, P. R., Li, Y., et al. (2001) Intravenous administration of human umbilical cord blood reduces behavioral deficits after stroke in rats. *Stroke* **32,** 2682–2688.

47. Willing, A. E., Lixian, J., Milliken, M., et al. (2003) Intravenous versus intrastriatal cord blood administration in a rodent model of stroke. *J. Neurosci. Res.* **73,** 296–307.

48. Ende, N., Weinstein, F., Chen, R., and Ende, M. (2000) Human umbilical cord blood effect on sod mice (amyotrophic lateral sclerosis). *Life Sci.* **67,** 53–59.

49. Ende, N., Czarneski, J., and Raveche, E. (1995) Effect of human cord blood transfer on survival and disease activity in MRL-lpr/lpr mice. *Clin. Immunol. Immunopathol.* **75,** 190–195.

50. Rowland, L. P. (1992) Amyotrophic lateral sclerosis and autoimmunity. *N. Engl. J. Med.* **327,** 1752–1753.

51. Garbuzova-Davis, S., Willing, A. E., Zigova, T., et al. (2003) Intravenous administration of human umbilical cord blood cells in a mouse model of amyo-

trophic lateral sclerosis: distribution, migration, and differentiation. *J. Hematother. Stem Cell Res.* **12,** 255–270.

52. Lu, D., Sanberg, P. R., Mahmood, A., et al. (2002) Intravenous administration of human umbilical cord blood reduces neurological deficit in the rat after traumatic brain injury. *Cell Transplant.* **11,** 275–281.

53. Chen, R. and Ende, N. (2000) The potential for the use of mononuclear cells from human umbilical cord blood in the treatment of amyotrophic lateral sclerosis in SOD1 mice. *J. Med.* **31,** 21–30.

54. Borlongan, C. V., Hadman, M., Davis Sanberg, C., and Sanberg, P. R. (2004) CNS entry of peripherally injected umbilical cord blood cells is not required for neuroprotection in stroke. *Stroke* **35,** 2385–2389.

55. Santa-Olalla, J. and Covarrubias, L. (1995) Epidermal growth factor (EGF), transforming growth factor-alpha (TGF-alpha), and basic fibroblast growth factor (bFGF) differentially influence neural precursor cells of mouse embryonic mesencephalon. *J. Neurosci. Res.* **42,** 172–183.

56. Ebadi, M., Bashir, R. M., Heidrick, M. L., et al. (1997) Neurotrophins and their receptors in nerve injury and repair. *Neurochem. Intl.* **30,** 347–374.

57. Ciccolini, F. and Svendsen, C. N. (1998) Fibroblast growth factor 2 (FGF-2) promotes acquisition of epidermal growth factor (EGF) responsiveness in mouse striatal precursor cells: identification of neural precursors responding to both EGF and FGF-2. *J. Neurosci.* **18,** 7869–7880.

58. Storch, A., Paul, G., Csete, M., et al. (2001) Long-term proliferation and dopaminergic differentiation of human mesencephalic neural precursor cells. *Exp. Neurol.* **170,** 317–325.

59. Abe, K. (2000) Therapeutic potential of neurotrophic factors and neural stem cells against ischemic brain injury. *J. Cereb. Blood Flow Metab.* **20,** 1393–1408.

60. Magy, L., Mertens, C., Avellana-Adalid, V., et al. (2003) Inducible expression of FGF2 by a rat oligodendrocyte precursor cell line promotes CNS myelination in vitro. *Exp. Neurol.* **184,** 912–922.

61. Justicia, C. and Planas, A. M. (1999) Transforming growth factor-alpha acting at the epidermal growth factor receptor reduces infarct volume after permanent middle cerebral artery occlusion in rats. *J. Cereb. Blood Flow Metab.* **19,** 128–132.

62. Sugimori, H., Speller, H., and Finklestein, S. P. (2001) Intravenous basic fibroblast growth factor produces a persistent reduction in infarct volume following permanent focal ischemia in rats. *Neurosci. Lett.* **300,** 13–16.

63. Bracci-Laudiero, L., Celestino, D., Starace, G., et al. (2003) CD34-positive cells in human umbilical cord blood express nerve growth factor and its specific receptor TrkA. *J. Neuroimmunol.* **136,** 130–139.

64. Rainsford, E. and Reen, D. J. (2002) Interleukin 10, produced in abundance by human newborn T cells, may be the regulator of increased tolerance associated with cord blood stem cell transplantation. *Br. J. Haematol.* **116,** 702–709.

65. Pomyje, J., Zivny, J., Sefc, L., et al. (2003) Expression of genes regulating angiogenesis in human circulating hematopoietic cord blood CD34+/CD133+ cells. *Eur. J. Haematol.* **70,** 143–150.

66. Gianni, A. M., Siena, S., Bregni, M., et al. (1989) Granulocyte-macrophage colony-stimulating factor to harvest circulating haemopoietic stem cells for autotransplantation. *Lancet* **2,** 580–585.

67. To, L. B., Roberts, M. M., Haylock, D. N., et al. (1992) Comparison of haematological recovery times and supportive care requirements of autologous recovery phase peripheral blood stem cell transplants, autologous bone marrow transplants and allogeneic bone marrow transplants. *Bone Marrow Transpl.* **9,** 277–284.

68. Willing, A. E., Vendrame, M., Mallery, J., et al. (2003) Mobilized peripheral blood cells administered intravenously produce functional recovery in stroke. *Cell Transplant.* **12,** 449–454.

69. Schabitz, W. R., Kollmar, R., Schwaninger, M., et al. (2003) Neuroprotective effect of granulocyte colony-stimulating factor after focal cerebral ischemia. *Stroke* **34,** 745–751.

70. Six, I., Gasan, G., Mura, E., and Bordet, R. (2003) Beneficial effect of pharmacological mobilization of bone marrow in experimental cerebral ischemia. *Eur. J. Pharmacol.* **458,** 327–328.

71. Mazzini, L., Fagioli, F., Boccaletti, R., et al. (2003) Stem cell therapy in amyotrophic lateral sclerosis: a methodological approach in humans. *Amyotroph Lateral Scler. Other Motor Neuron Disord.* **4,** 158–161.

72. Eglitis, M. A., Dawson, D., Park, K. W., and Mouradian, M. M. (1999) Targeting of marrow-derived astrocytes to the ischemic brain. *Neuroreport* **10,** 1289–1292.

73. Li, Y., Chopp, M., Chen, J., et al. (2000) Intrastriatal transplantation of bone marrow nonhematopoietic cells improves functional recovery after stroke in adult mice. *J. Cereb. Blood Flow Metab.* **20,** 1311–1319.

74. Weber-Nordt, R. M., Schott, E., Finke, J., et al. (1996) Umbilical cord blood: an alternative to the transplantation of bone marrow stem cells. *Cancer Treat Rev.* **22,** 381–391.

75. Laver, J. H., Abboud, M. R., Kawashima, I., et al. (1995) Characterization of c-kit expression by primitive hematopoietic progenitors in umbilical cord blood. *Exp. Hematol.* **23,** 1515–1519.

76. De Bruyn, D., Delforge, A., Lagneaux, L., and Bron, D. (2000) Characterization of CD34+ subsets derived from bone marrow, umbilical cord blood and mobilized peripheral blood after stem cell factor and interleukin 3 stimulation. *Bone Marrow Transpl.* **25,** 377–383.

77. Anzai, N., Lee, Y., Youn, B. S., et al. (2002) C-kit associated with the transmembrane 4 superfamily proteins constitutes a functionally distinct subunit in human hematopoietic progenitors. *Blood* **99,** 4413–4421.

78. Salven, P., Mustjoki, S., Alitalo, R., et al. (2003) VEGFR-3 and CD133 identify a population of CD34+ lymphatic/vascular endothelial precursor cells. *Blood* **101,** 168–172.

79. Encabo, A., Mateu, E., Carbonell-Uberos, F., and Minana, M. D. (2003) CD34+CD38- is a good predictive marker of cloning ability and expansion potential of CD34+ cord blood cells. *Transfusion* **43,** 383–389.

80. Ng, Y. Y., Bloem, A. C., van Kessel, B., et al. (2002) Selective in vitro expansion and efficient retroviral transduction of human CD34+ CD38- haematopoietic stem cells. *Br. J. Haematol.* **117,** 226–237.
81. Watt, S. M., Gschmeissner, S. E., and Bates, P. A. (1995) PECAM-1: its expression and function as a cell adhesion molecule on hemopoietic and endothelial cells. *Leuk. Lymphoma* **17,** 229–244.
82. Pranke, P., Failace, R. R., Allebrandt, W. F., et al. (2001) Hematologic and immunophenotypic characterization of human umbilical cord blood. *Acta Haematol.* **105,** 71–76.
83. Sorg, R. V., Andres, S., Kogler, G., et al. (2001) Phenotypic and functional comparison of monocytes from cord blood and granulocyte colony-stimulating factor-mobilized apheresis products. *Exp. Hematol.* **29,** 1289–1294.
84. Sanchez, M., Alfani, E., Migliaccio, A. R., et al. (2003) Amplification of T cells from human cord blood in serum-deprived culture stimulated with stem cell factor, interleukin-7 and interleukin-2. *Bone Marrow Transplant.* **31,** 713–723.
85. Chen, J. L., Li, Y., and Chopp, M. (2000) Intracerebral transplantation of bone marrow with BDNF after MCAo in rat. *Neuropharmacology* **39,** 711–716.
86. Ende, N., Chen, R., and Ende-Harris, D. (2001) Human umbilical cord blood cells ameliorate Alzheimer's disease in transgenic mice. *J. Med.* **32,** 241–247.
87. Ende, N. and Chen, R. (2002) Parkinson's disease mice and human umbilical cord blood. *J. Med.* **33,** 173–80.
88. Ende, N. and Chen, R. (2001) Human umbilical cord blood cells ameliorate Huntington's disease in transgenic mice. *J. Med.* **32,** 231–240.

14

Stem Cells and Regenerative Medicine

Commercial and Pharmaceutical Implications

L. Eduardo Cruz and Silvia P. Azevedo

ABSTRACT

Implantation was perhaps the first therapeutic tool in regenerative medicine. Since the first tissue implants were launched, the economical meaning of treating degenerative disease, and the influence in health care costs for the elderly population, has become more apparent *(1)*. It is also clear that tissue implant use fails, as the majority of implants do not integrate within the host tissue, die, or are no longer available after a period of time. Although implants may be mechanically durable and made of biologically inert materials, implants and their lack of biocompatibility accelerate the need for innovative means of regenerating loss or decayed tissue *(2,3)*. Alternatively, in the arena of the elderly, neurological and cardiac diseases will have major roles in the overall costs of health and consequent pharmaceutical, device, hospital, and clinical practices for the next 25 yr. The great therapeutic potential of stem cells in treating degenerative diseases can be rationalized, as cell therapy and tissue engineering may be considered the most relevant window of opportunity in health-related business for the dawn of the 21st century *(4)*. Will cell therapy be a tailor-made process? Will autologous cell transplantation become a standard medical procedure? Will off-the-shelf cell treatments soon be packaged, prescribed, and routinely used in the hospital environment?

Key Words: Implantation; stem cells; degenerative diseases.

INTRODUCTION

Autologous tailor-made cell therapy has established a new era in medicine. Positive reports on the benefits of cell therapy occur almost daily in both the mainstream media, as well as peer-reviewed journals *(5)*. Cardiac, immunological, neurological, and even endocrinological disorders are thought to benefit from cell therapy *(6)*. However, unless the product has

From: *Contemporary Neuroscience: Cell Therapy, Stem Cells, and Brain Repair*
Edited by: C. D. Sanberg and P. R. Sanberg © Humana Press Inc., Totowa, NJ

been standardized from a specific cell lineage(s), cell therapy will not likely become a routine procedure or significantly impact the health and social care system.

In terms of economical implications, cell therapy may be compared to biomaterials, bone substitutes, tissue engineering and/or artificial cells. All these issues originated in the beginning of the 1980s and are in their industry debut. Only a few products, or validated medical procedures, have actually reached the market. Overall, combining aspects of medical devices, pharmaceuticals, biotechnology, and tissue manipulations was considered a compromising subsector of the health industry for many years *(4)*. Therefore, start-up companies were the only business category dedicated to this field.

Cell and gene therapy follows a similar developmental path as tissue engineering. Cell transplantation has been widely studied in stroke *(6)* and cardiac disease *(7–10)*. However, for the approved use of either synthetic, engineered tissues or natural cells to be administered in the body, the potential biological product must prove to be safe, reproducible, validated, and standardized, which could be termed, "tissue and cell engineering for therapeutics." This industry would encompass cell- and tissue-engineered products, biomaterials, and biotechnologies, including allograft-derived tissue substitutes, cell collection, transportation, manipulation, testing, culturing, storing and thawing. The field of genetics would also be involved. There has never been such an amount of accumulated knowledge regarding this "new" living therapeutic entity. In all aspects, it comprises the challenge of bridging the safety gap, which remains constant whenever a chemical entity is administered to a patient. Safety, is an important challenge that accompanies the prospective use of embryonic stem cells *(11)*.

The majority of companies involved in cell therapy are small, highly sophisticated, and are privately owned and funded in terms of qualified staff. Although belonging to a small firm, these entrepreneurs must keep in mind that their innovative product will demand several millions of dollars and will take 8–10 yr to translate from the laboratory to the bedside.

In early 2000, owing to scientific and clinical evidence, cell therapy rapidly achieved a higher status in medical practice, and larger pharmaceutical groups turned their attention to this promising treatment option.

IMPACT OF INNOVATIVE TECHNOLOGIES IN HEALTH CARE

As a result of new medicine and technology, patients live longer and want to live better-quality lives *(12)*. The overall elderly population is increasing; thus, the health issues related to age also rise. Such diseases as diabetes, stroke, neurological disorders, cancer, and heart disease will be an even more important factor in health expenditures. Although stroke and neurological degenerative diseases are the second causes of death and disability world-wide, few new treatment options have been launched in the last decade *(13–15)*. However, this is not the case if the number of new drugs is evaluated for conditions like cancer, heart disease, diabetes, and even rare disorders (e.g., cystic fibrosis and sickle cell anemia). Neurological degenerative diseases might represent the most interesting field for cell therapy strategies *(16)*.

A great deal of time and investment is required for an innovation to become a generally accepted perception in health care *(17)*. Pharmaceutical companies have long acknowledged the simple fact that physicians—mainly key opinion leaders—will not exploit a new therapeutic tool unless it is proven safe and scientifically consistent.

In addition, health care interventions that improve clinical outcomes are achieved, but at significant cost. Cell therapy strategies will have to provide effective, positive results to be reimbursed and evaluated as an evidence-based practice to include in the management of acute and chronic diseases.

Innovative medicine, along with better education and sanitary and nutritional conditions, will help increase the total population of individuals over 85 yr of age who will live longer and better lives.

As the population continues to get older, neurological disorders may increase social costs *(8)*. Neurological disorders already affect approx 1.5 billion people worldwide and are responsible for most outstanding health expenses. Brain-related illness generates more health care–related costs, lost income, and overall economical implications than any other health condition, comprising an estimated $1.0 trillion annually worldwide and $350 billion annually in the United States. Of Americans over age 65, 20% have Alzheimer's disease, and half of those Americans above age 85 have Alzheimer's disease *(19)*.

According to the US Census Bureau (Fig. 1), it is predicted that the number of senior citizens (65–85 yr old) will more than double by year 2030, and the number of citizens in advanced senescence (>85 yr old) will be 2.5-fold higher. Individuals over age 85 will require 20-fold more medical assistance resources and twice as many hospitalization days. However, younger adults living in areas of high-population densities are more exposed to traumatic lesions. Although less significant in quantity, these lesions severely impact health costs and generate debilitated young people and/or individuals permanently incapacitated for normal productive lives *(20)*. In Brazil, with a population of more than 170 million people traumatic lesions are the leading cause of either hospitalization or death in the second and third decades of life *(21)*.

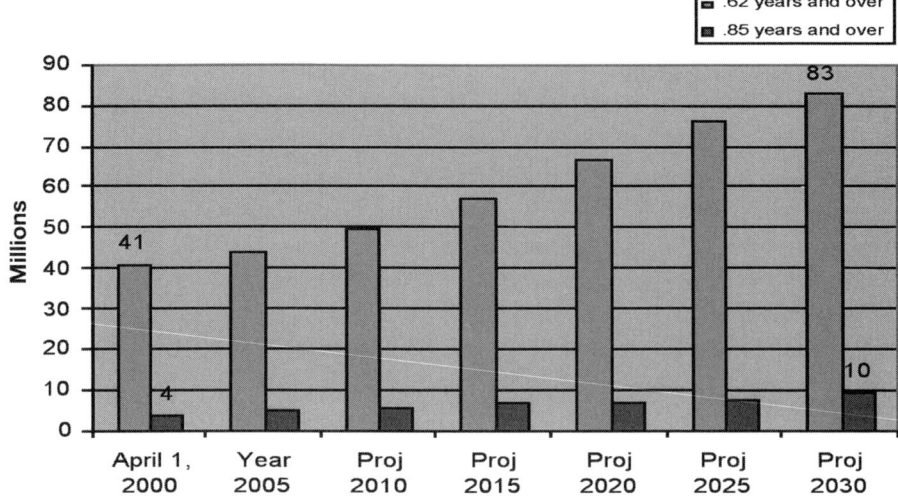

Fig. 1. The number of senior citizens will double by 2030.

The data above are in accordance with the World Health Organization forecasts. Thus, the number of individuals suffering from age-related neurodegenerative conditions, including cardiac disease, Alzheimer's disease, and Parkinson's disease (PD) will increase *(22)*. In the case of heart failure, there is a current transplant crisis *(23,24)*.

The emergence of biomimetics, biocompatible grafts, like the bone substitute with osteoblasts' inductive property *(25)*, and engineered heart valves *(26)*, forecast the materialization of new and safer tools of therapeutic significance in the near future *(27)*.

Human umbilical cord blood stem cells, once considered useful solely in cases of hematopoietic disorders, now appear to be a safe source of undifferentiated cell lineages of interesting plasticity and regenerative potential *(28–32)*.

In neurodegenerative disorders, cell therapy will more than likely focus on the cogniceuticals and sensoceuticals markets, i.e., memory and attention-related disorders, such as Alzheimer's disease and sensory decline and motor system illnesses, pain, retinal degenerative disorders, epilepsy, and Parkingson's disease (PD) *(18)*.

The aging of the baby boom generation, combined with new and improved treatments for neurodegenerative disorders will lead to an expansion of the present $30 billion neurodegenerative market.

Alzheimer's Disease

A study by the Lewin Group shows that the cost to Medicare in the United States regarding the treatment of Alzheimer's disease will soar from $31.9 billion in 2000 to $49.3 billion in 2010 *(19)*. In terms of health care expenses and lost wages of both patients and their caregivers, the cost of Alzheimer's disease nationwide is a staggering $80–100 billion per year. The annual cost of care for one Alzheimer's patient is $18,400 for mild symptoms, $30,100 for moderate symptoms, and $36,132 for advanced symptoms. From diagnosis to death, the average direct cost of caring for an Alzheimer's patient is $174,000. The average annual cost for nursing home care in the United States is approx $52,200 *(33)*.

Alzheimer's disease is an example of a neurological disorder with hardly any new treatment advances in neuropharmaceuticals. Available treatments offer only symptomatic relief, mainly for early stages of the disease. The market for Alzheimer's therapy is expected to grow from 16 million to 21 million patients by 2010 in the seven major pharmaceutical markets, which include the United States, France, Germany, Italy, Spain, United Kingdom, and Japan. Moreover, between 2005 and 2010, the drugs used to treat Alzheimer's disease could achieve sales over $2 billion *(19)*.

PD

Costs associated with PD include both direct and indirect expenses. Researchers recently determined that the US direct costs of PD are $16,634 per person/yr for people who are covered by both Medicare and a private insurance policy.

In 1997, researchers estimated that the annual US economic burden associated with PD was $25 billion. Through a recent multicenter collaboration, investigators calculated the direct medical costs associated with the treatment of PD among elderly people (average age, 72.7 years) who were covered by Medicare and private insurance. The researchers used a database of insurance claims information to identify 10,445 patients who met their criteria—patients who had had either two visits with a physician for the treatment of PD or one visit for treatment and two drug prescriptions to treat PD over an average of 675 d. Percentages of patients were as follows:

Hospitalized 52%
Admitted to a nursing home for short-term care 15.8%
Required medical care after a fall or injury 42.5%
Treated for dementia 12%
Treated with antidepressant medication 45.2%
Treated with antipsychotic medication 28%

The total average cost for treating each patient with PD in this study was $16,634, with hospitalization accounting for $9362 of that total and nursing home care for rehabilitation (<90-d stay) another $2282. Some patients incurred far greater expenses than others. The average yearly cost of medical care for a patient without PD who is covered by Medicare is $6711, which is less than half that of the population in this study. However, estimates could not be compared directly because, although the PD patients had access to private insurance, the non-PD patients did not, which may have affected the total expenditures in each group *(34)*.

Spinal Cord Injury

The average annual health care and living expenses, and the estimated lifetime costs, that are directly attributable to spinal cord injury vary greatly according to the severity of injury. The table below depicts expenses and severity of injury in spinal cord lesions.

Severity of injury	Average yearly expenses (In April 2005 dollars)		Estimated lifetime costs by age at injury (discounted at 2%)	
	First year	Each subsequent year	25 yr old	50 yr old
High tetraplegia (C1–C4)	$710,275	$127,227	$2,801,642	$1,649,342
Low tetraplegia (C5–C8)	$458,666	$52,114	$1,584,132	$1,003,192
Paraplegia	$259,531	$26,410	$936,088	$638,472
Incomplete motor function at any level	$209,324	$14,670	$624,441	$452,545

These figures do not include any indirect costs, such as losses in wages, fringe benefits, and productivity, which average $57,613 per year in March 2005 dollars but vary substantially based on education, severity of injury, and preinjury employment history *(20)*.

THE ROLE OF THE INDUSTRY

The Brain Industry

The brain industry, or neurotechnology, is already $100 billion per year in health care costs and therefore has a major role in economic development in upcoming decades. Approximately 350 public and private companies are researching, developing, and commercializing pharmaceuticals, biologics, medical devices, as well as diagnostic and surgical equipment for the treatment of neurological and psychiatric illness *(35)*.

Neurotechnology could be subdivided into neuropharmaceutical, neurodevice, and neurodiagnostic. Neuropharmaceutical can be further separated via its markets for cogniceutical, emoticeutical, sensoceutical, and neuro-nutraceutical, whereas neurodevice splits into markets for neuroprosthetic, neurostimulation, neurosurgical, and neurofeedback. Finally, neuro-diagnostic gives rise to neuroimaging, in vitro diagnostics, and neuro-informatics *(18)*;

The neuropharmaceutical sector alone reached sales of almost $90 billion in 2004, experiencing a growth rate of 13%, compared to 10% for overall pharmaceutical sales. Despite this outstanding performance, the progress in the development of new treatments, side effects, and low efficacy associated with the new drugs allow significant opportunity for innovative treatments.

The Cell Therapy Industry vs Academia

Stem cell therapy is an example of innovative technology, where the translation from laboratory to bedside is based on a completely new scheme of development, regulatory patterns, and scientific background, compared to the classical way of drug development in this last century. Although different viewpoints subsist in academia and industry, there is a need to accommodate strategies and perceive a common sense. Academia has long been the origin of basic knowledge of cell biology, whereas industry is responsible for spreading scientific novelties to benefit society in the large scale. The so-called "science or technology park" may be the bridge between academia and industry. Small in size, but large in critical knowledge, technology-based incorporations are generated in these "parks."

A "science and technology park" is defined by L. Sanz *(36)* as a space, physical or cybernetic, that is managed by a specialized professional team that provides value-added services. The team's main aim is to increase the competition of its region, or territory of influence, by stimulating a culture of quality and innovation among its associated businesses and knowledge-based institutions. Utilizing its sources, knowledge and technology can be transferred to companies and the market by actively fostering the creation of new and sustainable innovation-based companies through incubation and spin-off processes.

The pharmaceutical industry has evolved in such a way to provide full-scale research and development in-house, but small technology-based companies can put together or scale-up what is generated in university laboratories. Thus, private companies will have the most important role in further research of cellular therapeutics and will then help develop cell therapies in viable products. Small firms appear to be in a key position to transfer technology from this scientific base into the industry. A number of start-up companies have been created as spin-offs from academia in this sector. However, despite the recognized high quality of this research, the "commercialization gap" continues to exist.

One current issue affecting the relationship between academia and industry is the patenting of stem cell lines. This situation peaked when embryonic stem cells were almost banned for political reasons. However, many embryonic stem cell lines are patented, which supports ongoing market strategies. The stem cell patent issue is not simple. On one side, industry has to secure huge investments to support innovative research and development. The patent system is one method to protect this investment and ensure financial return. Patents also help technology transfer between academia and industry, and a strong intellectual property position is often necessary for start-up companies to secure venture capital investment and negotiate strategic partnerships with other industrial partners. On the other side, academia claims the need for free and open access to these cell lines for research purposes. Some scientists believe that human embryonic stem cell lines should not be patented at all.

Although much evidence of stem cell potential in animals has originated from academia, the development of these lines into clinical products requires industrial and commercial inputs. However, it is in no one's interest to push stem cells into clinical trials without the proper investigation of safety and efficacy.

Stem cell technology is at an early stage of development and presents an opportunity to develop harmonized guidelines on quality and safety issues based on adapted criteria to assess therapy efficacy. It also represents a chance to involve society at large in the decision-making process that will accompany the development of this technology. This approach should ensure the successful introduction of stem cell therapy *(37)*.

QUALITY AND SAFETY: REGULATORY ISSUES

Technical Aspects of Safety, Quality, and Standardization

Stem cells can be obtained from embryos, umbilical cord blood, and bone marrow; other sources are also being evaluated. Stem cells from the bone marrow are considered adult stem cells, whereas the umbilical cord blood stem cells are considered perinatal.

Although approval is required for all stem cell therapies, it is much easier to receive approval if the cells are autologous in nature. However, that may be changing in the near future. Regulatory agencies are issuing propositions to these issues that cell therapy regimens must be validated and quality-controlled (e.g., reproducibility and traceability). This will obviously cause significant upheaval for autologous cell treatments, even if with a minimal ex vivo manipulation. Cell therapy will not be accepted based only on good laboratory practices. Standardized protocols must be established, and regulatory agencies will certainly realize that, whatever the cell manipulation, if the procedure is routine, there must be a safe and validated protocol. Thus, each cell collection must be treated as if it were an official pharmaceutical production lot, with all the proper quality-control and assurance measures set in place. In addition, if cells or "samples" are more extensively manipulated in vitro, or when exogenous materials are used, standardized protocols, risk assessment, and quality-control measures should also be emphasized, as well as the foundations of safety, efficacy, and side-effect surveillance protocols.

It is important to note that a harmonized regulatory framework already exists at the EU level to handle the development, testing, and approval of biomedicine. However, neither the directives on medical devices nor medicinal products may be applicable in a practical manner. Therefore, specific harmonized regulatory frameworks may be needed for certain new biomedical materials, such as stem cells and stem cell products. Specific regulations shall become a reality for stem cell therapies and products in the near future.

A commonality in cell therapy regulation should be that the highest priority involves the proper balance between safety and potential quality of life for the patient. Regulatory issues should have enough flexibility to cope with the rapid evolution of the field, and the certification process should be streamlining because of the life cycle of the products, which are expected to be shorter than "regular medicine" in the first few years.

For a harmonious development of the market, both consumers and producers need a stable technical framework based on standards. These standards should include risk management and quality. Moreover, companies are not clear regarding legislation surrounding cell therapy. This position is difficult for young companies that need legal stability before they can receive investment funds. Cell therapy and tissue engineering are intriguing opportunities in health care. The amount, quality, management, and spread of information will make academia and industry find ways to facilitate cooperation that will lead to a faster launch of safe and beneficial cell therapeutic regimens.

CELL THERAPY: ECONOMICAL IMPLICATIONS

Although some investors claim that economical benefits from stem cell therapy are speculative *(38)*, there appears to be great promise. Published data on the potential for new stem cell-based therapies clearly suggest a prospective for new and powerful medicinal use.

More than 70 conditions have been identified as potential targets for stem cell therapy. These conditions vary from relatively rare conditions to cancer and heart disease, which afflicts millions and generates billions in spending. Some target conditions affect younger people, and other conditions primarily affect older individuals, which then tends to become more important in health care expenditures *(39)*.

To investigate the potential impact based on improvements in health care and spending changes, a relatively simple simulation model was developed by a group of researchers from Stanford and Princeton universities *(40)*. In this study, Drs. Laurence Baker and Bruce Deal presented the premise that stem cell researchers have the potential to accelerate the discovery of new advances and make possible the use of new therapies sooner than would otherwise have happened. Their diabetes simulating model is described below.

THE DIABETES SIMULATING MODEL

A hypothetical medical condition was constructed to roughly resemble type 1 (juvenile) diabetes—a condition that arises in childhood in which the body becomes unable to generate insulin, leading to sugar imbalances. People with type 1 diabetes can encounter a range of serious, ultimately life-threatening complications. Stem cell therapy appears to be effective for the treatment of type 1 diabetes.

An important potential benefit of stem cell therapy is a reduction in mortality for individuals with this condition. To quantify its possible effect, an example survival curve was designed for type 1 diabetes. For simplicity, it was assumed that there are 1000 new cases diagnosed among children per year, and that these diagnoses occur among children at age 10.

The survival curves were designed, beginning with the US population mortality rates by age for 2001, assuming that mortality rates for the simulated condition are seven times higher than that for the overall population (Fig. 2). This assumption is broadly consistent with the experience of some recent type 1 diabetes groups.

Figure 2 compares the survival curves for groups of 1000 individuals who are 10 yr old in 2000. The highest survival curve is for the baseline US population, and the lowest survival curve is for a population with the disease.

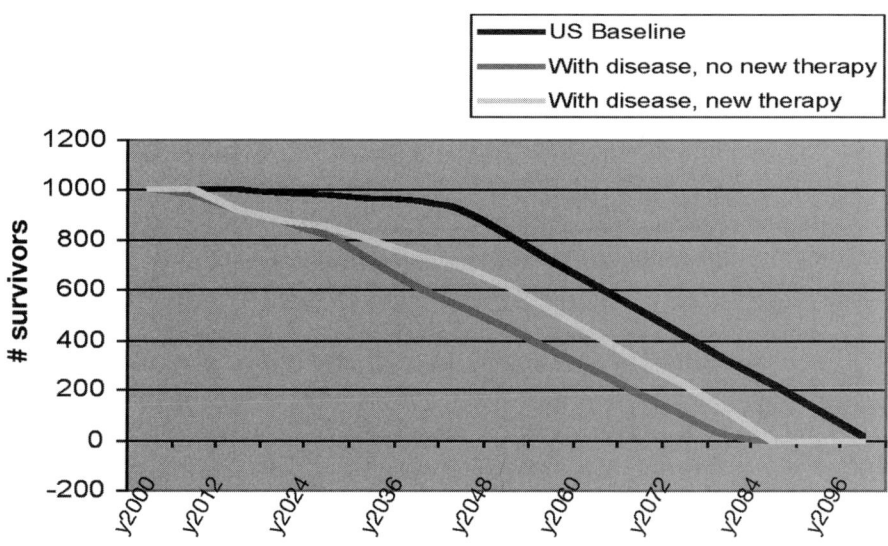

Fig. 2. Number of survivors along time.

Life expectancy in the US baseline group is age 77, and life expectancy in the group with the condition is age 54. The middle survival curve shows the case of a hypothetical new therapy introduced in 2020 that cuts the difference between the US baseline annual mortality rate and the hypothesized condition mortality rate by half. In the diabetes group, the introduction of this new therapy reduces mortality after 2020 and increases life expectancy nearly to age 60. Alternatively, the introduction of this new therapy in 2020 would produce 5307 additional life years in this group.

Figure 2 plots baseline on the observed mortality by age for the US population in 2001. The survival curve for the diabetes group without therapy is based on an assumed annual risk of mortality seven times higher than the US baseline. The curve for the diabetes group with a new therapy assumes a new therapy in 2020 reduces the difference between the US baseline mortality rate and the disease mortality rate by 50%.

Implications of accelerating the introduction of this therapy were also investigated. If a therapy were to become available 5 yr earlier, e.g., in 2015, life expectancy would increase by less than 5 mo, and there would be 473 additional life years for these individuals.

The introduction of a new therapy in 2015 or 2020 would benefit not only the individuals that were 10 yr old in 2000 but also those who were diagnosed before and after 2000. To estimate aggregate increases in life years, the impact of new therapies on individuals diagnosed each year from

1900 to 2040 was considered. A new therapy decreased the difference between baseline mortality and the assumed mortality rate among those with the condition by half. A new therapy in 2015, opposed to 2020, would generate 34,473 additional life years, counting all years and all affected individuals.

This is a substantial benefit. Although applying dollar values to life years can be controversial, many contemporary analyses of new medical technologies are evaluated around a standard that values 1 yr of life between $50,000 and $100,000. Using $75,000 per life year, 34,473 additional life years would be valued at nearly $2.6 billion.

Other potential benefits are also possible, including improvements in the quality of life. Successful therapies could enable individuals with type 1 diabetes to more easily engage in activities they enjoy and lower the burdens of medical visits. A common method of incorporating quality of life into economic analyses is to measure quality-adjusted life years (QALYs). The QALY adjusts for the quality of life by assigning a value, or health utility, for each year of life, with 0 representing death, and 1 representing full health. Although published health utilities for type 1 diabetes is not available, the literature on type 2 diabetes reports that health utilities in patients with type 2 without complications are not significantly different than those for the general population, but that health utilities can be substantially lower for those with complications. It was assumed that the burden of regular insulin injections would impose some loss of quality of life beyond that incurred by many type 2 diabetics, and it was also assumed that many people with type 1 diabetes will develop serious complications as well. Thus, the model of a year of life for type 1 diabetes is equivalent to 0.85 yr of full health. This is clearly an assumption but nonetheless is illustrative of confounds involving diabetes. In addition, the introduction of a new therapy that halved the impact of the condition in 2015, in contrast to 2020, would generate 48,617 additional QALYs, counting all years and all affected individuals. Valued at $75,000 per QALY, 48,617 additional life years would be valued at more than $3.6 billion.

Another potential benefit of a new therapy is improvement in the productivity of individuals. Individuals who live longer can contribute more goods and services to society. Reductions in the burden of illness could make those living with diabetes more productive as well. To illustrate this, we modeled a scenario in which individuals in our group could produce $33,000 worth of goods and services per year between ages 20 and 65 (if healthy) and produced 85% of this estimate if they had diabetes. Theoretically, a new therapy would keep people alive longer, thereby reducing the productivity loss associated with the condition by one half. In this case, a new therapy in 2015 would generate $1.1 billion in additional productivity, counting all groups in all years. The estimated potential benefits for this population that results from other sources (e.g., improved capacity for educational attainment if the burden of disease is reduced or the involvement of other related populations, such as parents or other family members who help care for children with this condition), was not included in this estimate. However, a full assessment would include these factors and possibly other sources of economic benefits.

These simulations can be used to evaluate the impact on the resources that society must devote to medical care. Thus, we investigated impacts on resources devoted to medical care other than the new therapy, then examined the role of resources associated with the therapy. Analyses of the costs associated with particular conditions can become quite involved. For these purposes here, we assumed that the relevant baseline cost figure in the population without the condition is approx $3000 per year. This is roughly consistent with annual health spending in the under age-65 population in the United States. Since individuals with the condition are comprised mainly of people younger than 65, this seems sensible. We also assumed that individuals with the condition spent, on average, an additional $4000 per year, so that their annual average spending is $7000 per year. Estimates of the incremental cost of type 1 diabetes are rare. However, several studies of the overall medical care costs attributable to diabetes (including type 1 and type 2) placed the average cost per person with diabetes between $1000 and $2000 per year in 1990 *(41)*. Inflation between 1990 and 2000 would increase these costs by approx 25%. A new analyses of health insurance claims for individuals under age 65 with type 1 diabetes was conducted and compared to individuals without diabetes, and results revealed a difference of more than $6000.

Using these figures, it can be estimated that the total amount of health care spending per person is $2000, provided that a new therapy is introduced in 2020 and lowers the difference between baseline population medical care costs and costs for those with the condition by half (i.e., reduces the incremental spending from $4000 to $2000). Figure 3 illustrates this impact. When the therapy is introduced in 2020, total annual costs fall from $312 to $223 million.

Note that in both cases, the estimates are based on a new therapy that cuts the difference between the US baseline annual health spending per person (assumed to be $3000) and annual spending in the population per person with the condition (assumed to be $7000) by 50%.

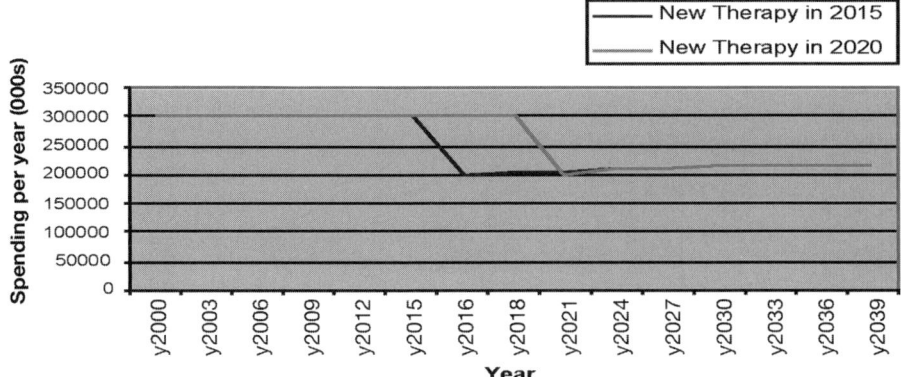

Fig. 3. Total medical spending in all cohorts, in scenarios that introduce a new therapy in 2015 and 2020.

After 2020, the dollars slowly increase because the new therapy also reduces mortality: more people are alive in any given year, leading to higher costs. If successful stem cell funding policies accelerate the launch of this new therapy to 2015, benefits are realized earlier, generating savings. Considering all costs in relevant groups, the acceleration of the hypothesized therapy would reduce nontherapy medical costs by $273 million in total, with the majority of this benefit affected by the lower spending amounts in the years between 2015 and 2020.

In addition, costs are associated with the therapy. To simulate this, the new therapy assumably would cost $2000 for the initial administration and would require annual maintenance treatment costs of $500 (Fig. 4). In the scenario where the therapy is introduced in 2020, there is a high one-time cost in 2020 of nearly $90 million to provide the therapy to all living persons with the condition. In subsequent years, therapy costs are incurred to provide the initial treatment to newly diagnosed persons and provide the maintenance therapy to previously diagnosed persons. If therapy were introduced in 2015 instead, therapy costs would be incurred earlier, and the therapy would be provided to more people. Thus, total costs for the therapy are higher with earlier introduction. In this simulation, total therapy costs are approx $136 million higher in total with a 2015 release, compared to 2020.

The nontherapy medical costs, combined with therapy costs, produced an estimate of net health care savings of approx $137 million in this scenario, counting all years and all affected individuals.

This simulation can be used to illustrate some important principles and the basic impact of marketing a new therapy 5 yr earlier than it otherwise

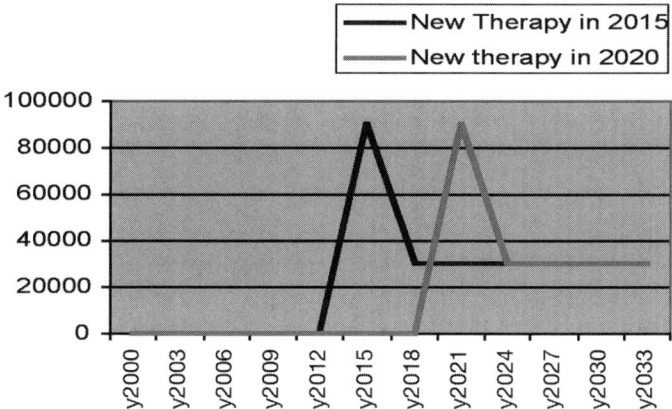

Fig. 4. Total therapy expenditure in all cohorts in scenarios that introduce a new therapy in 2015 and 2020. Note that in both cases, the estimates are based on a new therapy that is assumed to have a $2000 initial cost and to require maintenance costs of $500 per person/yr.

would have been. It can be tempting to evaluate the benefits of a new therapy relative to a hypothetical world with no new stem cell–based therapies. If one evaluated the impact of introducing a new therapy in 2015, compared to never having a new therapy, the benefits would be much larger. However, in this setting, it seems most plausible that most discoveries will eventually be made, and it is the timing that can be altered by stem cell therapy policies.

Another point is that therapies that produce larger improvements will result in lower costs. Varying the effect of the hypothesized therapy on mortality and on medical spending was studied. Some components of stem cell research focus discussion on potential therapies that will completely cure all patients with a given condition. If a complete cure is hypothesized, and mortality rates, QALYs, productivity, and health spending revert to US population baselines, the impact may be larger: in this simulation, it would generate more than 116,000 life years and 150,000 QALYs saved, as well as $364 million in health cost reductions. Although complete cures are not beyond belief, smaller successes could still lead to important improvements in health and cost reductions. Therapies that decrease the burden of symptoms, delay the onset of disease, or slow disease progression could still be beneficial, as could therapies that only work for a population subset with a given condition. As the impact of the therapy gets smaller, however, health benefits are reduced, and the costs of the therapy, relative to the benefits, can be an important factor. In this scenario, if a therapy with only a

10% reduction in mortality rates and medical spending occurs, use of the therapy would generate more than 5500 additional life years and 8000 QALYs, but it would generate a net increase in costs of $69 million, as costs for the therapy outpace reductions in other health care spending.

Yet, if the therapy itself is expensive, health care cost savings could be elusive, as indicated by Baker and Deal. The simulated benefits when the therapy is inexpensive and when it is costly was studied. At high levels of cost, the therapy expenses increase total health care spending, despite a reduction in on-therapy medical spending.

Varying the timing of therapy release is a key issue. In every case, introducing a new therapy at an earlier time results in more life years and lower spending. The benefits increase greatly with the time interval, and investments that accelerate achievement of a new therapy by a short time would produce only small benefits.

Finally, the simulation showed that characteristics of the condition matter. When this model was reestimated using simulated impacts for a condition designed to have characteristics similar to PD, using groups of 1000 individuals per year diagnosed at age 60, the mortality rates for these patients were double the population baseline rates. Based on analysis of insurance claims, the simulation estimated that baseline spending in similarly aged individuals without the disease averaged $6000 per person and those with the condition increased spending by about $4000 per year. The therapy cost was $2000 initially and $500 per year thereafter. In this setting, a therapy that reduced the difference between condition mortality and baseline mortality by half, and similarly reduced incremental nontherapy spending by half, introduced in 2015 instead of 2020, produced only 11,413 additional life years and total health cost savings of $1.4 million. These figures are much lower than the estimates for type 1 diabetes, because there is a shorter duration in which successful therapies can produce benefits in this older group, and the impact of the condition on mortality is not as large.

ECONOMICAL AND SOCIAL IMPACT OF CELL THERAPY INDUSTRY

It is difficult to precisely summarize potential gain from cell therapy, but some data from the biotechnology industry may be helpful. By December 31, 2003, the US biotechnology industry employed 198,300 people in 1473 biotechnology companies. This produced an average of 134 people per company. As there are a handful of relatively large biotechnology firms, as well as many smaller firms, the median firm likely has fewer than 134 employees. Thus, the economic impact is considered for attracting a 50-person biotechnology firm to a certain region.

Using a university research program as a guide, it is possible to anticipate that approximately one third of the budget of many research operations is spent on salaries, and employment data suggest that average salaries across all occupations in the industry are around $52,000 annually. This activity would have economic multiplier effects as well. Using a multiplier of 1.93, attracting a firm that spends $7.5 million per year on salaries and equipment could generate nearly $14.5 million in total new economic activity, positively impacting the region where the company develops its business.

CONCLUSIONS

Economic studies show that technological change is the major force driving long-run economic growth and rising living standards. Mandel *(38)* states, "if anything is going to bail us out of our long-term problems (energy, global warming, rising medical costs), it's going to be technological breakthroughs."

However, the technological path to growth is long and far more uncertain than it seems. In pharmaceuticals, an idea becomes a product in one out of 5000 circumstances. New technologies, even the most gifted ones, are more than difficult to get started in development and even harder to make commercially feasible. History is littered with great ideas that turn out to be dead ends, and there's no way of predicting which ones those will be. The ultimate impact of any new technology on the economy is fundamentally unpredictable.

Moreover, there is an increase demand for more rapid regulatory procedures among industry and academia. Still, one has to keep in mind that the race *(42)* for stem cell technology domain has originated some nonconformity, as seen the embryonic stem cell affair of 2005 *(43)*. Korean stem cell scientists are of outstanding importance. In a few weeks, one of the better-known scientists in the world was taken to the verge of complete disgrace. Even if most of his work is valid, he is a shattered man. If harmonization in cell therapy protocols were in place, perhaps such a catastrophe could have been prevented. The previous gap between academia, industry, and regulatory authority can no longer persist in a global economy. The good news is that the methods of collaborative research (e.g., Internet-based tools) can facilitate the process of finding out what may not be going well or as it was planned, and that is done in a planetary way. The information network is streamlining ethical principles in academia and industry. It should also be better utilized by regulatory authorities. The global information network will make scientific methods more transparent to society overall, as it is the public that sponsors academic work.

Although conservative in nature, there can be no conclusion other than there are substantial amounts of effective protocols showing the potential of cell and tissue engineering to better people's lives. Thus, for the young and the elderly, regenerative medicine will certainly have an important role in the Medicare industry. The action of which may involve independent or associated steps:

- Ex-vivo manipulation of embryonic, adult, or perinatal stem cells (bone marrow or umbilical cord blood), their expansion, differentiation, and potential integration into higher ordered structures that will be reintroduced in lesion regions and integrated in the process of regeneration (bioengineering or tissue engineering).
- Introduction into lesion regions of supramolecular structures similar to elements of the extracellular matrix and associated intercellular mediators, thereby facilitating the recruitment, expansion, and integration of populations of endogenous regenerative cells, promoting the repair of lesions or regeneration and renewal of degenerated tissues (Biomimetics).

The stem cell is a key element to regenerative medicine. Whether introduced off-the-shelf or from a previous collection (e.g., umbilical cord and bone marrow) or recruited from the patient *(44,45)*. Another element is biomaterial technology, in which a whole industry of diagnostic and even nutritional formulations will raise in parallel to this more individualized or focused medicine.

It seems that the younger the stem cell is (umbilical cord), the better that cell is; its plasticity or capability to regenerate is more disperse *(46)*. Several scientists and private companies are developing cell collection and characterization protocols *(47,48)*, and the protocol that will succeed will be proven to be safe, reproducible, and validated. Throughout history, medicine has sought to facilitate or improve the intrinsic self-renewing ability of biological systems. Stem cell, regenerative medicine, and cell and tissue engineering may be thought of as a window of opportunity in medical science and patient care.

In a logical sequence, validated autologous cell therapy procedures, alone or in combination with biomaterials, are expected to become the first medical routine in regenerative medicine and tissue and cell engineering for therapeutics. Then, the road may diverge in two ways, to the off-the-shelf autologous and allogeneic cells and to allograft and tissue substitutes of synthetic or natural derivation. Both will connect to pave the future of tissue and cell engineering for therapeutics—the combination therapy. The government and private institutions realize this potential and are setting milestones to meet the challenges that arise in stem cell research and its clinical application. Under the FDA guidelines, the critical path to the concept of cell products takes approx 10 yr to travel from discovery to patient use, and very few new proposed compounds for therapeutic use are approved *(49)*.

Although there is enormous promise for economic benefits of stem cell therapy, additional criteria must be satisfied to achieve effective, long-lasting repair of damaged tissues. An adequate number of cells must be produced for proper dosing. Cells must be able to differentiate into desired phenotypes. Cells must adopt appropriate three-dimensional structural support/scaffold and produce extracellular matrix. Produced cells must be structurally and mechanically compliant with the native cell. Cells must be able to successfully integrate with native cells and overcome the risk of immunological rejection. Finally, there should be minimal associated biological risks.

REFERENCES

1. Ratner, B. D., Hoffman, A. S., Schoen, F. J., and Lemmons, J. E., eds. (1996) Applications of materials in medicine and dentistry. In *Biomaterials Science: An Introduction to Materials in Medicine*. Academic Press, San Diego, pp. 283–388.
2. Spector, M. (1992) Biomaterials failure. *Orthop. Clin. North Am.* **23,** 211–217.
3. Anderson, J. M. (1994) Inflammation and the foreign body response. *Prob. Gen. Surg.* **11,** 101–107.
4. (2005) Bioengineered tissues: the science, the technology, and the industry. *Orthodontics Craniofacial Res.* **8,** 134.
5. Vats, A., Bielby, R. C., Tolley, N. S., et al. (2005) Stem cells. *Lancet* **366,** 592–602.
6. Borlongan, C. V., Fournier, C., Stahl, C. E., et al. (2006) Gene therapy, cell transplantation and stroke. *Front Biosci.* **11,** 1090–1101.
7. Grill, R., Gage, F. H., Murai, K., et al. (1997) Cellular delivery of neurotrophin-3 promotes corticospinal axonal growth and partial functional recovery after spinal cord injury. *J. Neurosci.* **17,** 5560–5572.
8. Henning, R. J., Abu-Ali, H., Balis, J. U., et al. (2004) Human umbilical cord blood mononuclear cells for the treatment of acute myocardial infarction. *Cell Transplant.* **13,** 729–739.
9. Martina, V., Jordan, C., Newcomb, J., et al. (2004) Infusion of human umbilical cord blood cells in a rat model of stroke dose-dependently rescues behavioral deficits and reduces infarct volume. *Stroke* **35,** 2390–2395.
10. Lindvall, O. and Kokaia, Z. (2004) Recovery and rehabilitation in stroke: stem cells. *Stroke* **35,** 2691–2694.
11. Newman, M. B., Misiuta, I., Willing, A. E., et al. (2005) Tumorigenicity issues of embryonic carcinoma-derived stem cells: relevance to surgical trials using NT2 and hNT neural cells. *Stem Cells Rev.* **14,** 29–43.

12. Lichtenberg, F. (2002) Benefits and costs of newer drugs: an update. *Natl. Bureau Economic Res.* 8996.
13. Heart and Stroke Foundation of Canada, Annual Report Card on the Health of Canadians, 2001.
14. Cowper, P. A. (2004) Economic effects of beta-blocker therapy in patients with heart failure. *Am. J. Med.* **116,** 104–111.
15. MedMarkets. MedMarket diligence, April 2005, vol. 4.
16. Medicines in Development for Heart Disease and Stroke. PhRMA Survey, America's Pharmaceutical Companies, 2005.
17. Mason, J., Freemantle, N., Nazareth, I., et al. (2001) When is it cost-effective to change the behavior of health professionals? *JAMA* **286,** 2988–2992.
18. The Neurotech Nexus, Institute for Global Futures (2005)
19. Available at: www.seniorjournal.com/NEWS/Alzheimer's/04-27-01AlzCostSoar. htm. December 2005.
20. The National Spinal Cord Injury Statistical Center, University of Alabama, Birmingham, 2005.
21. Cadernos do Brasil, Secretaria Executiva do Ministério da Saúde, SIM/ SINASC, 2004.
22. World Health Organization. Cardiovascular Disease Program, Geneva, Switzerland. www.who.int/cardiovascular_diseases/en.
23. Vilas-Boas, F., Feitosa, G. S., Soares, M. B. P., et al. (2004) Bone marrow cell transplantation to the myocardium of a patient with heart failure due to Chagas cardiomyopathy. A case report. *Arquivos Brasileiros de Cardiologia* **82,** 185–187.
24. Soares, M. B., Lima, R. S., Rocha, L. L., et al. (2004) Transplanted bone marrow cells repair heart tissue and reduce myocarditis in chronic chagasic mice. *Am. J. Pathol.* **164,** 441–447.
25. ExtraGraft. www.silvestrelabs.com.br. December 2005.
26. Simon, P. and Hoerstrup, M. D. (2002) Tissue engineering of functional trileaflet heart valves from human marrow stromal cells. *Circulation* **106,** I–143.
27. Ahsan, T. and Nerem, R. M. (2005) Bioengineered tissues: the science, the technology, and the industry. *Orthod. Craniofacial Res.* **8,** 134–140.
28. Broxmeyer, H. E., Douglas, G. W., Hangoc, G., et al. (1989) Human umbilical cord blood as a potential source of transplantable hematopoietic stem/progenitor cells. *Proc. Natl. Acad. Sci. USA* **86,** 3828–3832.
29. Jones, C. A. (1999) Umbilical cord blood transplantation. *Arch Dis. Child.* **81,** 282.
30. Knudtzon, S. (1974) In vitro growth of granulocyte colonies from circulating cells in human cord blood. *Blood* **43,** 357–361.
31. McGuckin, C. P., Forraz, N., Baradez, M. O., et al. (2005) Production of stem cells with embryonic characteristics from human umbilical cord blood. *Cell Prolif.* **38,** 245–255.
32. Sanberg, P. R., Willing, A. E., Garbuzova-Davis, S., et al. (2005) Umbilical cord blood-derived stem cells and brain repair. *Ann. NY Acad. Sci.* **1049,** 67–83.
33. American Health Assistance Foundation. Available at: www.ahaf.org. December 2005.

34. Orsini, L. S., Castelli-Haley, J., Kennedy, S., and Huse, D. M. (2004) Healthcare utilization and expenditures among privately insured patients with Parkinson's disease in the United States. Presented at the 8th International Congress of Parkinson's Disease and Movement Disorders, Rome, Italy, June 14–17, 2004. Published in *Mov. Disord.* **19(Suppl 9),** 145–146.
35. Johnson, S. (2005) The brain business: niche field of neurotechnology holds out great promise, but firms seeking to develop treatments take big risks. *Mercury News*, July 11, 2005, E1.
36. IASP International Association of Science Parks, December 2005. Available at: www.iasp.ws.
37. European Group on Ethics in Science and New Technologies. December 2005. Available at: europa.eu.int.
38. Mandel, M. *Business Week*, December 2005. Available at: www.businessweek.com/the_thread/economicsunbound.
39. American Heart association. Available at: www.strokecenter.org/prof/awards.htm.
40. Baker, L. and Deal, B. Some economic Implications of State Stem Cell Funding Programs. Available at: region.princeton.edu/media/pub/pub_main_10.pdf.
41. Javitt and Chiang, eds. (1995) Economic impact of diabetes. In *Diabetes in America*.
42. Orkin, S. H. and Morrison, S. J. (2002) Biomedicine: Stem-cell competition. *Nature* **418,** 25–27.
43. Scandal for Cloning Embryos: A Tragic Turn for Science. Available at: www.nytimes.com/2005/12/16/science.
44. Stewart, S., Pearson, S., and Horowitz, J. D. (1998) Effects of a home-based intervention among patients with congestive heart failure discharged from acute hospital care. *Arch. Intern. Med.* **158,** 1067–1072.
45. Orlic, D., Kajstura, J., Chimenti, S., et al. (2001) Bone marrow cells regenerate infarcted myocardium. *Nature* **410,** 701–705.
46. Verfaillie, C. M. (2000) Stem cell plasticity. *Graft* **3,** 296–298.
47. Gallacher, L., Murdoch, B., Wu, D. M., et al. (2000) Isolation and characterization of human CD34⁻Lin⁻ and CD34+Lin⁻ hematopoietic stem cells using cell surface markers AC133 and CD7. *Blood* **95,** 2813–2820.
48. Lei, M., Podell, E. R., Baumann, P., and Cech, T. R. (2003) DNA self-recognition in the structure of Pot1 bound to telomeric single-stranded DNA. *Nature* **426,** 198.
49. U. S. Food and Drug Administration Department of Health and Human Services. Last updated: October 21, 2005.

Index